ARTIFICIAL INTELLIGENCE
IN PROCESS FAULT DIAGNOSIS

ARTIFICIAL INTELLIGENCE IN PROCESS FAULT DIAGNOSIS

Methods for Plant Surveillance

Edited by

Richard J. Fickelscherer, PE

Department of Chemical and Biological Engineering
State University of New York at Buffalo
Buffalo, New York
USA

A Joint Publication of the American Institute of Chemical Engineers and John Wiley & Sons, Inc.
Published by John Wiley & Sons, Inc., Hoboken, New Jersey.
Published simultaneously in Canada.

For general information on our other products and services or for technical support, please contact our
Customer Care Department within the United States at (800) 762-2974, outside the United States at
(317) 572-3993 or fax (317) 572-4002.

Wiley also publishes its books in a variety of electronic formats. Some content that appears in print
may not be available in electronic formats. For more information about Wiley products, visit our web
site at www.wiley.com.

Library of Congress Cataloging-in-Publication Data is applied for
Hardback ISBN 9781119825890

Cover Design: Wiley
Cover Image: Courtesy Richard J. Fickelscherer and Daniel L. Chester

Set in 11/13pt Times LT Std by Straive, Pondicherry, India

DEDICATION

This book is solely dedicated to the late Bridget Fitzpatrick. She passed away unexpectedly way to soon and is deeply missed by all her fellow process control colleagues. She was a true ultimate professional in all her career endeavors.

CONTENTS

4   Operator Performance: Simulation and Automation   63

5   AI and Alarm Analytics for Failure Analysis and Prevention   85

10 Knowledge-Based Systems   **300**

LIST OF CONTRIBUTORS

Steven Apple
Global Director of Operator
Performance Services
Sneider Electric USA
Belton, TX
USA

Michael Baldea
McKetta Department of Chemical
Engineering
The University of Texas at Austin
Austin, TX
USA

Ron Besuijen
Center for Operator Performance
Dayton, OH
USA

Daniel L. Chester
Department of Computer and
Information Sciences (Retired)
University of Delaware
Newark, DE
USA

J. F. Davis
Office of Advanced Research
Computing and Department of
Chemical and Biomolecular
Engineering
UCLA
Los Angeles, CA
USA

Lieven Dubois
ISA 18, 88 and 101, Working
Group 8 of ISA 18.2,
Department of Alarm Management
and Training
Manage 4U
Leiden
Netherlands

M. J. Elsass
Department of Chemical
Engineering
University of Dayton
Dayton, OH
USA

Richard J. Fickelscherer, PE
Department of Chemical and
Biological Engineering
State University of New York at
Buffalo
Buffalo, NY
USA

Philippe Mack
PPITE SA
Liège
Belgium

Université de Liège
Liège
Belgium

Atique Malik
AI Control LLC
Edwardsville, IL
USA

Mark Nixon
Process Systems and Solutions
Emerson Automation Solutions
Round Rock, TX
USA

Perry Nordh PEng.
Honeywell Inc.
Calgary, Alberta
Canada

Rajan Rathinasabapathy
Department of Chemical and
Biomolecular Engineering
UCLA

Tech Services Phillips 66
Los Angeles, CA
USA

Ray C. Wang
McKetta Department of Chemical
Engineering
The University of Texas at Austin
Austin, TX
USA

Shu Xu
Process Systems and Solutions
Emerson Automation Solutions
Round Rock, TX
USA

FOREWORD

The idea of building machines that look like people and also behave like them has been a common theme in science fiction books and movies for many years now. Then there arose the digital computer era. When computers were first invented, they were machines that could manipulate symbols as well as numbers, such that early Artificial Intelligence (AI) researchers thought these computers were advanced enough to implement intelligent behavior. Those researchers were able to begin to program computers to do intelligent things like prove theorems of logic, solve word problems, discern calculus functions, play games, understand how to behave in restaurants, stack wooden blocks, diagnose diseases. They even developed rule-based systems that inevitably evolved into expert systems, which successfully found wide-spread, real-world practical usage.

Modern-day AI researchers are putting intelligent behavior in robots, making cars that drive themselves, playing games like Chess and Go at championship levels, making chatbots that carry on conversations and write essays. As a direct result, people are now unfortunately getting greatly more concerned about these machines eventually replacing them in the workplace, killer robots, fake news that might be generated by AI software, judges and lawyers being replaced by AI-programmed machines, etc. They are consequently now beginning to think about setting up safety regulations for AI programs to help ensure those programs do not potentially interfere with any of their basic human rights. Preempting such dangerous outcomes is currently being addressed as best as possible, considering that AI is such a rapidly evolving technology.

One human task that AI has long now tackled very effectively is the identification of things. This task is also known as classification or pattern recognition. Examples of it are facial recognition, recognizing animals in images, recognizing army tanks in images, and more recently, identifying possible operational faults in chemical and nuclear processing plants, which is the focus of this book. Correctly identifying such faults is the critical first step to making those plants run substantially safer for both humans and the environment, as well as to making them operate more efficiently. The capabilities of these automated process fault analyzers are rapidly improving beyond our limited human capabilities for always effectively doing such analysis unassisted. We are thus at the advent of a new era of human technological accomplishment, where AI will hopefully flourish for mankind's direct benefit. Its underlying potential appears to be unlimited. What is thus greatly needed now is for control engineers to implement these various particular diagnostic techniques and gather the resulting data that will hopefully build our enduring trust in those techniques.

DANIEL L. CHESTER, PHD

Newark, Delaware
May 2023

PREFACE

Automated process fault diagnosis is a very highly and widely studied sub-topic in the field of advanced process control theory. This has been true ever since computers were first introduced into process control automation. Continuous advances in computer technology and software programming capabilities, especially the rapid advances currently occurring in artificial intelligence (AI), have increasingly made actual applications of automated fault diagnosis more common throughout the processing industries. Where actually deployed, this technology has directly greatly helped to improve both process safety and operational efficiency in those participating industrial sectors.

The goal of this treatment is to be a comprehensive guide for those professionals actually implementing automated process fault diagnosis. It consequently may help them to more wisely choose and implement the best diagnostic methodology for their particular processing plant application. This should occur because our book consolidates the most recent advances in both theory and actual application practice of the various diverse diagnostic methodologies currently existing into a single comprehensive resource. An intended specific direct benefit of it therefore is to help facilitate the creation of computer programs capable of improving process safety and productivity, increasing process efficiency via reduced operating costs, and directly promoting better overall daily process operation.

Process Fault Analyzers are the resulting computer programs that can monitor real-time process plant operations to detect and identify the underlying cause(s) of process operating problems. Their motivation has been the enormous potential for improving these operations in terms of safety and productivity. Effective automated process fault analysis should help process operators: (i) prevent catastrophic operating disasters such as explosions, fires, meltdowns, toxic chemical releases; (ii) reduce downtime after emergency process shutdowns; (iii) eliminate unnecessary process shutdowns;

(iv) maintain better quality control of the desired process products; and (v) ultimately allow both higher process efficiency and production levels.

As mentioned, a diverse assortment of logically viable diagnostic strategies now exist for automating process fault analysis. However, such automation is currently not widely used within the processing industries. This is mainly due to one or more of these potential limitations: (i) prohibitively large development, verification, implementation, or maintenance costs of these programs; (ii) inability to operate a program based upon a given diagnostic strategy continuously on-line or in real time; (iii) lack of sufficient actual and/or high-fidelity simulated process fault situation data; and (iv) inability to model process behavior at the desired level of detail, thus leading to unreliable or highly ambiguous diagnoses. Subsequently, improved methods for more efficiently creating both highly robust and competent automated process fault analyzers are still being actively sought.

This treatment consequently details the present state-of-the-art and best practices of some of these currently employed diagnostic methodologies, with each of its chapters authored by practicing experts on that chapter's particular diagnostic method. Specifically, several viable alternative AI-based diagnostic methodologies' particulars are elaborated here. These methodologies, if they truly prove to be applicable, should successfully address many of the potential limitations previously enumerated. Our book thus presents a comprehensive overview of the current state of each of these methodologies/effective combinations, concentrating mostly on those methodologies directly utilizing AI programming techniques. It should therefore help ensure that the most appropriate choice of which of those particular techniques to use for a given desired target process application will become quite apparent. The comprehensive nature of this treatment for being both state-of-the-art and current best practice should thus make its highly technical material extremely relevant for immediate operational application throughout the processing industries.

RICHARD J. FICKELSCHERER, PE

ACKNOWLEDGMENTS

I would like to personally thank all the various chapter authors and co-authors for skillfully contributing to this present treatment. All are true experts in their particular fault diagnostic methodologies, as elaborated upon by them. I am especially grateful to Joe Alford, PE, for his enormous efforts, first of all in helping line up many of these experts and then reviewing and insightfully commenting on their original drafts of many of these chapters. His guidance in this book's creation was invaluable and essential to its current level of real-world usefulness. I believe I speak for all involved in creating this treatment and that it will ultimately benefit all individuals considering actually undertaking real-world automated process fault analysis projects. To those curious practitioners, we all wish you the best of luck.

Dr. Richard J. Fickelscherer, PE
Tonawanda, New York
falconeertech@verizon.net

MOTIVATIONS FOR AUTOMATING PROCESS FAULT ANALYSIS

Richard J. Fickelscherer, PE

Department of Chemical and Biological Engineering, State University of New York at Buffalo, Buffalo, New York, USA

OVERVIEW

This introductory chapter to our treatment establishes and briefly discusses the current motivations for directly automating process fault analysis. It begins by listing the various traditional methods currently employed for helping human operators perform more effective process fault management. Human limitations at actually performing such management are then enumerated. Also the advantages of human analysis versus those that computer automation currently possesses are compared. This comparison is made in order to identify the most feasible approaches/best possible pathways presently existing for actually automating process fault analysis. Please note that the literature references cited in this discussion (as well of those cited throughout the remainder of this treatment's various chapters and appendices written or co-written by me) are chosen because they are the earliest

Artificial Intelligence in Process Fault Diagnosis: Methods for Plant Surveillance,
First Edition. Edited by Richard J. Fickelscherer.
© 2024 John Wiley & Sons, Inc. Published 2024 by John Wiley & Sons, Inc.

mention of their respective particular observations that I have encountered in the open literature. These "old" citations are consequently meant to recognize and properly bestow the corresponding appropriate intellectual credit to those original pioneers in this continuing evolution of automated process fault analysis.

CHAPTER HIGHLIGHTS

- Discussion of the changing role of human operators in modern processing plants.
- Overviews of various traditional process fault management methods that are currently being employed by the processing industries in order to address these changes.
- Descriptions of various human limitations encountered when actually performing process fault management.
- Comparisons between human-based and current computer-based analysis advantages.
- Current major motivations for further developing automated process fault analysis.

1.1 INTRODUCTION

Economic competition within the chemical process industries (CPI) has directly led to the construction and operation of larger, highly integrated and more fully automated production plants. As a result, the primary functions performed by the human process operators in these plants have changed. An unfortunate consequence of these changes is that those operators' ability to perform process fault management has been diminished.[1] The underlying

---

[1]Dr. Joseph S. (Joe) Alford, PE, a retired control engineer, contends that this is not necessarily a universal situation based on his experience at Eli Lilly and Co. There, as computers took over the defined routine repetitive jobs that operators used to do in the manufacturing plants (e.g., manually walk the plant and collect data on clip boards), the company chose not to lay off operators, but rather to train them to use more of their natural intelligence (rather than just do rote jobs) and so trained them to be more involved in routine data analysis, process troubleshooting and fault management. So Lilly never directly laid off operators when automating their plants; instead they just used operators in more intelligent ways. This created a win-win situation for all involved. Operators liked their jobs more as that they felt their brains were being used and valued. Subsequently, process faults were better managed.

reasons behind this potentially highly dangerous situation and the various methods currently being used to counteract it are discussed here.

One continuing major trend driving the present modernization of the CPI has been the evermore increasing automation of all process control functions. The motivation for such automation is that it results in more accurately applying the best available process control strategies in a continuous, consistent, and dependable manner (Lefkowitz 1982; De Heer 1987). This automation has been made possible by advances in both computer technology and process control theory. Such advances have made automated control more economically feasible, reliable, and available (Lefkowitz 1982). Advancing process control computers have also provided a significant means for dealing with the diverse and complex information required to effectively operate a modern production plant (De Heer 1987). Together with continuing improvements in electronic instrumentation, these developments are directly allowing these plants to still operate effectively with considerably fewer human operators (Lefkowitz 1982).[2]

Another continuing trend for reducing CPI operating costs has been to maximize the availability of modern plants for production. This is typically accomplished by optimally scheduling the production runs and by minimizing the effects of unexpected production disruptions. A variety of methods are currently used to either eliminate or minimize the severity of unexpected production disruptions. None the less, as the complexity of the processing plants has increased, making these plants available for production has become much more difficult because the number of potential operating problems has also increased (Syrbe 1981). This tends to increase the frequency of unexpected production disruptions. Consequently, maximizing these plants availability for efficient process operation has become more dependent upon effectively managing their various potential operating problems (Linhou 1981).

---

[2]Joe Alford's experience has been that while it is true that many companies choose to operate plants with fewer operators as automation increases – it is not always the case. Operator jobs can be transformed so they have greater roles in FMEAs, alarm rationalization teams, process deviation investigations, real-time fault diagnosis, certain data analysis functions, helping schedule detailed operations, etc., such that they no longer have to fear for loss of their job but can look forward to their job becoming more interesting, challenging, and intellectually diverse.

1.2   THE CHANGING ROLE OF THE PROCESS OPERATORS IN PLANT OPERATIONS

The process operators' main task in plant operation is to continuously assess the process state (Lefkowitz 1982) and then, based upon that assessment, react appropriately. Process operators thus have three core primary responsibilities (Rijnsdorp 1986). The first is to monitor the performance of various control loops to make sure that the process is operating properly. Their second is to make adjustments to the process operating conditions whenever product quality or production efficiency fall outside predefined tolerance limits. The operators' third, and by far most important, responsibility is to avoid emergency situations if at all possible, and if not, properly respond to them.[3] This means effectively and reliably performing process fault management. Such management requires that the operators correctly detect, identify, and then implement the necessary counter actions required to eliminate the process fault or faults creating the emergency situation. If it is performed incorrectly, accidents can and have occurred on many occasions.[4]

---

[3]Lieven DuBois observes that alarm systems are currently set up to avoid safety functions, safety systems and emergency shut-down systems being activated; i.e., all these systems are designed to avoid emergency situations. But when activated they lead inevitably to down time, off-spec quality, energy losses, waste, flaring, etc. One of the key aspects in responding to alarms, according to EEMUA 191, is that the operator should diagnose the root cause of the alarmed situation, which means performing (process and equipment) fault management.

[4]Joe Alford observes that this is true for those plants that are generally continuous in nature: i.e., the above list of operator roles appears to be generally accurate in those plants. However, an estimated half of all manufacturing plants are batch in nature. In most batch plants, there are a large number of manual operations and thus these plants are more "semi-automated" than completely automated. So, for many batch plants, procedures are a combination of manual operations and automated operations. For example, in sterilizing a bioreactor, the overall control is automated, but still requires operators to manually check the large number of feed, sample, and other piping connected to the bioreactor with temp sticks or other temperature measurement devices to verify that they are achieving the desired sterilization temperature. In other batch operations, operators need to manually hook up different pieces of equipment to one another to prepare for an upcoming batch run or transfer of materials. In other unit operations, operators are changing the resin in chromatography separation columns in preparation for the next batch. And, with almost all batch processes, manual cleaning of equipment is needed after each batch. And the list goes on. The number of manual operations that operators must perform in many batch plants is such that the central control room is frequently not manned as the operators are out in the field performing manual operations. (Hence there is the critical need for remote alarming systems.)

The biggest change in the functions performed by the process operators has been directly caused by the increased automation of process control. Process operators now monitor and supervise, rather than manually control, process operations. Moreover, such functions are increasingly accomplished with interface technology designed to centralize control and information presentation (Visick 1986). As a result, their duties have become less interesting and their ability to manually control the process has diminished.[5] Both situations have increased the job dissatisfaction experienced by the process operators (Visick 1986). This has also directly diminished the operators' ability to perform process fault management.[6]

A second change in the functions performed by the operators has directly resulted from having fewer operators present in modern processing plants. Each operator has become responsible for a larger portion of the overall process system's production. This increases the risk of accidents because relatively fewer operators are available at any given time to notice the development of emergency situations or help prevent such situations from causing major accidents. Besides their increased risk, the potential severity of these possible accidents has also increased because larger quantities of reactive materials and energy are now being processed. This makes the operators' ability to perform effective process fault management much more critical for ensuring the safe operation of those plants.

One method used to help reduce the risk of a major accident has been the addition of emergency interlock systems to the overall process control strategy. Such systems are designed to automatically shut down the process during dangerous emergency situations, thereby reducing the likelihood of accidents occurring that could directly threaten human and environmental safety or damage the process equipment. Emergency interlock systems therefore help ensure that the process operation is safe during such emergencies by decreasing the effects of human error in such situations

---

[5]Joe Alford has not found this to be true in his experience; i.e., the operators' ability to manually do process control has not diminished as long as automated systems stick with simple single input, single output, PID feedback controllers. It is when control engineers start implementing algorithms like model predictive control that the operators' ability to do process control diminishes. But then again, he contends most process engineers can't effectively support model predictive controllers either. These controllers are sometimes just too complex and challenging to configure and properly support.

[6]Lieven DuBois believes that because automated and advanced control is built in or on top of basic process control systems, the role of the operator has changed. Nowadays an operator oversees more control loops, more measurements (instruments) and more equipment than ever before. It thus has become humanly impossible to learn and remember all possible faults and all potential consequences of such faults.

(Kohan 1984).[7] Eliminating any accidents also protects the operational integrity of the process system, which in turn allows it to be restarted more quickly after these automatic shutdowns.

However, the wide-spread use of emergency interlock systems has caused the operators' primary focus in plant operations to change from that of process safety to that of economic optimization (Lees 1981). In emergency situations, the operators are now more concerned with taking the corrective actions required for continuing to keep the process system operating rather than those which will safely shut it down. They rely upon the interlock system to handle any emergency shutdowns, trusting that it will take over once operating conditions become too dangerous to let production continue.

A potential problem with this strategy is that, in order to keep the process system operating, the operators may take actions that counteract the symptoms of a fault situation without correcting that situation itself (Goff 1985). Such behavior by the operators may cause them to inadvertently circumvent the protection of the emergency interlock system, thereby creating a situation which they falsely believe to be within that protection. Another potential problem of this strategy is that the emergency interlock system may fail, which again will create a situation in which the operators falsely believe that the process system is still protected by it. These potential problems can be reduced by: (i) prudently designing those interlock systems, (ii) being certain to add sufficient redundancy to detect critically dangerous situations (Kohan 1984), (iii) establishing formal policy by which particular interlocks can be bypassed during process operation (Kletz 1985), and (iv) adequately maintaining those interlock systems (Barclay 1988).[8]

In summary, the automation of the required process control actions and of emergency process shutdowns has shifted the operators' main activities away from direct process control to that of passive process monitoring. Moreover, this automation has also tended to shift their primary emphasis away from process safety to that of economic optimization. As a result of these changes, the operators' ability to always perform the most competent process fault management has been reduced. Unfortunately, this reduction

---

[7]Lieven DuBois notes that since then these systems have evolved drastically. However, some companies allow process operators to sometimes by-pass these interlock systems to keep the plant production running. These by-passes have directly led to disasters (e.g., BP Deep Water Horizon (2010)).

[8]From an ISA (International Society of Automation) perspective, interlocks are part of a broader paradigm known as "Safety Instrumented Systems" (SIS) which are governed by ISA, ANSI, and IEC standards (e.g., ANSI/ISA 84 on Safety Instrumented Systems).

has occurred during an era when such management has become more critical to both the safe and economical operation of the production plants. In response, various methods have been developed to help counteract this decline in the human operators' capability to perform effective process fault management.

1.3 TRADITIONAL METHODS FOR PERFORMING PROCESS FAULT MANAGEMENT

A variety of methods have been previously developed either to directly reduce the occurrence of process faults or to directly help the operators perform process fault management more effectively whenever it is required. The traditional methods currently utilized to reduce the occurrence of process faults include: (i) initially designing the process systems with greater operational safety in mind (e.g., by performing comprehensive FMEA and HAZOP studies), (ii) constructing process plants with higher quality, and therefore more reliable, process equipment, (iii) implementing comprehensive programs of preventative maintenance, and (iv) establishing and strictly following standard operating procedures and change control. The methods currently relied upon to directly help the operators perform process fault management include: (i) extensively training the operators in process fault management, (ii) increasing the effectiveness of alarm deployment strategies, (iii) designing better control consoles and man–machine interfaces, (iv) employing simple data analytics to, for example, discover cause/effect relationships, and (v) as a last resort safety measure, adding emergency interlock systems to the process control systems for the reasons previously discussed.

Despite these efforts, inadequate fault management continues to cause major accidents within the CPI. This is evident by the catastrophic accidents at Flixborough, England (1974), Bhopal, India (1984), Mexico City, Mexico (1984), and Sao Paulo, Brazil (1984) (Kletz 1985), Pasadena, Texas (1989), Texas City, Texas (2005), Jacksonville, Florida (2007), Port Wentworth, Georgia (2008), and Geismer, Louisiana (2013). It also represents a major problem for the Nuclear Power Industry, as is evident by the accidents at the power plants located in Three Mile Island, Pennsylvania (1979) and Chernobyl, Ukraine (1987). While the above disasters have been widely publicized, the vast majority of plant mishaps have not. As a result, the general lessons which could have been learned from these accidents are either never fully presented or are quickly forgotten (Kletz 1985). In fact, many accidents have been caused by the same mistakes being repeated over and over again.[9] The most general lesson that can be learned from the past incidents is that almost all plant accidents are preventable if the emergency situations preceding them are properly recognized and correctly acted upon.[10] At a minimum this requires properly recognizing those emergency situations. Unfortunately, as discussed next, even this does not guarantee that fault management will always be performed correctly.

1.4 LIMITATIONS OF HUMAN OPERATORS IN PERFORMING PROCESS FAULT MANAGEMENT

The preceding discussion has indicated that the various measures currently being taken to improve process fault management do not always guarantee successful results: accidents still occur. One reason for this is that some of

[9]In addition to the above recognized incidents, Steve Apple points out that there are very many that have not been officially reported or noticed, i.e., there are a far greater number of un-reported near-misses. He believes the non-reporting is really a function of the litigious society in which we are living. What large manufacturer wants to notify the public that they almost blew up the neighborhood a couple of times last week? And they even go to great lengths to hide these incidents internally. As a result, tribal learning is lost, as these events are not recorded and learned from unless they are terribly harrowing. And future events that could have been avoided are left to the fate of the operator's ability to diagnose from very little information or knowledge of previous occurrences. For instance, years ago in dealing with a large company who had an incident in Bhopal, India (just for example), he discovered new company policies in place. He was trying to gather plant data for empirical modeling efforts. Their established policy as a result of that incident was that they were ordered to destroy any process data more than 90 days old. This is not currently still the case, but it lasted for several years until process engineers were able to prove the historical value of that data was greater than the risk of it remaining available.

As one final point on this topic, Steve Apple has actually been in control rooms where operators were bypassing safety systems on start-up or shut down. In fact, they were purposely bypassing systems designed to protect against unit failure for limited periods of time because "that's how they had always done it." The catastrophe at Texas City, Texas (2005) is just one example of this sort of risky practice. The operators had traditionally allowed a level limit to be bypassed on starting up the unit. When the second sensor that told them it was time to reduce the flow to that unit failed, they were captured in a "point of no return" situation that they could not recover from. The extensive re-design and instrumentation that would have been necessary to avoid this situation was painful enough that engineers simply turned their heads and allowed this practice to continue. It was not an everyday event- just on startup and they assumed it had a backup sensor to prevent failure. Boom!

[10]According to Duncan A. Rowan, PE, a retired DuPont forensic investigator, the majority of the catastrophic accidents he investigated during his career were caused by simple single fault situations that were misinterpreted and improperly responded to by process operating personnel.

these measures are not always properly implemented nor adequately maintained. Even if they are, these measures alone still do not provide the operators with sufficient support in all emergency situations. Moreover, it is extremely doubtful that the measures guaranteed to provide such perfect support can ever be developed. Humans just have certain inherent limitations that cause their performance as process operators to be potentially unreliable.

One of these limitations is known as **"vigilance decrement."** Studies have shown that humans do not perform monitoring tasks very well. The number of things which go unnoticed increases the longer a human performs a given monitoring task (Eberts 1985). With process operators, this phenomenon results directly from fatigue and boredom associated with control room duties in modern production plants. Automation has left process operators with fewer control functions to perform. This leads to both greater job de-skilling and dissatisfaction. That in turn causes boredom which could lead to inattention. Since an inattentive operator will probably not have an accurate, up-to-date cognitive model of the present process state when confronted with an emergency situation, they may mistakenly base their decisions upon an inaccurate model. Studies have also shown that the quality of a decision depends upon the amount of time the decision maker has. In an emergency situation, an inattentive operator will usually be forced to gather data and make their decisions in less time than if they had been paying full attention. Both of these situations will increase the likelihood of human error. Counteracting this limitation requires a means for relentlessly monitoring and correctly determining the actual process state. Since the agent performing that surveillance would always be aware of this true state, such an agent would maximize the time available to the decision maker when process operating problems arose.

Another limitation of human operators is a phenomenon called **"mind set"** (Kletz 1985), which is also known as **"cognitive lockup"** and **"cognitive narrowing"** (Sheridan 1981), **"tunnel vision"** (Lees 1983), and **"the point of no return"** (Rasmussen 1981). Sometimes when an operator becomes sufficiently certain as to the cause of abnormal process behavior, they becomes exclusively committed to that particular hypothesis and acts upon it accordingly. This commitment continues regardless of any additional evidence that they receive which refutes that hypothesis or which makes alternative hypotheses more plausible. In most of the cases, this additional evidence is actively ignored by the operator until it is too late for him to initiate the proper corrective actions (Sheridan 1981). Moreover, the longer the operator observes that the response of the system is not as they would expect, the harder they tries to force it to be so (Sheridan 1981). Counteracting this limitation requires a means for examining all the available evidence in a rational, unbiased manner so that all plausible fault hypotheses consistent with that evidence can be derived. These hypotheses would have to be ranked according to how well they explained the observed process behavior, and this ranking would have to be updated as new evidence became available.

A third human limitation is the phenomenon known as **"cognitive overload."** Even when the detection of system failures is automatic, the sheer number of alarms in the first few minutes of a major process failure can bewilder the process operators (Sheridan 1981).[11] Rapid transition of the process state may also do this, especially if those operators have not experienced a similar situation and have not been told what to expect (Linhou 1981). Both situations greatly increase the levels of stress experienced by those operators (Fortin et al. 1983). Under stressful situations, humans lose information processing capability. A direct consequence of this loss is that the operator may not be able to quickly determine the true process state and formulate the appropriate corrective response (Dellner 1981). Counteracting this limitation requires a means for rapidly, rationally, and consistently correctly determining the true process state, regardless of how abnormal it is or how quickly it is changing. Such an analysis would thus help focus the operator's attention on the most likely causes of the observed process behavior, rather than having them attempt to imagine all of the possible causes of such behavior.[12]

---

[11]Lieven DuBois points out that since alarm management standards have been accepted (by regulators), the number of alarms occurring during abnormal situations have been steadily decreased by using different techniques. Some of these improvements are discussed in greater detail in Chapters 3–5.

[12]In support of the above, a major factor then in many, if not most, emergency situations is from the resulting information overload presented to the operator. Joe Alford contends that this overload originates directly from the way automation engineers configure the overall control system (i.e., not following best practices). Information overload has been cited as a factor in most of the major accidents mentioned earlier. The point here is that poor performance in dealing with emergency situations is often not the fault of the operator – as operators are often confronted with a myriad of alarms and alerts and other information messages, often not prioritized – so they have little chance of quickly sorting through the mountain of information, blinking lights, and blaring horns that occur when abnormal situations are occurring. ISA/ANSI standards such as 18.2 (Alarm Management) and 101 (Man Machine Interfaces) are tools that should be utilized by engineers to create a more optimal environment for helping operators manage emergency situations.

Thus, one of the unintended consequences of increased automation in processing plants is the ease and low cost (essentially free) of adding alarms, alerts, and information messages for operators. Consequently, control engineers end up just configuring lots and lots of them. So, when dealing with managing abnormal situations, more focus needs to be directed to the control engineers that are making a bad situation worse by the huge amount of information they have their control system designs generate during an incident, often including many nuisance alarms.

A fourth limitation of human operators is that the situation confronting them may require knowledge that is either beyond their ability to understand (Goff 1985), that is, outside the knowledge that they have gained from their experience and training (Kletz 1985), or that they have forgotten (Kletz 1985). Although operators are generally competent individuals, they typically do not fully understand the underlying fundamental principles involved in the process system's design and operation (Kletz 1985). Such knowledge is required so that the operators are more capable of flexible and analytical thought during emergency situations. This creates the somewhat paradoxical situation of the need for highly trained personnel to operate "automated" plants (Visick 1986). Counteracting this limitation requires a medium in which all pertinent information about both the process system's normal and abnormal operation can be permanently stored and quickly retrieved. It also requires a method for determining which of that information is relevant to the solution of the problem currently confronting the process operator.

The final human limitation is that, even in the best of situations, humans make errors. Despite efforts intended to reduce such errors, human errors can never be totally eliminated. Sheridan (1981) eloquently states the reason why:

> Human errors are woven into the fabric of human behavior, in that, while not intending to make any errors, people make implicit and explicit decisions, based upon what they have been taught and what they have experienced, which then determines error tendencies.

He adds:

> The results of the human error may be subsequent machine errors, or it may embarrass, fluster, frighten, or confuse the person so that he is more likely to make additional errors himself.

Counteracting this limitation requires a means for storing the correct solutions to operating problems confronted in the past, correctly classifying the current plant situation as one of those problems if applicable, and then instantiating the appropriate stored solution with the current process state information. This would enable all the proper analyses performed in the past to be efficiently reused in a systematic manner, thereby eliminating the need to recreate them each time they are required. It should also decrease the chances that the wrong analysis would be used or that the correct analysis would be used improperly.

1.5   THE ROLE OF AUTOMATED PROCESS FAULT ANALYSIS

Unfortunately, the various traditional measures currently being taken to help operators perform process fault management have not been able to provide them the support that they need to totally eliminate process accidents. Typically, these accidents have had very simple origins (Kletz 1985; Lieberman 1985). The reason that they still occur is because the number of possible process failures which need to be considered and the amount of process information which must be analyzed commonly exceed those that an operator can always effectively cope within emergency situations.

Furthermore, this situation probably cannot be counteracted by further additional investments in these traditional measures: many of them have already been exploited to nearly their full potential. Thus, in order to further improve process safety, additional process fault management methods need to be developed and successfully deployed to directly help address this problem.

As to be extensively elaborated further throughout the remainder of this treatment, the most attractive, but currently greatly underutilized, approach for helping the operators perform process fault management is to automate process fault analysis, that is, to automate the underlying reasoning required to determine the cause or causes of abnormal process behavior. Not surprisingly, various logically viable diagnostic strategies for automating fault diagnosis in chemical and nuclear process plants have been proposed for nearly as long as computers have been used in process control. However, for a variety of reasons discussed in the next chapter, at the present time, the potential of existing process control computers to analyze real-time process information for such purposes is still relatively unexploited by the CPI (Venkatasubramanian 2001).

Automated process fault analysis should be used to augment, not replace, human capabilities in process fault management. Consider the current relative strengths and weaknesses of computer analysis compared with human analysis. Computers can outperform humans in doing numerous, precise and rapid calculations, and in making associative and inferential judgments (Sheridan 1981). Computers furthermore have potentially unlimited and almost completely infallible memories compared to humans, and, furthermore, can be readily networked together to powerfully enhance their capabilities. They are also just starting to begin to "learn" in complex environments. On the other hand, humans are better at those functions which cannot be standardized. Humans readily exploit commonsense reasoning very effectively in decision making, whereas computers do not, beyond that which can be expressed as the heuristics human experts rely their analyses upon (Alford et al. 1999a,b). They are also better at decision making that has not been adequately formalized (i.e., creative thought). Humans likewise perform better pattern recognition in co-ordinations that involve the integration of a great many factors whose subtleties or non-quantifiable attributes defy computer implementation (Lefkowitz 1982). In order to ultimately achieve successful plant deployments, these various advantages in current computational capabilities always need to be kept in mind and more fully exploited when designing actual automated process fault analyzers.

Currently, the computer offers a means to rapidly analyze process information in a systematic and predetermined manner. If such analysis is already being done by the operators, automating it would free them to perform other functions. If it is not being done, it could be because the operators either do not have sufficient time or the capabilities required to perform it. In either case, properly automating such analysis should make the information reaching the operators more meaningful (De Heer 1987). Thus, the main advantage of deploying real-time, on-line process fault analysis is to dramatically reduce the cognitive load on the process operators (Laffey et al. 1988; DuBois et al. 2010). This would subsequently allow them to concentrate on those analyses which still require human judgments to perform.

As a final observation, it is possible that many of these current problems being encountered in achieving more effective process fault management could be counteracted by replacing operators with more highly trained process engineers.[13,14] The higher wages paid to these individuals would be

---

[13]Note: Joe Alford speculates one advantage of replacing operators with engineers is that it might finally force engineers to mend their ways and do a better job of alarm and HMI management. They would have to actually experience first-hand the results of their zealous configuration of so much information in process control computers. Another issue he observed in most of the plant automation audits he did over several years was that engineers did not inactivate alarms/alerts when they were not relevant, leading to many nuisance alarms and subsequent loss of respect of the alarm system by operators. So, again, it is great to highlight some of the inherent weaknesses of operators in dealing with process faults, but a significant part of the problem is how control engineers configure control systems.

[14]As Steve Apple further points out, one of the greatest problems with current HMI's is that most are designed by engineers with little consideration of human factors. The ultimate human factors today are video games. The average kid can pick one up and play with very little instruction. The average HMI actually being used on process units? – not so much. Most HMI's are designed to follow the P&ID's that were developed for the process. And most engineers understand them, and think they actually are instructional for operators who have very little understanding of thermodynamics, fluid flow, heat transfer, etc. So we as engineers design them for OUR understanding, expecting everybody to see what seems intuitively obvious to us. High performance situational awareness HMI's try to understand how humans collect and recognize data instantaneously. So, much of our goals in redesigning control systems is to actually REMOVE information that is not pertinent- both in the form of process data and alarms as well as fancy looking graphics that just take up real-estate. A perfect HMI would lead any user/operator to the same conclusion and ultimately the same resolution to the problem (so, unfortunately, there is yet no perfect HMI).

offset by their more efficient operation of the given process system. None-the-less, since engineers are subject to the same human limitations as process operators are, doing this would not eliminate the need to directly address the fore-mentioned currently existing problems that humans exhibit in performing highly effective process fault management. Therefore, it is our overwhelming contention that automating process fault analysis currently represents the best remaining relatively unexploited means presently available for directly addressing these problems head-on. Doing so should both bolster and compliment the process operators' innate capabilities at performing effective process fault management whenever required.

REFERENCES

Alford, J.S., Cairney, C., Higgs, R. et al. (1999a). Real rewards from artificial intelligence. *Intech* (April ed.) 52–55.

Alford, J.S., Cairney, C., Higgs, R. et al. (1999b). Online expert-system applications: use in fermentation plants. *Intech* (July ed.) 50–54.

Barclay, D.A. (1988). Protecting process safety interlocks. *Chemical Engineering Progress* 84 (2): 20–24.

De Heer, L.E. (1987). Plant scale process monitoring and control systems: eighteen years and counting. In: *Proceedings of the First International Conference on Foundations of Computer Aided Process Operations* (ed. G.V. Reklaitis and H.D. Spriggs), 33–66. New York: Elsevier Science Publishers Inc.

Dellner, W.J. (1981). The user's role in automated fault detection and system recovery. In: *Human Detection and Diagnosis of System Failures* (ed. J. Rasmussen and W.B. Rouse), 487–499. New York: Plenum.

DuBois, L., Foret, J., Mack, P., and Ryckaert, L. (2010). Advanced logic for alarm and event processing: methods to reduce cognitive load for control room operators. *IFAC Proceedings Volumes* 43 (13): 158–163.

Eberts, R.E. (1985). Cognitive skills and process control. *Chemical Engineering Progress* 81: 30–34.

Fortin, D.A., Rooney, T.B., and Bristol, H. (1983). Of christmas trees and sweaty palms. In: *Proceedings of the Ninth Annual Advanced Control Conference*, 49–54. West Lafayette: Indiana.

Goff, K.W. (1985). Artificial intelligence in process control. *Mechanical Engineering* (October ed.) 53–57.

Kletz, T.A. (1985). *What Went Wrong?, Case Histories of Process Plant Disasters*. Houston, TX: Gulf Publishing Co.

Kohan, D. (1984). The design of interlocks and alarms. *Chemical Engineering Magazine* (May ed.) 73–80.

Laffey, T.J., Cox, P.A., Schmidt, J.L. et al. (1988). Real-time knowledge based systems. *AI Magazine* 27.

Lees, F.P. (1981). Computer support for diagnostic tasks in the process industries. In: *Human Detection and Diagnosis of System Failures* (ed. J. Rasmussen and W. Rouse), 369–388. New York: Plenum Press.

Lees, F.P. (1983). Process computer alarm and disturbance analysis: review of the state of the art. *Computers and Chemical Engineering* 7 (6): 669–694.

Lefkowitz, I. (1982). Hierarchical control in large scale industrial systems. In: *Studies in Management Science and Systems*, vol. 7, 65–98. New York: North-Holland Publishing Co.

Lieberman, N.P. (1985). *Troubleshooting Process Operations, Tulsa.* Oklahoma: PennWell Publishing Co.

Linhou, D.A. (1981). Aiding process plant operators in fault finding and corrective action. In: *Human Detection and Diagnosis of System Failures* (ed. J. Rasmussen and W.B. Rouse), 501–522. New York: Plenum Press.

Rasmussen, J. (1981). Models of mental strategies in process plant diagnosis. In: *Human Detection and Diagnosis of System Failures* (ed. J. Rasmussen and W.B. Rouse), 251–258. New York: Plenum.

Rijnsdorp, J.E. (1986). The man-machine interface. *Chemistry and Industry* (May ed.) 304–309.

Sheridan, T.B. (1981). Understanding human error and aiding human diagnostic behaviour in nuclear power plants. In: *Human Detection and Diagnosis of System Failures* (ed. J. Rasmussen and W.B. Rouse), 19–35. New York: Plenum.

Syrbe, M. (1981). Automatic error detection and error recording of a distributed fault-tolerant process computer system. In: *Human Detection and Diagnosis of System Failures* (ed. J. Rasmussen and W.B. Rouse), 475–486. New York: Plenum.

Visick, D. (1986). Human operators and their role in automated plant. *Chemistry and Industry* 199–203.

Venkatasubramanian, V., "Process fault detection and diagnosis: past, present, and future, Proceedings of CHEMFAS4, Seoul, Korea, 2001, pp. 3–15.

2

VARIOUS PROCESS FAULT DIAGNOSTIC METHODOLOGIES

Richard J. Fickelscherer, PE

Department of Chemical and Biological Engineering, State University of New York at Buffalo, Buffalo, New York, USA

OVERVIEW

This chapter introduces each of the various process fault diagnostic methodologies that are further elaborated throughout this treatment. The current status of each of these different strategies, especially those which are actually being deployed within the processing industries, is overviewed here. The overriding emphasis of this overview has been placed upon describing those methodologies whose underlying algorithms directly utilize artificial intelligence (AI) programming techniques. The continuing improvement of this rapidly maturing technology is one major reason for such automation now being more readily adopted by industry. Also, other critical reasons for this adoption have been the simultaneous rapid development of measurement capabilities (e.g., smart digital devices) and computing power (e.g., cloud computing). As their actual performance relentlessly improves,

Artificial Intelligence in Process Fault Diagnosis: Methods for Plant Surveillance,
First Edition. Edited by Richard J. Fickelscherer.
© 2024 John Wiley & Sons, Inc. Published 2024 by John Wiley & Sons, Inc.

AI programs for automating process fault diagnosis are becoming dramatically more robust and competent. Subsequently, their adoption by industry is proving to be both more wide spread and cost effective, directly overcoming some of the previous major drawbacks to such programs' usages in actual process applications. Forthwith, their usages are now becoming more prevalent throughout the processing industries. This is occurring directly because their actual capabilities at improving process performance in daily operations are now being more fully realized.

CHAPTER HIGHLIGHTS

- Overviews of the various process diagnostic strategies that are currently being deployed by the processing industries.
- Emphasis on those particular strategies requiring or benefiting from artificial intelligence (AI) software programming algorithms and methodologies.
- Current state-of-the-art and optimized best programming practices of those corresponding AI software implementations.
- Elaborations directly facilitating better informed choices about the most appropriate diagnostic strategy selection for a given desired target process system application.

2.1 INTRODUCTION

Automating process fault diagnosis directly addresses a paramount industrial safety issue and also promotes more efficient process operations. Recent technology R&D advances, especially within the field of artificial intelligence (AI), makes such automation continuously more relevant throughout the entire processing industries. The foreseeable future only bodes well for the increasingly successful use and further advancement of these various real-time software technologies in actual industrial processes. Practical reasons for thus referring to this present treatment when creating actual process fault analyzers include:

(1) Proven technology on actual industrial process applications.
(2) Current state-of-the-art of the various viable AI-based diagnostic methodologies' advances.
(3) Comprehensive overview of optimized best current practices in applied process fault diagnosis.

(4) Easier comparison of alternative AI-based diagnostic strategies for determining the most effective one/hybrid combination for each specific industrial application.

Subsequently, our present treatment stresses results of actual successful applications of process fault diagnosis rather than just unproven developments in its underlying theory. It is therefore an attempt at a comprehensive elaboration of the current state-of-the-art and best practice for each of the various AI-based diagnostic methodologies/effective combinations now being deployed in industry. This ensures that it will be very relevant to those operations personnel currently trying to actually apply such technology.

2.2 VARIOUS ALTERNATIVE DIAGNOSTIC STRATEGIES OVERVIEW

The various diagnostic methodologies detailed in this treatment include many of the following possible logically viable strategies (after Mouzakitis 2013; Asif et al. 2020):

(1) Model based: (A) Quantitative—(i) analytical redundancy; (ii) parity space; (iii) state variable estimation and parameter estimation; (iv) diagnostic observers.

(2) Model based: (B) Qualitative—(i) abstraction hierarchy; (ii) fault trees; (iii) sign-directed graphs; (iv) Fuzzy systems.

(3) Hardware based: hardware redundancy; voting techniques; limit checking; frequency analysis.

(4) History based: Fuzzy logic; artificial neural networks; artificial immune systems; statistical methods: (i) principal component analysis (PCA); (ii) linear and multiple regression; (iii) polynomial regression; (iv) partial least squares regression (PLSR); expert systems; pattern recognition; alarm analysis.

As is plainly indicated here, a diverse variety of logically viable diagnostic strategies now exist for automating process fault diagnosis. Despite this situation, unfortunately, such automation is currently still not being widely used throughout the processing industries. In order to help rectify this present reality, active research and development efforts attempting to further improve these various diagnostic strategies are relentlessly advancing. This ever progressing R&D effort has increasingly centered on both improving and effectively combining together those various methodologies into more robust and generalized approaches. Our present treatment mainly emphasizes only those techniques actually proven as effective industrially implemented diagnostic approaches, and which also directly utilize mature, routine-specific AI programming techniques.[1,2]

[1]The following is Joe Alford's experience in actually deploying industrial AI technology. Forthwith, the limited use of these AI technologies is currently not all due to the present development state of these technologies themselves, but rather to the challenges of integrating these tools into existing automation systems and also validating and supporting the resulting multi-vendor systems. In some cases, the AI product comes with its own PLC computer and then the challenge is digitally interfacing this with a plant's Distributed Control System (DCS) and integrating the data collection, alarm management, and human-machine interface (HMI) with that of the DCS. Also, as with most multivendor systems, when one vendor upgrades their software (which often occurs on a roughly annual basis) the challenge to the customer is in retesting/revalidating the resulting overall system to make sure all applications still work together properly. This normally requires some down time of the automation system and can be time consuming and labor intensive to do. The holdup is therefore not the many user friendly AI and data analysis tools available on the market. It is rather getting them integrated with other computers into a larger real-time automation environment that is often the challenge. For example, a frequent challenge is that the data generation format of a given AI tool is often different than that of the plant historian- yet the goal is to have all process and lab data in a single data base for data mining and analysis purposes. Another common challenge is that most AI tools and control systems have their own short term data base to capture short term trends, alarm records for the current manufacturing lot, etc. However, comprehensive fault detection often requires real-time access to the plant historian which is usually an off-line long term repository of historical data. Integrating real-time access to all the relevant sources of data in determining an approaching or real fault can thus be extremely challenging.

[2]Steve Apple believes one of the BEST uses of process fault diagnostics not mentioned here is the ability to obtain early prediction of faults, thus avoiding them entirely. That has been the main goal of much of the AI in the past- predictive analytics. He has used expert systems, Logic Trees, Neural Nets, Bayesian nets, Chaotic Systems Theory, Fuzzy Logic- you name it, Steve has tried them all. So the idea with these for faults diagnosis is early fault prediction. This can be done by lining up the data from fault diagnosis and monitoring the variables that indicate path to failure. Or even path to process loss, etc. Depends on what you consider to be a "fault". The idea is to use reliable sensors to replicate what is normally relegated to lab analysis or unreliable sensors in the process. As an example, Bentley Nevada for years used the Gensysm G2 engine to build a system that took all their rotating equipment failure data and made predictive models that gave early warnings of rotating equipment failure. Several others he has seen had models that were based on hours of service, and other types of measurable parameters around known equipment.

As Joe Alford pointed out, these systems often suffer from being "different". i. e., different user inter face with its own learning, different console containing the system, etc. Additionally, this led to two damning factors. The first was repetitive information from different systems (e.g. "I can already see that on my control system"). The second was the problem of false positives. While a user can handle sometimes missing a prediction, they cannot handle when it makes a prediction that does not pan out. And consequently they will ignore it's predictions in the future- like the boy who cried wolf. Steve further believes Joe Alford is also spot on with his analysis of interfacial difficulties and maintenance thereof.

Overviews of the distinguishing logics of some of the various above diagnostic strategies and their possible effective combinations are briefly discussed below. Specifically, the particular strategies described here are: (1) fault tree analysis, (2) alarm analysis, (3) decision tables, (4) sign-directed graphs, (5) history-based statistical methods, (6) methods using qualitative models, (7) methods using quantitative models, (8) methods using artificial neural networks (ANNs), (9) methods using artificial immune system (AIS) strategies, and (10) methods using knowledge-based systems (KBSs). Most of current development statuses of these various strategies, along with their corresponding actual industrial employments, are then more fully elaborated throughout this treatment.

The appendix to this overview chapter briefly describes failure modes and effects analysis (FMEA), a commonly used procedure for identifying and numerically rating potential process faults and their direct consequences. FMEA is typically performed during its target process system's original design or retrofit planning phase. This normally allows for the complete preemptive elimination of many of these possible operational pitfalls before the commencement of on-line process production. Appendix 2.A was written by Dr. Joseph S. Alford, PE, currently an actual industrial FMEA facilitator.

2.2.1   Fault Tree Analysis

Fault tree analysis is a method that was developed in the early 1960s to analyze the safety of ICBM launch control systems. Currently, the method is widely used to analyze the safety of many complex systems, including those used in chemical and nuclear processing plants. Using Boolean logic, the technique describes the relationships that exist between the various possible basic events in a particular system (e.g., such as system component failures, human errors, etc.) and the final outcomes of those events (e.g., fires, explosions, etc.). The final outcomes are normally referred to as top events. Based upon the relationships and the probabilities associated with the occurrence of the various basic events (referred to as those events' failure rates), fault trees can be used to estimate the probability associated with the occurrence of each of these top events. Fault tree analysis is thus consequently more suited for analyzing process system designs for potential operational safety problems rather than for diagnosing process faults in real-time. Despite this basic mismatch in its underlying orientation, Teague and Powers (1978) were first to present an algorithm that combines a priori estimates of failure rates with real time process data in order to determine a quantitative basis for ordering the sequence in which potential fault hypotheses are examined. Getting accurate failure rate data however still remains a major limitation of the approach. Regardless, diagnostic strategies based upon fault tree analysis continue to be proposed (Ruijters and Stoelinga 2015), many being centered about Fuzzy logic reasoning techniques (Ren and Kong 2011).[3]

2.2.2 Alarm Analysis

Alarm analysis is multifaceted,[4] including the diagnoses of process faults by associating those fault situations with the corresponding pattern and order of occurrence of process alarms in the target process system. These relationships can be used to create structured representations called alarm trees. Prime cause alarms (i.e., those most closely associated with the actual process fault situations) are stored at lower levels in the resulting alarm tree representations, while the other "effect alarms" are stored at higher levels. These alarm trees thus establish the priority relationships that exist between the various process alarms. Their overall goal is to suppress irrelevant alarms, thereby helping the process operators to focus their attention only upon the most highly prioritized critical alarms. Alarm analysis, in part, thus attempts to intelligently reduce the amount of information that process operators have to interpret during major process upsets. It therefore directly addresses the problem of cognitive overload commonly experienced by them during the so-called associated alarm floods that frequently occur during emergency situations.

Chapter 3, written by Lieven DuBois, briefly discusses the current standards regarding optimal alarm system design strategies, their effective usage, and consequently the directly resulting improvements in overall process operations management from their application. It also outlines how advanced automated fault diagnosis should be effectively integrated into optimal alarm deployments.

[3]Steve Apple contends that the chemical processing industries replicate fault trees most normally in their PHA's (Process Hazards Analysis). These are required by OSHA 1910 for all potentially hazardous operations in the plant. PHA groups meet regularly to update and design PHA standards. These studies show the logical conclusion of any failures in the plant- what happens when this valve closes accidentally, or if it fails open, and a complete rundown to the natural conclusion of any hazardous situation that may come as a result of any control system activity in the plant. They estimate maximum pressures, relief valve settings, maximum temperatures, and then also what is released, or what explodes, and set up mitigation routes for any situation. As you might expect, PHA's often produce a whole new raft of requested alarms. Some of which may be valid and some which may not according to the rules of the alarm philosophy requirements of the plant.

[4]Reference ANSI/ISA 18.2, Management of Alarms for the Process Industries.

Chapter 4, written by Ron Besuijen, discusses the results of actual industrial studies in which effectively automating alarm analysis dramatically reduced the cognitive load of process operators during simulated fault situations. Furthermore, it details the improvement in operator training in complex procedures benefiting from the combination of high fidelity process simulations with operational AI programs performing emergency shut downs. This directly allows the operators to prominently make both better informed and timely decisions in their corresponding process systems' operations during actual fault situations. It further emphasizes the necessary interplay between AI automation and human capabilities, with the computer-aided process supervision directly supplementing and supporting those remaining human capabilities which cannot yet be effectively automated.

Alarm analysis thus directly supports effective alarm management by proving insight into the underlying process conditions affecting performance. Such analysis can help match abnormal conditions with the most likely pattern of process alarms in the target process system. In most cases, multiple possible patterns exist for each alarm under consideration. Many of the patterns are interactive and may have a natural hierarchy of process dependence. Prime cause alarms (i.e., those most closely associated with the actual process fault situations) are potentially more important to identify since they are more closely related to the root cause of the condition. Other "effect alarms" are related to the upset propagating, which results in secondary alarms.

Identification of secondary (i.e., effect) alarms that can potentially be suppressed is key to managing the flow of data. Effective data management directs the operator's attention more directly to the root cause condition. The overall goal is to suppress redundant, symptomatic, consequential, nuisance, or correlated alarms, thereby helping the process operators to focus their attention only upon the more critical alarms.

Alarm analysis consequently attempts to intelligently refine the alarm data into more actionable information that provides guidance during major process upsets. As discussed, this helps to directly reduce the cognitive overload commonly experienced by the process operators during many emergency situations.

Chapter 5 was originally meant to discuss the current state of alarm analysis including both traditional and emerging methods. It was to be written by Bridget Fitzpatrick. Regrettably, Bridget passed away before its completion. Briefly, the chapter would have presented an overview of process engineering, data analytics, advanced process modeling, and modern learning methods. These modern learning methods were to include both machine learning and AI. Methods to be reviewed were to include example data structures and presentation methods.

Traditional methods were to focus on:

- Use of first principles to determine likely root causes;
- Use of simple data analytics to identify redundant, symptomatic, consequential, or correlated alarms with heuristic thresholds;
- Use of data analytics to identify problem areas with chronic alarm shelving;
- Use of data analytics to identify problem areas with upsets related to transitions out of suppression and out of service states.

Emerging methods were to include:

- Use of advanced modeling to determine if remediation of the identified prime cause alarm scenarios will resolve the underlying issue;
- Use of automated machine learning to identify both likely cause and the common successful interventions from review of process data and operator actions;
- Use of deep reinforcement learning to match prime cause alarm scenarios with operating conditions and to identify process data segments without successful matching for further review.

We, along with Bridget's many fellow process control colleagues throughout the world, are all saddened by her sudden and untimely passing. She will be greatly missed. This book is dedicated in her memory.

Fortunately, Steve Apple graciously volunteered to attempt to create a substitute chapter based on the above outline of Bridget's proposed chapter. They were close colleagues for many years and intimately worked together on several major industrial process control projects. Forthwith, the following was his reply when requested to participate in this endeavor to complete Bridget's contribution to this treatment:

"Sorry we lost somebody so dear to us all. She was a special and brilliant person – a definite encourager and mentor to all those whose lives she touched. I will miss her greatly".

The first appendix to this chapter (Appendix 5A) details a typical example of the underlying logic, written as object-oriented program pseudo-code, required for performing the overall process state identification normally necessary specifically in both effective alarm analysis and process fault diagnosis. This pseudo-code was originally created by Richard Fickelscherer and Daniel Chester for an actual persulfate process system application on behalf of the FMC Corporation. Appendix 5.B details how state identification is now actually handled in our KBS shell, FALCONEER™IV.

2.2.3 Decision Tables

Decision tables (also commonly known as fault dictionaries) are derived from the cause–effect relationships that exist between process fault situations and observed process variables. For a given fault situation, each observed process variable is assigned a value of low, normal, or high depending upon how that fault situation affects it. Matching the resulting patterns of current process responses to those predicted thus indicates the possible underlying fault(s). The key advantages of decision tables are that they are relatively easy to derive, implement, and understand. The information that they contain is also highly structured, which facilitates the examination of every possible combination of patterns. In turn, the patterns represented within the tables can easily be converted into simple Boolean logic expressions for computer implementation.

The disadvantages of this method include: (1) it does not take into account the possibility of sensor failure or process noise; (2) it does not take into account unsteady state process operating conditions (such as the transients which almost always accompany a given severe process fault situation); and (3) the tolerance limits on the sensor variables may need to be different for the various possible fault situations. Another major disadvantage of this method, as is also true with alarm analysis, is that anything that even slightly alters the patterns of variable response could radically alter the resulting fault diagnosis (i.e., the method by itself is not very robust, even with Fuzzy set logic (Dash et al. 2003)). Furthermore, even if the patterns do not get distorted, the level of discrimination between the various possible process fault situations is typically not very good (i.e., the resulting fault hypotheses have low diagnostic resolution). Consequently, because of their limited usefulness, in actual applications decision tables have been coupled with other diagnostic strategies, chiefly KBSs.[5]

---

[5]Joe Alford observes that decision tables and many other fault diagnostic techniques normally require some form of data pre-treatment, i. e., data validation, before the applicable fault technique can be effectively applied. For example, data filtering is often used to minimize the "noise" component of data. If-then-else rules are then often used to flag (and eliminate) invalid data points. Such pre-treating is usually necessary because some small percentage of incoming data is invalid for any of a number of reasons. Flagging these data points (required for some real time applications) and eliminating them from the analysis is important to do to minimize some of the enumerated problems with decision tables.

2.2.4 Sign-Directed Graphs

Sign-directed graphs follow the same logic of decision tables, but are a more succinct and crisper method for representing all of the possible patterns that can result from each process fault situation (Kramer and Palowitch 1987). Nodes in these diagrams represent the observed process variables, while the arcs in them represent the causal relationships between those variables. The fundamental premise underlying this approach is that the origin node of a fault must be linked in the graph to all of the observed consequences of that fault. As with decision tables, the advantages of this method are that it is relatively easy to use, implement, and understand. Updated and improved algorithms for implementing these graphs are continuously becoming available (Liu et al. 2016), with this methodology now allowing multiple fault diagnosis (Yu and Lee 1991) and being extended to also perform PCA (Vedam and Venkatasubramian 1999).

2.2.5 History-Based Statistical Methods

History-based monitoring strategies in general are normally best suited for those processes for which explicit mathematical models are difficult to construct due to the lack of available information or sufficient knowledge pertaining to the target process (Das et al. 2012). PCA is one such dimensional reduction technique and has been used extensively in process fault diagnosis since the early 1990s. PCA is directly suited to handle noisy and ill-conditioned data sets, as is also PLSR. Advantages of using these techniques, along with their improvements and extensions, are still being very actively investigated (Gertler et al. 1999; Chiang et al. 2000a,b; Russel et al. 2000; Qin 2003, 2012; Isom and LaBarre 2011; Dunia et al. 2013; Wu et al. 2014).

Chapter 6 discusses the current state of these various statistical approaches and their possible effective extensions. Specifically, one such exciting extension – time-explicit Kiviat diagrams (Wang et al. 2015) – has recently been improved by Shu Xu, Mark Nixon, Ray Wang, and Michael Baldea. These diagrams place axes (the principal components from PCA) radially around a center point and data samples are rendered as closed polygons. "Normal operating regions" are calculated based on historical data and then used for fault detection. This clever and original approach depicts process data in a lower dimensional space to facilitate visualization and fault detection for all three possible types of chemical processes: continuous, batch, and periodic. In addition, the case studies described here demonstrate the superior performance of the proposed approach in quickly detecting faults while maintaining low type I/II error rates.

The appendix to this chapter (i.e., Appendix 6.A), written by Richard Fickelscherer and Daniel Chester, presents a very flexible and rigorous strategy for continuously performing on-line virtual statistical process control (SPC). Such SPC directly compliments standard process fault diagnosis by establishing which process sensors or calculated process performance parameters are currently under control or not. Its usage was included as an optional configuration choice for any of the various process variables actively being monitored in any actual FALCONEER™ IV applications (Fickelscherer and Chester 2013). Using Virtual SPC process monitoring directly allows for tighter enforced operational targets to be more readily and immediately achieved by process personnel.

2.2.6   Diagnostic Strategies Based upon Qualitative Models

A diverse variety of diagnostic strategies based upon qualitative models of dynamic process behavior have been suggested, being extended to PCA (Maurya et al. 2005) and methodologies to automatically create these models from sign-directed graphs (Wang et al. 2005; Gao et al. 2010). Such qualitative differential equations are called confluences and lie at the center of a reasoning approach called qualitative physics. Correctly describing physical systems with confluences is a procedure known as envisionment. Although the solutions generated by confluences are inherently ambiguous, Forbus (1987) suggests that such descriptions more closely follow the reasoning employed by expert problem solvers and thus lead to reasoning strategies that are easier for non-experts to follow. This assertion was supported by actual knowledge engineering experiments conducted with DuPont process experts during the original FALCON KBS Project (Fickelscherer 1990).

As indicated above, the largest inherently debilitating problem with employing confluences to describe process systems is that such modeling eliminates too much useful diagnostic information, thereby causing spurious diagnoses to be generated. In order to try to eliminate these incorrect results, Kramer and Oyeleye (1987) were first to suggest a method for combining confluences with sign-directed graphs. Combining qualitative model diagnostic methods with Order of Magnitude Reasoning strategies have also been suggested by many authors as a means to help reduce the inherent ambiguity of the various diagnostic methodologies based upon qualitative modeling.

Chapter 7 discusses the current state of one particular diagnostic approach employing qualitative models. It was written by Rajan Rathinasabapathy, Michael J. Elsass and James F. Davis. Briefly, Chapter 7 introduces smart manufacturing (SM). SM is the strategic investment in people, technology, and practice that enables manufacturers to extract significantly increased value from their existing assets and resources. Assessing operational health and taking action early is a proactive functional capability in SM. It has particular relevance and value when applied to process lines in which causes of problem behaviors and their manifestations can be masked or separated in time and location because of integrated effects. They call this Process Health and Early Detection. Significant attention has been paid to the health and maintenance of the individual devices that make up a process operation.

In this chapter, they develop and demonstrate how SM, AI, data, and modeling are integrated for explainable, normal and abnormal unit, and line operational assessments. A key development is an intermediate flow sheet based diagnostic, called *diagnostic localization* that allows earlier actions to be taken before fault root causes are fully discernable or manifest. These capabilities are associated with qualitative modeling, symptom and topographical matching, and search with time and change discretized. Additionally, they address ways to discern when the model itself is incomplete. Why various modeling approaches are selected and how they are integrated to meet these requirements for human-in-the-loop diagnostic localization are illustrated with a test bed application implemented on an Ethylene plant, focusing on how an algorithm called Causal Link Assessment (CLA) executes. They summarize the CLA algorithm here and also include several examples to provide additional detail.

2.2.7 Diagnostic Strategies Based upon Quantitative Models

Overcoming the diagnostic ambiguity inherent in all of the strategies discussed previously requires additional diagnostic information. For process fault analysis, such information is usually available in the form of quantitative input/output, parity equations or state space models derived from the first principles underlying the target process system's normal operating behavior. In general, this so-called deep knowledge of the process system not only improves the fault analyzer's ability to better discriminate between the various possible fault situations (i.e., better possible diagnostic resolution), it also typically increases the sensitivity at which the fault analyzer can diagnose each of those fault situations. The key feature common to most diagnostic strategies based upon quantitative models is that they use redundant sources of information (i.e., analytical redundancy) to detect discrepancies. This redundancy typically arises from having both calculated and measured values of given process variables available. Residual differences in the quantitative models' evaluations with real-time process data directly indicate the presence of process fault situations.

In general, the major drawback with basing diagnostic strategies upon quantitative models lies in the fact that enough sufficiently accurate models are sometimes difficult, if not impossible, to derive.[6] Such models also require greater real-time computer resources to solve and analyze.

Many quantitative diagnostic strategies are based upon estimation of state variables and unmeasured process parameters, parity equations, and diagnostic observers. These techniques use accurate models to estimate process variables and model parameters directly from measured process variables. These estimates are then compared via statistical tests to reference values obtained under normal operating conditions. Significant discrepancies in these comparisons indicate the presence of process faults. Park and Himmelblau (1983) and King and Gilles (1986) were first to describe methods for extending these techniques to non-linear models. Fathi et al. (1993a,b) extended these techniques to KBS approaches. The other advances in these techniques have been surveyed by many authors (e.g., Das et al. 2012). However, it is doubtful these techniques will ever be used to diagnose highly disruptive process faults (e.g., pump failures, etc.), or to monitor process operation during process startups, emergency process shutdowns, etc.

Many other alternative quantitative diagnostic strategies also rely upon the availability of analytically redundant information. A method for detecting measurement errors and process leaks through redundancy classification was first presented by Mah and Stanley (1976). Stephanopoulos and Romagnoli (1980) also used these principles to develop a serial elimination algorithm for diagnosing measurement errors. Kretsovalis and Mah (1987) presented a more general algorithm for performing redundancy classification. Gertler and Singer (1985) also developed a method based upon such redundant information that ensures that the resulting unique fault diagnoses will always be correct. However, this method does not optimize the diagnostic sensitivity of the resulting fault analyzer for each possible fault situation, nor does it handle multiple fault situations. Other strategies for incorporating quantitative models into process fault diagnosis have been presented by many authors (Fickelscherer and Chester 2013, 2016). Qin and Li (2001) have extended these techniques by calculating Exponentially Weighted Moving Averages (EWMAs) on errors in the models' residuals and signals with the underlying fault magnitudes being estimated directly

---

[6]Sometimes this directly results from not having sufficient available process sensor measurements currently being taken in the target process system. Prudently and judiciously adding strategic unique and/or redundant sensors can sometimes directly rectify this model shortage [Fickelscherer & Chester 2013].

from those residuals. Lastly, formal theories of fault diagnosis based upon first principle based models have been previously proposed by both Kramer (1987) and Reiter (1987).

If a sufficient number of accurate first principle quantitative models of normal process operation can be derived for a given target process, we contend that the Method of Minimal Evidence (MOME) Fuzzy logic diagnostic strategy as codified in our FALCONEER™ IV computer program is the best means for accomplishing optimal process fault diagnosis (Fickelscherer and Chester 2013). This claim is based upon the extensive research and development we have carried out while trying to automate process fault analysis in many real-world applications. Evaluating and analyzing accurate quantitative models of normal process operation simply just provides the most unimpeachable source of process knowledge from which to vigilantly monitor current process operations. This judgment has been repeatedly confirmed by the demonstrated highly competent performance of the resulting FALCONEER™ IV programs in many actual complex process applications.

Chapter 8, written by Richard Fickelscherer and Daniel Chester, fully describes our optimized quantitative model-based diagnostic methodology and its associated patented Fuzzy logic software algorithm (Chester et al. 2008).[7] This algorithm directly simplifies the very difficult problem of automated process fault diagnosis into the much easier, and incrementally solvable, problem of process modeling. It thus allows anyone capable of doing such modeling leverage of the same process knowledge required to originally design a given target process to then further intelligently operate that process. The currently required effort to successfully accomplish this feat has now been substantially reduced by over two orders of magnitude compared to past similar attempts at automating process fault analysis, with even better results to date. Specifically, where applicable, fully operational and highly competent process fault analyzers require less than a person-month to create and validate for most moderately complex process systems. It consequently directly allows fundamental process knowledge of normal operation to be continuously utilized whenever the process is operating, with the program immediately detecting and diagnosing any abnormal process operating conditions and then sounding timely intelligent alerts to the appropriate process operators and engineers.

The resulting intelligent process auditing and supervision can consequently directly further reduce operating costs and substantially help dramatically improve both the safety and efficiency of overall process production.

---

[7]The actual Visual Basic program pseudo-code implementing this patented algorithm is presented in Appendix 8.A of this chapter.

Process data already currently being actively monitored can thus be more completely analyzed to further extract all the pertinent and timely process diagnostic information contained within that data. This refined diagnostic information can therefore potentially help the process operators to more fully optimize, in real-time, current process operations throughout all of the processing industries, well beyond their current unmonitored levels, for any of those particular process systems which can be adequately modeled in sufficient detail. This chapter's underlying diagnostic strategy and associated patented algorithm thus directly demonstrates yet another facet of the possible benefits from the use of quantitative process modeling for optimizing a given process operations' safety and efficiency. Some of these benefits are summarized in Appendix 8.B.

2.2.8 Artificial Neural Network Strategies

Good discussions of ANNs and summaries of their early application to process fault analysis are given by both Himmelblau (1992) and Venkatasubramian and Chan (1989). ANN methods originally evolved in the field of AI. Briefly, ANNs are trained to identify various process faults by altering the connection weights within the neural network via recursive back-propagation calculations. They are advantageous for process fault analysis because an a priori model of the target process operation is not required. Watanabe et al. (1989) demonstrated competent performance with a two tiered neural network, the first level identifying the fault while the second estimates its magnitude. Kramer (1991) was first to extend ANN methods to perform non-linear PCA.

A major problem limiting application of ANN's is that adequate training data (i.e., data collected and/or accurately simulated during each particular target fault situation at a sufficient variety of possible fault magnitudes/rates of occurrence) is typically unavailable for all the faults of interest. Furthermore, even if such sufficient fault data is available, supervised ANN training is computationally intensive[8] and highly dependent upon the underlying organization of the number of nodes and number of node layers utilized.[9] If properly constructed and trained, however, such networks are

---

[8]Shu Xu believes data availability and computational power should not be an issue nowadays thanks to the development of pervasive sensing and cloud-computing technology.

[9]The results of these training calculations produce "black box," non-linear models which, due to their hidden layers, cannot offer explanations of their results (Angeli & Chatzinikolaou 2004). Developing effective ANN configurations is widely described as a "Black Art" (Russel & Norwig 1995).

capable of effectively interpolating current patterns in the process data to correctly identify fault hypotheses within their domain.[10,11,12]

Chapter 9 discusses the current state of ANN applications in process fault diagnosis. It was written by Lieven DuBois and Philippe Mack.

[10]Joe Alford has had significant success at Eli Lilly in using neural nets to model process behavior and faults, using validated historical data and these were "quantitative" models, albeit "black box" models. That is, a user inputs all the available historical data as inputs to the neural net and then lets the automated model fitting algorithms within the tool determine (i.e., learn) which inputs are important and which ones are not, with the technique also computing the weighting factors for each input of each node in the net. This results in a quantitative model which can be configured/programmed into a real-time process control environment for real-time use. Assuming sufficient historical data is available, existing neural net products (e.g., Neural Ware, Process Insights) allow the development of a quantitative model (whether linear or non-linear) to be fairly quick and easy. The technique also has a huge benefit in quickly telling a user if the input data available is able to quantitatively predict the output variable of interest. If so, the user can use any of a number of techniques to create the model, including perhaps first principles- or they can choose to stay with a neural net representation of the model. If a cause-effect relationship is not revealed with the initial attempt at a model using a neural net technique, the user saves all the time that would otherwise be wasted in trying to create a first principles or other form of model. That is, the available input data is not able to be quantitatively predictive of the output variable of interest: this can otherwise take a lot of time to conclude if using various versions of first principle equations to attempt the model.

[11]Joe Alford further notes that ANNs are not advertized for or good at extrapolating. They are thus only reliable at interpolating within the boundaries of the data sets in which they were trained. Consequently, they cannot be expected to accurately predict an output when the input data they are using is outside the envelope of training data.

[12]Steve Apple's response to Joe Alford's above ANN's comments. ANN's can also suffer from interpolation issues depending on the sparsity of data in the region where the algorithm was trained. In regions within the overall data envelope where there is little data available, they can sometimes over-predict a result that might not be reasonable if more data were available. These issues can be somewhat solved by a couple of methods. Mark A. Kramer showed in the late 1990's that ANN's can be "bound" by the inclusion of PLS (Partial Least Squares) analysis. In other words, using the combination of ANN and PLS (or even linear or least squares) models describes areas of both internal data sparsity, and extrapolation. Whereas the ANN tends to contain a lot of non-linearity once it exits areas of data clouds, PLS (or even linear or least squares) tends to bound the solution to a directed line to the next data cloud, or even outside the data region.

The second commonly used method of bounding the solution is by the inclusion/insertion of data from first principles models. To use this method, regions of sparse process (empirical) data are supplemented by the "filling in" of simulated data for those process conditions by running case studies in a properly tested and certified first principles model. Of course, first principle models are not known for their match to actual plant data, so some sort of data reconciliation is necessary to align the two models. This method has been proven successful for both interpolation and extrapolation in practice. An example is in polymer plants where some empirical data is just not reasonable to collect, as it would ruin the finished product. So, you must fill it in. In the past, he used Harmon Ray's reactor models to augment data with good results, but this was not extended for performing failure analysis yet.

2.2.9 Artificial Immune System Strategies

Artificial immune system (AIS) strategies, as with ANN's, also forgo the need to have detailed process models. Again, then, AIS strategies also require having enough actual and/or accurately simulated fault data to create sufficient "anti-bodies" to detect and diagnose their targeted faults (Ghosh and Srinivasan 2011; Dai and Zhao 2011). They are also, again, highly computationally complex and resource intensive. Zhoa et al. (2014) have developed methods to extend AIS to batch processing and combine AIS with ANN's to more efficiently monitor fault-free steady state process operations. To address insufficient specific process fault data, Shu and Zhao (2016) have leveraged process fault data from similar processes to create "vaccines" which extend the fault set covered in the target process system. Unlike supervised ANN's however, AIS strategies are structured to allow them to independently learn of new "anti-bodies" when confronted with novel faults (i.e., perform unsupervised learning to detect and diagnose such unaccounted for novel faults outside the original training data).

2.2.10 Knowledge-Based System Strategies

The knowledge-based system (also commonly referred to as the expert system) approach to problem solving is a programming paradigm that also originally evolved within the field of AI. In reality, this approach encompasses a vast collection of diverse problem-solving strategies.[13] The common characteristic uniting all of these various KBS strategies is that, when it comes to problem solving, each unequivocally stresses the dominance of domain-specific knowledge over the particular problem-solving algorithm being employed. This programming strategy has found wide spread support because it has been successfully applied to solve a variety of extremely difficult problems in many diverse application domains which previously could only be solved by human experts.

The biggest advantage of KBSs in process fault diagnosis applications lies in their unlimited flexibility when it comes to combining domain knowledge obtained from diverse sources, especially the experiential heuristic knowledge commonly used by domain experts. Many specific successful strategies for

---

[13]Possible Knowledge Representation constructs in KBS applications include: frames, objects, production rules, Semantic Networks, lists, scripts, trees, graphs, Fuzzy sets, first-order logic, Horn clauses and functions; various possible solution inference search strategies include: data driven, goal driven, hybrid, depth first, breadth first, heuristic, hill climbing, recursion, resolution, blind and exhaustive; possible uncertainty calculations include: certainty factors, probability, Demster-Shafer Theory, Dynamic Baysian Networks, Boolean logic and Fuzzy logic.

combining the various sources of the domain knowledge used in fault diagnosis can be found throughout the open literature. Details on the performance of diagnostic knowledge base systems developed for actual process system applications are also abundantly available there (e.g., Alford et al. 1999a,b).[14]

Chapter 10 discusses the current state of KBS applications to process fault diagnosis. It was also written by Lieven DuBois and Philippe Mack. Its Appendix 10.A highlights an actual diagnostic KBS application Lieven DuBois created for compressor trip prediction.

For completeness, Chapter 11 recounts many of the lessons learned during the very first real-world industrial KBS application project in process fault analysis, also known as the FALCON Project. This project was a joint venture in the mid-1980s between the University of Delaware, DuPont and the Foxboro Company. It developed the original **FALCON** (Fault AnaLysis CONsultant) System, a pioneering real-time, on-line KBS. FALCON very successfully performed process fault detection and diagnosis in an actual Adipic Acid plant then owned and operated by DuPont in Victoria, Texas. This experience led to the formulation of the generalized quantitative model-based diagnostic strategy called the Method of Minimal Evidence (MOME) (Fickelscherer 1990, 1994).

The FMC Corporation later on hired Richard Fickelscherer as a consultant to create another hand-compiled process fault analyzer directly based on MOME, called FALCONEER (FALCON via Engineering Equation Residuals), for their then Electrolytic Persulfate plant in Tonawanda, New York (Skotte et al. 2001, Fickelscherer et al. 2003). After this KBS application also proved successful, he collaborated with Daniel Chester to help him convert the MOME diagnostic strategy into an easy to use, standardized Fuzzy logic based compiler program, called FALCONEER™ IV (Fickelscherer et al. 2005, Fickelscherer and Chester 2013). This Fuzzy logic algorithm incorporating the MOME diagnostic strategy substantially reduces the total effort required to effectively both create and maintain competent process fault analyzers and has since been patented (Chester et al. 2008). Chapter 8 and Appendices 8.A and 8.B describe this algorithm

---

[14]These articles describe how Eli Lilly developed their real-time Expert System extensions (using Gensym's G2 real-time expert system shell) of their fermentation manufacturing plant automation systems. Lilly's on-line KBS applications in their fermentation pilot and manufacturing plants were the first such applications within the entire pharmaceutical industry; among other things they also performed real-time fault detection and diagnosis. The Lilly KBSs received national awards from both AIChE and ISA. It is also very noteworthy that such systems fell under the umbrella of FDA scrutiny and consequently had to be formally validated. Doing such (as required by cGMPS) when applying unproven AI technology for plants making human medicines was not a trivial task.

in sufficient detail so that others can also independently create such optimal automated process fault analyzers for themselves.[15]

2.2.11 Role of the Process Operators in Automated Fault Detection and Diagnosis

The ultimate end users of fault analysis software are the actual process operators.[16] Their acceptance and trust is paramount in both a resultantly successful and sustainable fault analyzer application. Consequently, their associated human needs have to be understood and properly addressed directly by these programs.

---

[15]The overriding goal of each chapter in this treatment is to also make it possible to do likewise for its particular alternative diagnostic strategy or effective combination of strategies.

[16]One last general observation made by Steve Apple. While he was at Gensym, they introduced thousands of systems around the world with Expert Systems and AI deployed. This work directly generated thousands of papers that were delivered at societies all over the globe directly identifying millions of dollars in savings with successive projects. And yet- a few years after installation, he would go back to visit those sites- even the ones with the largest successes- and discover the project was no longer active. They only stayed installed in sites where they were the only answer for the problem. The key problem was that the system was different. It had a different GUI, it was on a different server, and it needed a full-time caretaker to keep it maintained and running. He encountered the same results with Neural net/fuzzy logic/etc. systems he installed with Pavilion. After a couple of years, the keyboard was dusty, except in installations where the solution was integrated as part of the control system.

One operator explained it to him quite well. They stated: "I am expected to use this system here- a Honeywell DCS- and I am expected to mind it 24 hours a day. It gives me alarms and shows me trends. To use the "smart" system, I have to go to another terminal, sign in, remember how to use it, and get answers that are sometimes questionable. Lots of the stuff it tells me repeat things that I could already ascertain from the existing control system, so it was redundant to what I already knew. And after working with it for a while, I sort of began to understand what it would tell me anyway. And when it lost connection- nobody had the knowledge of how to fix it: that guy had left the company."

A friend told Steve once- a man would rather have a problem he cannot solve than a solution he does not understand. So, unless you can keep a highly educated person on site at a facility, say, in Port Arthur, Texas, and pay him well enough to stick around, it's difficult to realize sustainability for these sorts of applications, unless, there is some sort of corporate support, or ongoing service contract with the solution provider. The path to future sustainability therefore lies in integration with the existing control system, and making information available to the operator as a pop-up or the likes. Anything outside this control system suffers the same problems. And it seems that lately cyber security issues are throwing an additional wrench in the works. Cross-sharing of information and external access are getting to be extremely difficult. The problem now is that everything is Microsoft; i.e., it used to be they were all proprietary systems- which afforded a high level of security just due to the obscure nature of the data coding and encryption.

Consequently, what was once a difficult interface between disparate collection data systems is now greatly simplified. But the access to it is blocked for that very reason. Don't get Steve wrong- he continues to try to push these solutions- but he's currently just hitting a different kind of headwind than previously encountered.

These application's original implementation and long-term sustainability thus requires early and comprehensive continuing efforts by their developers. Chapter 12, written by Perry Nordh, PE, addresses these issues directly. While Chapter 13, written by Rajan Rathinasabapathy, Atique Malik, and Richard Fickelscherer, discusses some of these issues as they pertain to the more encompassing field of actual industrial applications of advanced process control.

2.3 DIAGNOSTIC METHODOLOGY CHOICE CONCLUSIONS

The common theme underlying all of these various viable AI-based diagnostic methodologies and their effective combinations is that they try to identify process faults from patterns of diagnostic evidence (i.e., symptoms) that occur during those faults. All are each logically sound and thus potentially valid strategies for performing such pattern matching. However, even if implemented perfectly, their various specific methodology limitations regarding possible diagnostic resolution and diagnostic sensitivity hinder their usefulness in many actual process applications: there is just no universal applicability inherent in any particular diagnostic strategy.

This situation has led to the continuing search for the most optimal diagnostic methodology/effective combinations, in terms of both actual correct performance and overall cost effectiveness, so that automating process fault diagnosis will become more commonplace throughout the processing industries. The current clear consensus amongst most active researchers in this field is that hybrid combinations of the various diagnostic strategies actually provide the most promising avenues for achieving this (Das et al. 2012): that is, unfortunately no single particular diagnostic methodology is both practical and effective for every potential industrial application. In reality, each such application has its own set of difficult challenges to be overcome by specifying and utilizing the most appropriate hybrid diagnostic strategy. The overriding goal of this book, consequently, is to help developers identify that best combination and then further guide them in successfully implementing it for their particular target application.[17]

---

[17]As mentioned in an earlier comment, Joe Alford contends a major reason that automated fault analysis techniques are not more widely used is due to challenges in interfacing and integrating such tools and techniques initially into real time automation systems, and then validating and supporting such multi-vendor systems. When used in real-time, other challenges also exist with fault diagnosis techniques such as the need for data validation algorithms – since incoming data needs to be automatically checked as valid before being sent to algorithms - which is a task that has normally historically done off-line, sometimes manually, before applying a data analysis tool. Getting all the various sources of data (process, lab, historian, etc.) in a compatible format is also often still quite a challenge.

REFERENCES

Alford, J.S., Cairney, C., Higgs, R. et al. (1999a). Real rewards from artificial intelligence. *Intech* (April) 52–55.

Alford, J.S., Cairney, C., Higgs, R. et al. (1999b). Online expert-system applications: use in fermentation plants. *Intech* (July) 50–54.

Angeli, C. and Chatzinikolaou, A. (2004). On-line fault detection techniques for technical systems: a survey. *International Journal of Computer Science & Applications* 1 (1): 12–30.

Asif, A.M., Liu, Y., Lakhan, M.N. et al. (2020). Recent advances in nuclear power plant for fault detection and diagnosis. *Journal of Critical Reviews* 7 (18): 4340–4349.

Chester, D. L., L. Daniels, R. J. Fickelscherer, and D. H. Lenz, United States Patent No.: US 7,451,003, "Method and System of Monitoring, Sensor Validation and Predictive Fault Analysis", 2008.

Chiang, L.H., Russel, E.L., and Braatz, R.D. (2000a). Fault diagnosis in chemical processes using fisher discriminant analysis, discriminant partial least squares, and principal component analysis. *Chemometrics and Intelligent Laboratory Systems* 50: 243–252.

Chiang, L.H., Russel, E.L., and Braatz, R.D. (2000b). *Fault detection and diagnosis in industrial systems*. Springer Science & Business Media.

Dai, Y. and Zhao, J. (2011). Fault diagnosis of batch chemical processes using a dynamic time warping (DTW)-based artificial immune system. *Industrial & Engineering Chemistry Research* 50: 4534–4544.

Das, A., Maiti, J., and Banerjee, R.N. (2012). Process monitoring and fault detection strategies: a review. *International Journal of Quality & Reliability Management* 29 (7): 720–752.

Dash, S., Rengaswamy, R., and Venkatasubramanian, V. (2003). Fuzzy-logic based trend classification for fault diagnosis of chemical processes. *Computers and Chemical Engineering* 27: 347–362.

Dunia, R., Edgar, T.F., and Nixon, M. (2013). Process monitoring using principal components in parallel coordinates. *AIChE Journal* 59 (2): 445–456.

Fathi, Z., Ramirez, W.F., and Korbicz, J. (1993a). Analytical and knowledge-based redundancy for fault diagnosis in process plants. *AIChE Journal* 39 (1): 42–56.

Fathi, Z., Ramirez, W.F., Tavares, A.P. et al. (1993b). Symbolic reasoning and quantitative analysis for fault detection and isolation in process plants. *Engineering Applications of Artificial Intelligence* 6 (3): 203–218.

Fickelscherer, R. J., Automated Process Fault Analysis, Ph. D. Thesis, Department of Chemical Engineering, University of Delaware, Newark, DE., 1990.

Fickelscherer, R.J. (1994). A generalized approach to model-based process fault analysis. In: *Proceedings of the 2nd International Conference on Foundations of Computer-Aided Process Operations* (ed. D.W.T. Rippin, J.C. Hale, and J.F. Davis), 451–456. Austin, TX: CACHE.

Fickelscherer, R.J. and Chester, D.L. (2013). *Optimal Automated Process Fault Analysis*. New York: AIChE/John Wiley and Sons, Inc.

Fickelscherer, R.J. and Chester, D.L. (2016). Automated quantitative model-based diagnostic protocol via assumption state differences. *Computers and Chemical Engineering* 90: 94–110.

Fickelscherer, R. J., D. H. Lenz, and D. L. Chester, "Intelligent Process Supervision via Automated Data Validation and Fault Analysis: Results of Actual CPI Applications," Paper 115d, AIChE Spring National Meeting, New Orleans, LA, 2003.

Fickelscherer, R.J., Lenz, D.H., and Chester, D.L. (2005). Fuzzy logic clarifies operations. *InTech* (October) 53–57.

Forbus, K.D. (1987). Intelligent Computer-Aided Engineering. In: *AAAI Workshop on Artificial Intelligence in Process Engineering*. New York, NY: Columbia University.

Gao, D., Wu, C., Zhang, B., and Ma, X. (2010). Sign directed graph and qualitative trend analysis based fault diagnosis in chemical industry. *Chinese Journal of Chemical Engineering* 18 (2): 265–276.

Gertler, J., and D. Singer, "Augmented models for statistical fault isolation in complex dynamic systems," American Control Conference, Boston, MA, 1985, pp. 317–322.

Gertler, J., Li, W., Huang, Y., and McAvoy, T. (1999). Isolation enhanced principal component analysis. *AIChE Journal* 45 (2): 323–334.

Ghosh, K. and Srinivasan, R. (2011). Immune-system-inspired approach to process monitoring and fault diagnosis. *Industrial and Engineering Chemistry Research* 50: 1637–1651.

Himmelblau, D.M. (1992). Use of artificial neural networks to monitor faults and for troubleshooting in the process industries. In: *Proceedings of the On-Line Fault Detection and Supervision in the Chemical Process Industries* (ed. P.S. Dhurjati), 144–149. Newark, Delaware: IFAC Symposium.

Isom, J.D. and LaBarre, R.E. (2011). Process fault detection, isolation, and reconstruction by principal component pursuit. *American Control Conference San Francisco, CA* 2011: 238–243.

King, R., and E. D. Gilles, "Early Detection of Hazardous States in Chemical Reactors," IFAC Kyoto Workshop on Fault Detection and Safety in Chemical Plants, Kyoto, Japan, 1986, pp. 137–143.

Kramer, M.A. (1987). Malfunction diagnosis using quantitative models and non-Boolean reasoning in expert systems. *AIChE Journal* 33 (1): 130–147.

Kramer, M.A. (1991). Nonlinear principal component analysis using auto-associative neural networks. *AIChE Journal* 37 (2): 233–243.

Kramer, M. A., and O. O. Oyeleye, "Qualitative Simulation of Chemical Process Plants," IFAC Tenth World Congress, Munich, Germany, 1987.

Kramer, M.A. and Palowitch, B.L. (1987). A rule-based approach to fault diagnosis using the signed directed graph. *AIChE Journal* 33 (7): 1067–1078.

Kretsovalis, A. and Mah, R.S.H. (1987). Observability and redundancy classification in multicomponent process networks. *AIChE Journal* 33 (1): 70–82.

Liu, Y., Wu, G., Xie, C. et al. (2016). A fault diagnosis method based on sign directed graph and matrix for nuclear power plants. *Nuclear Engineering and Design* 297: 166–174.

Mah, R.S., Stanley, G.M., and Downing, D.M. (1976). Reconciliation and rectification of process flow and inventory data. *Industrial and Engineering Chemistry Process Design and Development* 15 (1): 175–183.

Maurya, M., Rengaswamy, R., and Venkatasubramanian, V. (2005). Fault diagnosis by qualitative trend analysis of the principal components. *Chemical Engineering Research and Design* 89 (9): 1122–1132.

Mouzakitis, A. (2013). Classification of fault diagnostic methods for control systems. *Measurement and Control* 46 (10): 1–15.

Park, S. and Himmelblau, D.M. (1983). Fault detection and diagnosis via parameter estimation in lumped dynamic systems. *Industrial and Engineering Chemistry Process Design and Development* 22 (3): 482–487.

Qin, S.J. and Li, W.H. (2001). Detection and identification of faulty sensors in dynamic processes. *AIChE Journal* 47 (7): 1581–1593.

Qin, S.J. (2003). Statistical process monitoring: basics and beyond. *Journal of Chemometrics: A Journal of the Chemometrics Society* 17 (8-9): 480–502.

Qin, S.J. (2012). Survey on data-driven industrial process monitoring and diagnosis. *Annual Reviews in Control* 36 (2): 220–234.

Reiter, R. (1987). A theory of diagnosis from first principles. *Artificial Intelligence* 32: 57–95.

Ren, Y. and L. Kong, "Fuzzy Multi-State Fault Tree Analysis based on Fuzzy Expert System," Proceedings of the 9th International Conference on Reliability, Maintainability and Safety (ICRMS), IEEE, 2011, pp. 920–925.

Ruijters, E. and Stoelinga, M. (2015). Fault tree analysis: a survey of the state-of-the-art in modeling, analysis and tools. *Computer Science Review* 15–16: 29–62.

Russel, E.L., Chiang, L.H., and Braatz, R.D. (2000). Fault detection in industrial processes using canonical variate analysis and dynamic principal component analysis. *Chemometrics and Intelligent Laboratory Systems* 51: 81–93.

Russell, S.J. and Norwig, P. (1995). *Artificial Intelligence: A Modern Approach*, 583–585. New York: Prentice Hall International, Inc.

Shu, Y. and Zhao, J. (2016). Fault diagnosis of chemical processes using artificial immune system with vaccine transplant. *Industrial & Engineering Chemistry Research* 55: 3360–3371.

Skotte, R., D. Lenz, R. Fickelscherer, W. An, D. Lapham III, C. Lymburner, J. Kaylor, D. Baptiste, M. Pinsky, F. Gani, and S. B. Jørgensen, "Advanced Process Control with Innovation for an Integrated Electrochemical Process," AIChE Spring National Meeting, Houston, TX, 2001.

Stephanopoulos, G. and Romagnoli, J.A. (1980). A general approach to classify operational parameters and rectify measurement errors for complex chemical processes. In: *Computer Applications to Chemical Engineering*, 153–174. American Chemical Society.

Teague, T.L. and Powers, G.J. (1978). *Diagnosis Procedures from Fault Tree Analysis*. Carnegie Mellon University, unpublished manuscript,.

Vedam, H. and Venkatasubramian, V. (1999). PCA-SDG based process monitoring and fault diagnosis. *Control Engineering Practice* 7: 903–917.

Venkatasubramian, V. and Chan, K. (1989). A neural network methodology for process fault diagnosis. *AIChE Journal* 35 (12): 1993–2002.

Wang, B., Cao, W., Ma, L., and Zhang, J. (2005). Fault diagnosis approach based on qualitative model of sign directed graph and reasoning rules. *Fuzzy Systems and Knowledge Discovery* 339–343.

Wang, R.C., Edgar, T.F., Baldea, M. et al. (2015). Process fault detection using time-explicit Kiviat diagrams. *AIChE Journal* 61 (12): 4277–4293.

Watanabe, K., Matsuura, I., Abe, M. et al. (1989). Incipient fault diagnosis of chemical processes via artificial neural networks. *AIChE Journal* 35 (11): 1803–1812.

Wu, F., Yin, S., and Karimi, H.R. (2014). Fault detection and diagnosis in process data using support vector machines. *Journal of Applied Mathematics* 1–9.

Yu, C. and Lee, C. (1991). Fault diagnosis based on qualitative/quantitative process knowledge. *AIChE Journal* 37 (4): 617–628.

Zhoa, J., Shu, Y., Zhu, J., and Dai, Y. (2014). An online fault diagnosis strategy for full operating cycles of chemical processes. *Industrial & Engineering Chemistry Research* 53: 5015–5027.

2.A

FAILURE MODES AND EFFECTS ANALYSIS

Joseph S. Alford PE, Eli Lilly (retired)

West Layfette, Indiana, USA

2.A.1 INTRODUCTION

Presented in this appendix is a brief summary of failure modes and effects analysis (FMEA). FMEA is an off-line technique; probably the most widely and commonly used one by industry in identifying possible faults and then directly driving actions (often process redesign actions). It is thus useful for reducing the severity of consequences of failures, reducing their potential frequency of occurrence, and/or increasing their immediate detectability. Conducting an FMEA analysis during original process design or before retrofits (or even at other times) can have a major beneficial impact on the complexity of future on-line techniques that plant personnel may want to implement. That is, conducting an initial off-line FMEA may end up eliminating many of the possible process failures that future on-line techniques might otherwise have to deal with.

Artificial Intelligence in Process Fault Diagnosis: Methods for Plant Surveillance,
First Edition. Edited by Richard J. Fickelscherer.
© 2024 John Wiley & Sons, Inc. Published 2024 by John Wiley & Sons, Inc.

2.A.2   FMEA PROCEDURE

Designing and operating an industrial plant can involve thousands of components, many of which can fail and cause an upset to the plant. A plant can also involve many formal operational procedures, whether automated or manual, many of which, if not executed properly, can also cause an upset to the plant. The literature contains many examples of failures of components and/or operational procedures in industrial plants that have caused massive equipment destruction and/or significant loss of life (e.g., Three-Mile island 1979; Chernobyl 1987; Texas Oil Refinery 2005; BP Deep Water Horizon 2010; Bhopal 1984). In anticipation of the possibility of such failures and as a consequence of ones actually occurring, a number of failure mode detection and analysis techniques have been developed in past decades to assist plant designers and operations management in reviewing potential and actual failures. Perhaps the most common of these techniques is FMEA. This is an inductive reasoning (forward logic) single point of failure analysis and is a core activity in reliability engineering, safety engineering, and quality engineering. It is an off-line semi-quantitative technique, normally initially used during original plant design, to list potential failures and then score each one with respect to the potential severity of consequences, potential frequency of occurrence, and detectability. The three scores are multiplied together to provide a RPN (Risk Priority Number) metric. This metric is compared to an acceptability threshold (usually set by the FMEA team or plant management), with all failures with a RPN above the threshold then driving plant redesign or other remediation work in order to reduce the RPN so that it is less than the threshold number.

FMEAs, while commonly used for plant or product design, are also commonly used in other industrial plant paradigms, such as evaluating potential failures with existing plant procedures, computers and other systems, and also plant support services (e.g., maintenance activities, utilities, and shipping/receiving operations).

FMEAs exist in a few different flavors, with, for example, one form known as FMECA (failure modes, effects, and criticality analysis), and another form only focusing on two of the above three mentioned criteria. Details of the different forms of FMEA are beyond the scope of this Introduction, which will only focus on FMEA general principles.

As noted, conducting a FMEA during process design has an objective of reducing the frequency and magnitude of potential failures and increasing their on-line detectability. Perhaps some potential failures can even be preemptively eliminated. This is important in that the reduction of significant failure modes (as a result of executing an FMEA) will likely reduce the extent and complexity of whatever on-line failure detection and analysis algorithms are implemented in the final automated manufacturing plant operations. Overall this treatment mainly deals primarily with on-line, real-time fault detection and diagnosis techniques, some of which may be challenging to implement. So, minimizing the breadth of potential failures in a plant (e.g., via a preliminary FMEA) cannot help but simplify the on-line fault detection work to follow.

The following is a general review of the FMEA technique in a little more detail, including an example.

FMEA is a structured way of determining failure modes of highest risk to a plant. The method also prompts ways to analyze and reduce risk.

- *Failure modes* are any errors or defects in a process, design, or product. They can be potential or actual.
- *Effects analysis* refers to studying the consequences of those failures.

The reasons for conducting FMEA include

"After the fact" learning from a plant failure is costly and time consuming.

Anticipating failures in advance and adjusting the design of the plant, product, service, or system, accordingly, can:

- reduce the consequences (i.e., severity) of a failure,
- reduce the frequency of its occurrence,
- improve the ability to quickly detect the failure.

Quantifying a failure mode involves a focus on:

Severity of consequences of failure (S).
Frequency of occurrence of failure (F).
Detectability of failure (D).

Notes: Consequences can be of various forms (e.g., loss of life, personal injury, equipment damage, release of a toxic compound to the environment, reduced product yield, and QC rejection of product lot).

Frequency is how often the failure might occur. Manufacturer's often publish "mean time between failure" data for their products than can help in estimating frequency metrics.

Detectability usually refers to how quickly operators will become aware of a failure, once it occurs (e.g., immediately, within minutes, within an hour, etc.). However, some applications can include "advance warning of a probable failure" which offers the potential of avoiding the failure.

A matrix is defined by the FMEA team in which categories of severity, frequency, and detectability are defined. Categories are assigned a number (e.g., 1–5, or 1–10) depending on the number of defined categories.

Each failure is evaluated and scored by the FMEA team, calling in plant subject matter experts if needed.

RPN (risk priority number) is then calculated; RPN = S*F*D.

A threshold value of RPN is set by the plant team/management.

All failures with an RPN greater than the threshold are the target of remediation/redesign efforts to reduce the RPN.

Example: An FMEA is performed on a pharmaceutical tablet fill-finish assembly/inspection line including ingredient hoppers, blender, conveyors, and a tableting machine.

Five categories for S, F, and D are defined, so each mode will be scored with a number from 1 to 5.

Severity Score Range/Definitions

5—some equipment destroyed, mfg. line out of service
for >1 month, injuries possible, and/or "out of spec" product produced.

4—Assembly line out of service 1–4 weeks, some product rejection likely.

3—Assembly line down 2–7 days, some product rejection likely.

2—Assembly line down for 1 day, some product rejection likely.

1—current production run put on temporary hold until situation
corrected (likely less than one day), no product rejection likely.

Frequency score range/definitions—examples:

5—possibility of occurrence—once/week.

4—possibility of occurrence—once/month.

3—possibility of occurrence—once/six months.

2—possibility of occurrence—once/year.

1—possibility of occurrence—once/five or more years.

Detectability score range/definitions—examples:

5—failure not known for more than one day after event.

4—failure not known for several hours after event.

3—failure not known for several minutes after event.

2—failure not known for up to 1 min after event.

1—failure known to operators immediately.

The following table is then completed by the FMEA team:

Failure mode	Cause	Impact	S	F	Detect	D	RPN
Wrong compound in blender bin	Operator error	Product, quality	4	2	Lab. assay	4	32
Tablet press speed-run away	Comm. from computer	Product quality	2	4	RPM alarm	1	8
Roller compact vac. pump stop	Mechanical problem	Product quality	2	2	Out of Spec dissolution lab test	4	16

Please Note: integer values for S, F, and D scores are not required; fractions are OK to use (they are sometimes needed, e.g., when different members of the FMEA team disagree on scores).

FMEA team establishes a threshold value as to what RPN scores will result in remediation attention.

Team decides on a value of 20.

Therefore, all failures with a value of RPN >=20 will result in remediation attention (e.g., re-design) in order to try and reduce RPN < 20.

Failures with RPN < 20 will not usually result in remediation efforts—such failures are deemed as acceptable risk to the process.

Follow-up action:

FMEA follow-up focus is on a wrong compound in the hopper (RPN = 32).

Team decides to install an NIR instrument on the blender. This reduces the detectability score from 4 to 2, which reduces the RPN from 32 to 16.

The new RPN represents acceptable risk since it is less than 20. An operator might still put a wrong ingredient in the hopper, but it will now be quickly detected which will then stop the manufacturing line process and minimize the product that would otherwise have to be thrown away. Note: having a second operator manually check and verify any and all additions to the hopper, might be another way of reducing the RPN, this time by reducing the probable frequency of occurrence.

2.A.3   CONCLUSION

This example suggests that by conducting a preliminary FMEA could directly identify potentially very serious faults before the fact. Consequently, more instrumentation may be added to the process or other procedures or design changes can be implemented. Such modifications should be helpful not only in reducing failures, but in increasing the probable effectiveness of whatever on-line fault detection and diagnosis techniques are subsequently implemented (the focus of the remainder of this treatment).

3

ALARM MANAGEMENT AND FAULT DETECTION

Lieven Dubois

ISA 18,88 and 101, Working Group 8 of ISA 18.2, Department of Alarm Management and Training, Manage 4U, Leiden, Netherlands

CHAPTER HIGHLIGHTS

- Appropriate current PFD International Standards.
- Detailed Alarm Management Lifecycle specification.

ABBREVIATIONS USED

BPCS   basic process control system
DCS   distributed control system (Wikipedia 2023a)
EEMUA   the Engineering Equipment and Materials Users Association
HMI   human machine interface
IEC   International Electrotechnical Commission
ISA   International Society of Automation

In remembrance of Bridget Fitzpatrick.

Artificial Intelligence in Process Fault Diagnosis: Methods for Plant Surveillance,
First Edition. Edited by Richard J. Fickelscherer.
© 2024 John Wiley & Sons, Inc. Published 2024 by John Wiley & Sons, Inc.

OPC   open platform communications, formerly object linked
           embedding (OLE) for process control (Wikipedia 2023b)
PLC   programmable logic controllers
SCADA   supervisory control and data acquisition

3.1   INTRODUCTION

Alarm management was a term introduced around the millennium, after the
publication of the first guideline for alarm systems – *A Guide to Design,
Management, and Procurement* – also known as the Engineering Equipment
and Materials Users Association (EEMUA) Publication 191 (EEMUA 2013).

The guideline was a result of an investigation by the UK Health and
Safety Executive after an explosion and fire at the Pembroke refinery in
Milford Haven in Wales in 1994. The incident investigation concluded that
the then present alarm system did not help the operators in performing their
tasks, during the time leading up to the explosion. This was caused due to
an overload of alarm messages (285 per minute leading up to the explosion)
and the poor information contained in those alarms.

In 2009, the first standard for alarm management was published (ISA
18.2 Committee 2009) followed by an international standard, International
Electrotechnical Commission (IEC) 62682, in 2014 that was based on the
ISA/ANSI 18.2 standard (IEC 2014). The ISA/ANSI 18.2 standard was
then formally updated in 2016 (ISA 18.2 Committee 2016).

In this chapter, it is not the purpose to explain all aspects of alarm management (AM) but explain the relationship between AM guidelines and
standards and on-line fault diagnosis.

3.2   APPLICABLE DEFINITIONS AND GUIDELINES

The ANSI/ISA 18.2:2016 standard and the IEC 62682 standard define an
alarm as an:

> audible and/or visible means of indicating to the operator an equipment
> malfunction, process deviation, or abnormal condition requiring a
> timely response (ISA 18.2 Committee 2016).

The reader can distinguish several important elements in this definition:

- There must be a means to draw the attention of an operator.
- There must be an abnormal situation caused by an equipment malfunc-
  tion, process deviation, or other abnormal condition.
- The operator must undertake a timely response.

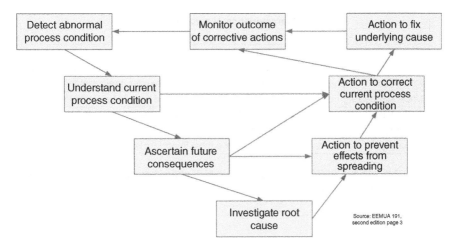

Figure 3.1 The components of the operator response start from the top left block (EEMUA 2013).

A timely response is of importance, as an alarm needs to occur early enough, if possible, to allow corrective action to be effective. A response is usually a human action that the automation system cannot undertake (Figure 3.1).

EEMUA 191 sketches the operator response as follows:

Usually, it is the alarm system that annunciates the abnormal condition, but consequently, the operator needs to become aware of it. This is usually done by means of visual, auditive, or any other means of getting the attention of the operator to the annunciated situation. The operator must understand the current process condition, ascertain current and possible future consequences, undertake actions to correct the current process conditions, take actions to prevent the effects from spreading, investigate the root cause, and monitor the outcome of all actions taken.

For the purpose of this book, the focus is on on-line identifying, investigating, and fixing the root cause of the abnormal situation.

Therefore, the alarm message should contain diagnostic information. EEMUA 191 (EEMUA 2013) specifies this as a characteristic of a good alarm (Figure 3.2).

Identifying the root cause as part of the transmission of an event to the operator alarm console is of interest and has an added value.

To fully understand the added value, we need to look at what the human brain is doing when receiving and reading an alarm message text.

Figure 3.3 will illustrate what is happening (Wickens et al. 2004).

Characteristics of a good alarm	
Relevant	i.e. not spurious or of low operational value
Unique	i.e. not duplicating another alarm
Timely	i.e. not long before any response is needed or too late to do anything
Prioritised	i.e. indicating the importance that the operator deals with the problem
Understandable	i.e. having a message which is clear and easy to understand
Diagnostic	i.e. identifying the problem that has occurred
Advisory	i.e. indicative of the action to be taken
Focusing	i.e. drawing attention to the most important issues

Figure 3.2 The EEMUA 191 recommended characteristics of a good alarm (Table 1, Third Edition, page 4).

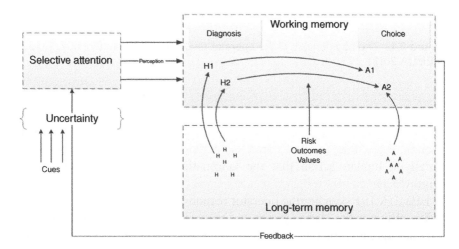

Figure 3.3 Schematic of evaluation process (from Human Factors Engineering, figure 7_2 on page 163).

When the alarm message is read and it has the attention of the operator, the diagnosis process starts in the brain. From the long-term memory different hypothesis (H1, H2, …, Hn) are drawn (downloaded), each with a corresponding action (A1, A2, … An). How many long-term hypotheses are stored in the brain of the person depends on experience, training, and/or general knowledge. Which hypotheses are selected for evaluation depends on cues or clues observed. It is also uncertain whether all hypotheses are evaluated, due to stress caused by many factors, among which information overload (the next alarm already appearing) and the requirement to undertake action before the consequence of not responding or too late response occurs. While evaluating all hypotheses, the operator needs to consider the potential risks related to each corresponding action. Eventually, the operator needs to make a choice. They will perform an action (or have an action performed, e.g., sending a technician into the plant or in the field), and they will be in uncertainty whether the chosen action will be effective, until the process or situation returns to normal, and the alarm is cleared (removed from the alarm list, returned to normal).

While the human brain is known to be faulty (all kind of biases), computers are known to be able to evaluate swift and adequate all known hypotheses, evaluate all available cues, calculate risks, and propose the best possible solution.

There are some other human aspects to consider. The less experience a person has, the less hypotheses will be available in the long-term memory. The older and thus more experienced the person gets, the longer it takes to access and retrieve all hypotheses.

And there is also the nature of alarm systems to consider. The better the alarm system is designed and the better the plant is engineered, the smaller the chance that high impact, high risk alarms will occur. Which results in less experience for newer staff.

Training on simulators or digital twins (see Chapter 10 – Knowledge-Based Systems, section Digital Twins) can correct some of the human aspects, but not all. Adding diagnostic information to the alarm message will do so for sure.

3.3   THE ALARM MANAGEMENT LIFE CYCLE

3.3.1   Introduction

ANSI/ISA 18.2 introduced at its publication in 2009 a life cycle model for adequate and sustainable management of alarm systems for the process industries (ISA 18.2 Committee 2009). IEC adopted this model in its IEC 62682 standard, published in 2014 (IEC 2014). The next subsections will briefly describe the application of these work processes for diagnostic alarms.

3.3.2   Life Cycle Model

See Figure 3.4.

3.3.3   Alarm Philosophy

Plants or organizations that implement alarm management according to the standard should create and maintain a so-called "alarm philosophy." The alarm philosophy specifies the objectives of the alarm system and the objectives of the different work processes. A document (not shown in the diagram below)

Life cycle model

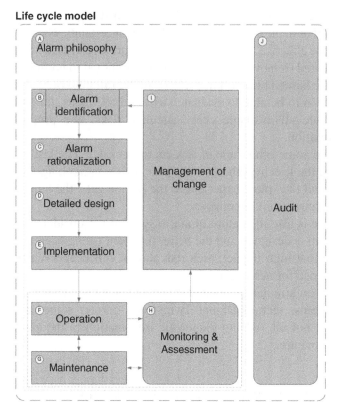

Figure 3.4   Alarm management life cycle from ISA 18.2 and IEC 62682 (ISA 18.2 Committee 2016) (IEC 2014).

that is a subset of the Philosophy stage of Alarm Management and discussed in the ANSI/ISA 18.2 and IEC standard is an Alarm System Requirements Specification (ASRS) which is consistent with the alarm philosophy but typically goes into more detail than the philosophy, may be specific to individual control systems, and is an important source document in driving detailed system design and testing. The ASRS is often developed as a portion of the overall control system "functional requirements."

3.3.4   Alarm Identification

Alarm identification is a work process with one or more subwork processes where potential alarms are identified. If a diagnostic application is used to generate alarms, then this work process should be mentioned in the alarm philosophy.

Identifying a potential alarm is different from providing additional information to an existing alarm. If the diagnostic application is only called up from existing alarms (see alarm presentation), then it is not a subwork process of this alarm management life cycle stage.

3.3.5   Alarm Rationalization

The standard recommends that any suggested (identified) operator alarm should be put through the rationalization process. This process will evaluate if the suggested alarm meets all the criteria of an operator alarm according to the site's alarm philosophy and will also quantify some of the alarm's attributes (e.g., setpoint, priority).

This process is also applied when an existing alarm, including a diagnostic alarm, would be subject to change because either it is a nuisance alarm, or no alarm response is defined, or the alarm is not documented properly (i.e., according to the specifications of the alarm philosophy).

At the end of the rationalization process, any retained alarm is classified, prioritized, and properly documented. The required information is then passed on to the next work process (detailed) alarm design.

When during alarm rationalization it is decided that the suggested alarm does not meet all the criteria, it can be used as some process notification to the operator, the maintenance engineer, the asset manager, or any other role the notification is intended for. It is recommended to document the reason why such notification is not withheld as an alarm.

3.3.6   Alarm Design

At this stage any alarm should be designed (i.e., integrated in the operator's human machine interface – HMI) and configured in the basic process control system (BPCS) or in any supervisory control system that is used in the plant.

Alarms are for control room operators. If the information from the diagnostic application is not for the operator, then it should never be annunciated as an alarm. Such notifications can go elsewhere, such as to the maintenance service, contractor, quality department, plant utilities department, and so on. The information could be presented in the HMI of the operator, so they can observe something is going on, but it needs no action from their side. Such presentation can be in a color change or an icon that appears next to the representation of the device that has a fault. If the operator has time and a fault diagnostic application is properly integrated with the HMI of the operator, then the operator might be able to click on it to see what is going on.

3.3.7 Implementation

After design an alarm needs to be implemented, i.e., tested, and the operator(s) should get training on this new, diagnostic alarm system.

3.3.8 Operation

After proper testing and training, the (diagnostic) alarm is put in operation. It becomes part of the alarm system.

When an alarm becomes a nuisance, e.g., when chattering, fleeting, or stale, it can be shelved by the operator, i.e., put on a stack and no longer be annunciated. It is the audible annunciation that often causes the stress for the operator, because it is interrupting their mental tasks.

3.3.9 Maintenance

When an alarm is not working properly or the equipment the alarm relates to is under maintenance, the alarm can be put out-of-service and under maintenance. As soon as the related equipment is repaired or fixed and/or the alarm is fixed – the criteria to repair an alarm at this stage is defined in the alarm philosophy – it is put back into service.

3.3.10 Monitoring and Assessment

Any alarm system shall be monitored and assessed against the criteria set forth in the alarm philosophy. At this stage, so-called nuisance alarms are identified. Nuisance alarms are:

- Fleeting alarms: alarms that appear only for a short time and disappear without operator response.
- Chattering alarms: alarms that appear and disappear cyclic with or without operator response.
- Stale alarms: alarms that remain in the alarm display for a long time.

The alarm philosophy specifies criteria for fleeting, chattering, and stale alarms. It also specifies if such alarms can be repaired (in the maintenance stage) or need to go through rationalization (and through management of change).

3.3.11 Management of Change

Management of change applies to all work processes listed left from the work process in the life cycle figure. Management of change procedures shall keep track of any changes in the alarm configuration (or alarm attributes) at each stage (rationalization, design, implementation). Management of change keeps track of all changes, authorizations to change, and passing to different stages. Usually, plant procedures are applied as well as the selected plant tools to track such changes.

3.3.12   Audit

The audit process is a work process that is performed at a frequency specified in the alarm philosophy (annual, biennial, triennial, …). It is a process whereby all work processes are reviewed; to check if they are properly implemented, operators are interviewed to evaluate the usefulness of the alarm system and related procedures and the information is compared with eventually new standards and practices. The audit process provides input to the alarm philosophy. For example, the alarm rates specified in the alarm philosophy can be adjusted to further reduce the workload the alarm system imposes on the operator.

3.4   GENERATION OF DIAGNOSTIC INFORMATION

3.4.1   Introduction

Diagnostic logic can be:

- Part of the sensor or actuator, a so-called intelligent device or smart sensor.
- Part of the BPCS.
- A separate application connected to the BPCS and the HMI, providing diagnostics upon demand.
- A separate application connected to the BPCS, generating alarms (Figure 3.5).

3.4.2   As Part of the Basic Process Control System

Basic process control systems include programmable logic controllers (PLC), distributed control systems (DCSs), supervisory control, and data acquisition systems. During the last two decades most of the lower-level control equipment (smart sensors and intelligent actuators) has a CPU or a programmable logic chip on board in which diagnostic logic can be programmed (see Chapter 10 – Knowledge-Based Systems, section Software as a Knowledge-Based System). It is also observed that such intelligent devices no longer communicate with the BPCS using analog or digital electrical signals, but

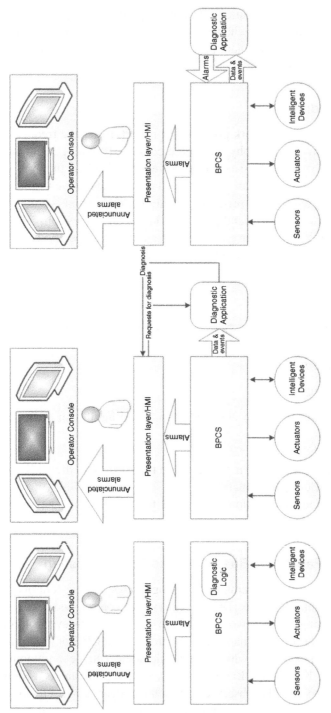

Figure 3.5 Different ways to embed diagnostics in alarms.

communicate through a "bus" connection, transferring already digitized values, but also device status information. Such device status information can contain diagnostic information that can be used by the BPCS to be included in some process notification. ISA 18.2 TR8 describes different types of process notifications for different roles in plant management (ISA 18.2 WG8 n.d.). Whether or not such information is used for alarm purposes is subject of alarm rationalization. Whether or not such information is part of the alarm message depends on the alarm design principles (alarm design and presentation).

The logic builders (programming capabilities) of the BPCS could be used to pick up the information from intelligent devices and/or combine this with analog and digital values it receives from other sources, thus generating diagnostic information that could be part of an alarm. Typically, changing the alarm logic requires not only proper rationalization and design but also strict management of change (MOC). In some industries, it is not allowed to change the logic of the BPCS after validation of such logic by regulating authorities.

3.4.3 As a Separate Application

It is in many situations easier to develop diagnostic information in an application separate from and connected with the BPCS. Apart from integration issues, which seem to decrease over the last decades due to the use of open platform communications (OPC) and data historians, this holds the advantage of using knowledge-based systems or apply machine learning for this purpose.

The following ways of integration can be envisioned:

- The diagnostic application generates alarms and sends them to the BPCS for display in the "regular" alarm display of the operator console (left example in Figure 3.5).
- The diagnostic application monitors process data and events and sends diagnostic information to the operator (in the HMI) only upon demand (middle example in Figure 3.5).
- If available, the alarm help function in the BPCS can provide diagnostic information (as pictured in the middle example).

3.5 PRESENTATION OF THE DIAGNOSTIC INFORMATION

3.5.1 Introduction

There are several ways to include diagnostic information in an alarm message:

- As part of the alarm text or alarm help.
- As a link to the fault diagnostic application.
- As a notification icon (status indication) in the operator's HMI.

- On a separate list.
- A combination of all.

3.5.2   As Part of the Alarm Text

Many alarm messages are composed as follows:

Timestamp I priority I tag in alarm I type of alarm I alarm text

which usually results in something like this:

20220222 20:22:22 I High I P101T01 I PVHI I *Pump 101 Temperature high*

Very often the alarm text portion of the message informs no more than what is just in front of it: a text description of the tag name and a text description of the alarm type. No wonder that such poor messages do not assist the operator in performing all the required actions. This was a consequence of the limited space on monitors and printers at the early stages of DCS (80–132 characters).

The text of the above message could be improved with additional information and a call for action, such as: *Temperature Pump 101 more than 81° during last 10 minutes. Expected 40 minutes until trip. Check inflow, outflow, and filter.*

With such information, at least the operator knows what items they should check.

An on-line fault diagnosis application could rule out which of the three possible causes is unlikely to be the real root cause. The alarm message might then look like: *Temperature Pump 101 more than 81° during last 10 minutes. Expected 40 minutes until trip. Filter clogged. Ask maintenance to replace or clean filter immediately.*

3.5.3   As a Link to the Diagnostic Application

When possible the alarm design could add hyperlinks to this alarm text, such that the operator can click on a hyperlink to pull up the relevant additional information that has been configured/programmed into the control system. Example:

Temperature *Pump 101* more than *81° during last 10 minutes.* Expected 40 minutes until trip. *Filter clogged.* Ask maintenance to replace or clean filter immediately.

In the above example, three hyperlinks (i.e., the three parts of the message that are underlined) could be made available to call upon a knowledge base or expert system that provides further information:

- Clicking on Pump 101 would show a detailed representation of the pump on an operator console display. Such representation is usually part of the functionality of the control system vendor's or supervisory control and data acquisition (SCADA) HMI software.
- Clicking on the temperature link would show a trend graph of the temperature of the pump, with the setpoint (the desired value), the alarm limit, the persistence time, and an extrapolation to the trip point, at a predefined place in the operator's HMI (ISA 101 Committee 2015). Such functionality is often part of the functionality of the control system vendor's or SCADA HMI software.
- Clicking on "filter clogged" would bring up the diagnostic information on how the diagnosis was achieved, at a predefined area in the operator's HMI. If this would be plain text, an operator might not read it, given the time involved and urgency to take action. While identification of the root cause is clearly valuable for the operator and is often practical to do, an on-line text description of how the root cause was identified is usually not part of the control system vendor's or SCADA HMI software.
- Clicking on the help function in an alarm message, when available, could also provide the diagnostic information required for the alarm.

Note: with respect to the example alarm message just described, it can be valuable for a control system to provide spare fields as well as flexibility and expandability in the format of alarm messages.

For example, some vendors provide a few spare fields in their alarm and data record message formats for customers to use as desired (and some vendors do not). For example, for batch processes, it is valuable for subsequent data mining and analysis purposes for the alarm/data record to include the batch lot number and relevant batch step. It is also important that the text part of an alarm/data record allow for enough characters to fully describe the event. A common criticism of alarm records is that they are too cryptic, due to an insufficient number of alphanumeric characters allowed in the alarm/data message or record. Expanded, more meaningful alarm messages not only make data analysis more efficient but can also reduce the need for operators to navigate to different HMI screens to find more information.

3.5.4   As an Indication in the HMI

If the diagnosis or fault detection does not require a timely operator action, the resulting notification can be displayed in the operator HMI as an alert or a notice. TR8 of ISA 18.2 provides guidelines on how to distinguish these types of notifications (ISA 18.2 WG8 n.d.). The NAMUR Recommendation NE 107 on Self-Monitoring and Diagnosis of Field Devices distinguishes models of Diagnosis of Field Devices (NAMUR 2017) and proposes graphic illustrations of status signals (Figure 3.6).

It might be that the proposed colors conflict with the colors chosen in the style guide of the HMI philosophy (ISA 101 Committee 2015) or the colors reserved for alarms in the alarm philosophy (ISA 18.2 Committee 2016). If so, colors must be chosen according to the appropriate philosophy. The HMI of the operator (subject of ISA 101) is nowadays also used by other roles in the plant.

If the capabilities of the HMI allow the use of different colors for different roles, the problem could be solved as follows (Figure 3.7).

Failed	Out of Specification	Maintenance Required	Check Function
⊗	⚠?	◆	▽🔧
⊗	⚠?	◇	▽🔧
⬤	▲	◆	▽
Failure	Out of Specification	Maintenance Required	Check Function

Figure 3.6   Graphical representation of status signals according to Table 4 of NE 107.

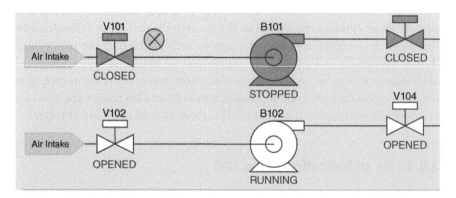

Figure 3.7   Status indication of a failed actuator using the graphical symbol of NE 107 for the operator role.

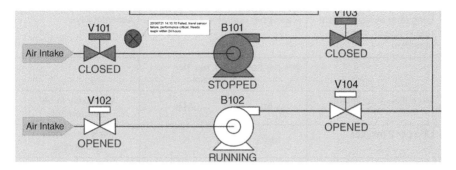

Figure 3.8 Status indication of a failed actuator using the graphical symbol of NE 107 for the maintenance role.

In the above example, the upper compressor/ventilation line is closed due to a failing valve, which is indicated by the circle with the cross next to it. If the operator would click on it, they would see the diagnostic event causing this problem. No alarm is annunciated. No alarm colors are used (Figure 3.8).

In the above example, the graphical symbol next to the valve is in red, indicating an action is required by maintenance. If the maintenance technician would click on it, they can read the diagnostic event or fault.

The use of auxiliary graphical symbols rather than coloring the entire graphical representation of the device is strongly recommended. If the entire valve would be red inside, the user cannot distinguish anymore whether the valve is in closed or open position.

3.6   INFORMATION RATES

3.6.1   Introduction

The alarm system guideline (EEMUA 2013) was built on research work and interviews with plant operators in continuous industries in the late nineties of the past century (Figures 3.9 and 3.10).

It was derived that an alarm every 10 minutes would be acceptable during normal operating circumstances, while less than 10 per 10 minutes could be manageable in abnormal situations, as in non-rationalized alarm systems many consequential alarms after a trip or shutdown are not suppressed.

For understanding alarms in general, alarm management practitioners often talk about remediation alarms and mitigation alarms.

Remediation alarms are the ones that occur before the trip or shutdown. They should inform the operator to prevent the trip or shutdown from happening (remediate the situation).

Long term average alarm rate in steady state operation	Alarms per operator		Acceptability
	No. per hour	No. per 10 minutes	
More than 1 per minute	> 60	> 10	Very likely to be unacceptable
One per 2 minutes	30	5	Likely to be over demanding
One per 5 minutes	12	2	Manageable
Less than 1 per 10 minutes	< 6	< 1	Very likely to be acceptable

Figure 3.9   Table 14 from EEMUA 191 Third Edition.

Number of alarms displayed in 10 minutes following a major plant upset	Acceptability
More than 100	Definitely excessive and very likely to lead to the operator abandoning use of the system
20–100	Hard to cope with
Under 10	Should be manageable – but may be difficult if several alarms require a complex operator response

Figure 3.10   Table 15 of from EEMUA 191 Third Edition alarms after a major plant upset.

Mitigation alarms are the ones that occur after the trip or shutdown or when the hazard has been released (e.g., when the gas is detected or the tank is overflowing) These alarms should assist the operator in mitigating the effects of the abnormal situation.

In some industries, such as food and beverage, where such safety systems and emergency shutdown systems do not exist, or are only available on machine level, unattended alarm messages can lead to even worse problems, such as contaminated products and damage of the market image.

Typically, diagnostic information, resulting from real-time, on-line diagnosis, should be added to *remediation* alarms.

3.6.2   Nuisance Alarms

3.6.2.1   Chattering Alarms   When letting a diagnostic application (or any source) generate (operator) alarms, great concern should be taken not to create nuisance alarms such as repeating (or chattering) alarms (or to generate alarms during batch process phases for which the alarm is not relevant). Alarms typically create an audible signal (a noise) that is stressful for operators, especially when the alarm is a nuisance alarm. A repeating alarm on the same subject will be shelved, ignored, and hence will be useless.

3.6.2.2 Fleeting Alarms Fleeting alarms are the ones that would appear and quickly disappear because some of its true values or states are no longer valid. Fleeting alarms desensitize the operator.

Consequently, when designing diagnostic alarms attention should be given on persistency, i.e., its validity and truth values should be persistent enough (valid for a significant amount of time) to be alarmed and thus not fleeting.

3.6.2.3 Stale Alarms A stale alarm is an alarm that remains in the operator display for hours, usually meaning that the operator cannot take an appropriate action, or that no action is defined, or that the message is not meant for the operator but for another role in the plant.

When designing a diagnostic alarm, attention should be given to the role of the receiver (is the information meant for the operator or for somebody else), the action to undertake (is it something that the operator can handle or not), and the time the action should be taken (is the corrective action within an adequate time span, not too early).

REFERENCES

EEMUA (2013). *Publication 191: Alarm Systems – A Guide to design, management and procurement.* London, United Kingdom: The Engineering Equipment and Materials Users' Association.

IEC (2014, 10). IEC 62682 Management of alarms systems for the process industries. In: *Management of Alarms Systems for the Process Industries.* Geneva: IEC.

ISA 101 Committee (2015). *ANSI/ISA-101.01-2015 Human Machines Interfaces for Process Automation Systems.* 64. Research Triangle Park 27709, North Carolina, USA: International Society of Automation.

ISA 18.2 Committee (2009). *ANSI/ISA-18.2-2009 Management of Alarm Systems for the Process Industries.* Research Triangle Park, North Carolina, 27709: ISA.

ISA 18.2 Committee (2016, March 17). *ANSI/ISA 18.2-2016 Management of Alarm Systems for the Process Industries.* Research Triangle Park, North Carolina, 27709: ISA.

ISA 18.2 WG8 (n.d.). *ANSI/ISA 18.2:TR8 Guidelines for Non-alarm Notifications.* Raleigh: ISA.

NAMUR (2017). *NE107 Self-Monitoring and Diagnosis of Field Devices.* Leverkusen: NAMUR.

Wickens, C.D., Lee, J.D., Liu, Y., and Gordon, S.E. (2004). *An Introduction to Human Factors Engineering*, seconde (ed. L. Jewell). Upper Saddle River, New Jersey, USA: Pearson International Edition.

Wikipedia. (2023a, 04 23). en.wikipedia.org/wiki/DCS. Retrieved from en.wikipedia.org: https://en.wikipedia.org/wiki/Distributed_control_system

Wikipedia. (2023b, 04 23). en.wikipedia.org/wiki/OPC. Retrieved from en.wikipedia.org: https://en.wikipedia.org/wiki/Open_Platform_Communications

4

OPERATOR PERFORMANCE: SIMULATION AND AUTOMATION

Ron Besuijen

Center for Operator Performance, Dayton, OH, USA

CHAPTER HIGHLIGHTS

- Smart alarming studies to improve operator alarm management.
- Integration of AI Based methods in such management.
- Improvements of operator performance with Safe Park applications.
- Advantages of implementing high-fidelity process simulators.

4.1 BACKGROUND

Modern control systems in the petrochemical industry can deliver an increasing amount of information and automated support for the process operators. It is not unusual for a panel operator to have more than 5000 data

Special recognition for chapter structure and editing: David Strobhar, PE, Chief Human Factors Engineer, Beville Operator Performance Specialists, www.beville.com.

Artificial Intelligence in Process Fault Diagnosis: Methods for Plant Surveillance,
First Edition. Edited by Richard J. Fickelscherer.
© 2024 John Wiley & Sons, Inc. Published 2024 by John Wiley & Sons, Inc.

points available on the console. The author knows of a console that has almost 13,000 data points. These will include transmitters used for pressure, flow, temperature, level, conductivity, vibration, thrust, valve position, and differential pressure. Analyzers continually measuring the components of the process. Areas of the plant are monitored by infrared fire and gas detectors.

There are also notifications for errors of non-process data, including analyzers, logic controllers (programmable logic controller [PLC], etc.), control systems, electrical systems, fire water, and buildings (temperature, heating, ventilation & air conditioning [HVAC] smoke). There are several hundred alarms possible with all these systems combined.

Although they are not used as often as in the past, level, flow, temperature switches that only indicate if a certain condition has been met. Because the reliability of these switches has been low and they are difficult to test, they are being replaced by transmission of analog values.

For any control system to work, including those that incorporate artificial intelligence (AI), the reliability of the sensors is critical. Inaccurate or wrong data leads to erratic or potentially erroneous control. Preventive maintenance is required to periodically validate the accuracy of these instruments. It is not uncommon to see a few flow transmitters indicating a flow when the process is out of service.

Modern safety systems have increased the number of position indicators and transmitters exponentially. One example is a heater that has 13 pressure transmitters on a fuel gas system to ensure all the safety and isolation valves are functioning properly before any of the burners are ignited.

The use of fault tolerant safety systems with a two out of three voting triples the instruments on every shutdown variable. To ensure the reliability of the safety systems redundant transmitters are being used. For example, to initiate a trip on a steam drum, three level transmitters will be used where two of the three must agree to initiate an action. In this design it is easy to detect if one of the transmitters fail and then the transmitter can be bypassed. The voting for the trip condition now becomes two out of two. This design will also reduce the probability of nuisance trips as a faulty transmitter can easily be compared to two other transmitters. The increase in reliability and fault tolerance also results in increased complexity.

Although much of this data will not be required for most of the responses required by operations, it does highlight the importance of how this information is presented and the need to remove some of the actions required by operations during upset conditions. Automation can reduce the noise in the signals the operator uses in troubleshooting and provide them more time for troubleshooting.

With the increase in data points is usually an increase in the number of alarms and notifications. On one cracking furnace the number of alarms was increased three and one-half times after a process safety management upgrade. With the majority being high priority alarms.

In this chapter three tools to improve fault detection, or the response to faults, will be covered. The first is smart alarming that is aimed at reducing the number of alarms for a given event. The second is the use of safe park applications which place the process in a predetermined condition that can provide the operator with the time to respond to secondary issues and determine the cause of the problem. The third tool is the use of a process simulator and other benefits that can be seen from a simulator program. A study that utilized these three tools is then discussed.

How pattern recognition is used in fault detection is covered and an example of why this is important. The last topics will be the importance of human-centered AI (HCAI) and AI mental models.

4.2 AUTOMATION

4.2.1 Smart Alarming

Alarm rationalization is a common practice in the petrochemical industry. However, this is typically viewed from a steady-state operation to give the panel operator enough information to prevent an outage or escalation of a problem. The intent is to avoid repetition of alarms to allow the panel operator to quickly sort through them and determine the cause of the upset. This can be very effective when there is a deviation of a transmitter, failure of a control valve, or loss of a pump.

With only two to three alarms the operator can zero in on a problem and initiate an appropriate response and minimize the upset. An Australian military study showed that in a command-control task (what console operators do), interruptions can degrade subsequent situation awareness and task performance for up to 40 seconds (Loft et al. 2015).

Alarm rationalizations are effective for smaller events. During a large event, like a plant outage or trip, the number of alarms can exceed 100 in a few minutes even with alarm rationalization.

During these events, operations are required to safely isolate and secure the process. This task can take some time, limiting the Panel Operators ability to address the alarm flood that occurs. An alarm notifying the panel operator of an event that develops during the outage can easily be missed, allowing the event to escalate. This could be a control valve passing (i.e., leaking) that has been closed or a process loss of containment (i.e., spill/release).

This type of event occurred in a petrochemical facility during a complete outage when a pressure control valve to flare was passing and sub-cooled a distillation tower. The low-pressure alarm was lost in the several pages of alarms. The additional flaring was hidden within the normal flaring required during an outage.

Programs can be developed to reduce the number of alarms during large events to a manageable level for the Panel Operator. A program can be used to disable or change the trip point of an alarm. A study of which alarms are known to come in during the specific event can be used determine which alarms are expected and therefore will not cause an additional hazard if they are disabled or altered.

If the number of alarms is reduced to a small number for the course of the event, then the Panel Operator can quickly scan the alarm summary to look for anomalies that are not normal for that type of outage. Detection and identification of problems, particularly independent failures, are improved with less visual noise in the system.

4.2.2 Safe Park Applications

Isolating a petrochemical facility can involve over 100 process changes required of the panel operator that must occur in a short period of time. Performing this task will limit the panel operators ability to respond to secondary problems or communicate with other affected units.

Applications have been created that make many of the known process changes once the new state of the plant has been identified, often called safe park. These applications also speed up the process of isolating the facility, maintaining inventory and limiting the risk of creating off-specification inventory. Both will benefit the start-up of the facility and reduce the amount of flaring required with its associated environmental impact.

Safe-park applications can be initiated automatically or be left to the operator to initiate. While an operator will take longer to decide to implement a complex action such as safe park than a simple control change, the resulting automation will result in a faster and more accurate response than the operator could achieve alone. Use of automation to perform these sorts of complex tasks utilizes the human performance principle known as the decision complexity advantage (Wickens 1992).

The graphic interface is as important as the actions of a safe park application. A well setup graphic will allow the operator to easily monitor the actions to ensure they are occurring and allow the operator to check-off that the additional responses are completed for the event. This would include actions required by the field operator such as manually isolating valves or shutting down pumps, etc. All the actions required in the procedure for the event will be included on this graphic. Also included are any prerequisites may also be required for some of the steps. A pump or compressor may need to be shutdown by operations before related systems can be isolated by the safe park application (e.g., the reflux pump will need to be shutdown before the reflux valve is closed).

Testing these applications on a simulator allows the engineers to test the communication with the control system and to ensure the program will function in a dynamic environment. I have also tested these applications when the plant was shutdown for maintenance in case there are communication differences between the live system and the simulator. Isolating the control valves first is a recommended step.

Having a safe park application fail halfway through an event can be worse than not having the tool. This would leave the operator trying to figure what steps are left, or worse, having the operator believe the application is working and not responding at all.

Applications can also be developed to assist operations in responding to predictable process upsets. For example, when there is a loss of feed to the process in a short period of time the controllers may have difficulty maintaining a steady flow. The tuning is typically set up for steady state operation with the intent to keep a consistent flow to the distillation towers. This allows for the multivariable controllers on the towers to optimize as close as possible to the product specification. For a sudden reduction of feed an application can calculate the required feed reduction through the remainder of the process. A new flow set-point can be ramped in, or first placed in "auto" if it is in a cascade loop. The cascade loop can then be re-established after the ramp is completed.

This same concept can be used on a heater or furnace to reduce the process flow. A calculated amount of fuel and feed can be cut in that order. This reduction could take much longer if the normal control system is used. If there are several heaters in the process the total feed through the process can be reduced quite quickly without having to trip the heater. Rapid cooling of the process coils due to a trip can cause the process coils to plug depending on the type of process involved.

Rapid rate reduction strategies can also assist with meeting environmental guidelines to reduce flaring. Some flare systems are unable to achieve complete combustion at higher flow rates.

4.3   SIMULATION

Process simulators are typically thought of as an operations training tool. With fewer process upsets, the opportunity for operators to practice troubleshooting is becoming rarer. Like any skill, troubleshooting requires practice to maintain proficiency.

There are benefits other than just training that can be gained from a simulator system. The closer the simulator system is to the actual process the more benefits that can be realized. Utilizing the same control software and hardware allows testing of software updates on the simulator before being implemented in the actual process unit. This also allows the system engineers to practice updates and repairs on a system that does not affect the operating unit. Basic controllers and advanced control systems can be tested and tuned. A complete control system interface upgrade or graphic updates can be tested.

Safety system logic can be tested in a dynamic environment that may highlight issues that do not show up in a standard logic test. A high-fidelity simulator can be used to develop and test a safe park application. This would be very difficult to achieve in a facility with high on-stream times. These applications require extensive testing to ensure all the actions occur or no unwanted consequences happen. Smart alarming also requires similar testing.

Process control and safety systems are becoming more complex as they increase the efficiency of production and protection of the equipment. This may seem like it may require less training of the operations personnel. Even though they can be quite beneficial, the unexpected consequence has been that operations require more training as these systems are more difficult to troubleshoot when they have a problem.

A simulator training program allows operations to experience process upsets in a dynamic environment. Procedures are a good tool to assist with this, however they are very linear, and the process is very multi-variable. Although there may have been one cause, there can be many disturbances occurring at the same time. Working through upsets in a simulated environment allows operators the opportunity to develop the mental processes (e.g., sense making, decision making) required to sort through data and prioritize the actions required.

4.4   RESEARCH

Research was conducted by the Centre for Operator Performance to validate the benefits of smart alarming and safe park applications (Simonson et al. 2022). To date, faith rather than proven results has underpinned the use of these tools. Alarm management has been shown to reduce the number of alarms and provide better prioritization. Upset automation has been shown to cut the number of control actions necessary during an upset.

The goal of the research was to measure operator and plant performance in response to an abnormal situation. The use of actual operators and a high-fidelity simulator would ensure the results reflect real-life as much as possible. Human factors researchers, Richard Simonson, Joseph Keebler, Elizabeth Blickensderfer from the Embry-Riddle Aeronautical University, Daytona Beach, Fla., developed the experimental design. This added to the confidence that the outcome would accurately indicate the effectiveness of alarm management and upset automation.

4.4.1   Method

The study focused on the indoor control-panel operating tasks for the finishing side of ethylene manufacturing. It used a high-fidelity dynamic simulation of the cold side of the process, modeled by software from CORYS, Grenoble, France.

4.4.1.1   *Simulated Scenario and Independent Variables*   The simulation began with a malfunction of the propylene refrigeration compressor unit (i.e., "601 trip"); hence, the plant was in "upset" mode and not steady state. When the refrigeration compressor unit trips, the control room

operator must isolate the plant, which includes closing valves, shutting off pumps, and flaring (i.e., burning off excess gases).

As the operator was isolating the plant, a valve malfunction would occur. This involved an open valve that the operator closed as part of the standard procedures following the 601 trip. The issue was that the operator's display showed the valve as closed but it remained open in the (simulated) plant. The operator's task was to detect the valve malfunction while continuing to resolve the trip of the 601 compressor. Accomplishing this required approximately 100 process moves. The simulation specialist administered the modeled scenarios and acted as a role player when necessary (e.g., responding as the "cracking panel operator" and "outside operator" during the scenarios).

4.4.1.2 Alarm Design The study included two types of alarm design. No rationalization (which yielded approximately 250 alarms) and smart/state-based alarms (which gave approximately 5 alarms).

4.4.1.3 Automation The study also incorporated two levels of automation: presence or absence of the safe park application. With safe park, operators would perform approximately eight moves in responding to the simulated events. Without the safe park application, operators would carry out approximately 120 moves in response to the simulated events.

4.4.1.4 Study Conditions Each operator experienced four separate, simulated scenarios:

	No rationalization	Smart alarm
Safe park automation app	Scenario A	Scenario B
No safe park automation app	Scenario C	Scenario D

To avoid learning effects, the study used four different scenarios (one per study condition) with the malfunctioning valve differing in each. The sequence of experiencing the scenarios was randomized for each participant.

4.4.1.5 Measures Each operator's perceived mental workload level and the flare released were measured. A subjective questionnaire, NASATLX, measured the perceived mental workload. Every operator completed the NASA-TLX after each simulated scenario, resulting in four unique workload scores per operator (one per scenario). The amount of flare (i.e., the amount of excess chemicals released) was recorded by the simulator's historian function. In the actual plant, after a 601 trip an application reduces furnace feed by 22% within three minutes. For consistency in this study, the feed rate was left at that level, although in a non-simulated 601 trip event, operations would have cut the feed rate further.

The total flare released was collected by importing historian data into a spreadsheet in one-minute averages. The averages then were totaled over the period of time between the onset of the first alarm and when the operator identified the anomaly.

4.4.1.6 Procedure   Upon arriving for the simulation day, each operator completed the scenarios one by one. The order of the scenarios for each operator varied, being randomly assigned, to avoid order effects. Each scenario lasted between 10 and 45 minutes. After finishing a scenario, the operator completed the NASA-TLX questionnaire. The simulation specialist then debriefed the operator about the scenario and the operator was given a short break. This process repeated until the operator had gone through all four scenarios.

4.4.2 Testing and Results

Eleven finishing-side operators currently employed at an ethylene facility in Canada participated. All the operators were shift workers on a five-week rotating schedule who were slated for their required simulation-based training. They were active panel operators and familiar with the equipment, processes and procedures of the particular plant and specific to their job. The operators experience levels ranged from several months to 30 years; the average was about 7 years.

The operators were informed in advance that the scenarios involved a propylene refrigeration compressor trip, which is documented in an emergency procedure, and that a secondary valve failure also would occur. The details of the valve failures were not disclosed. Each operator completed all four scenarios on a single day.

4.4.3 Operator Performance

To ensure consistency, a script was created for implementing each step at the same time for all the scenarios. This included all the required steps by the cracking panel operator and field operator as well as changes to the simulator. The simulator trainer performed the steps and gathered the results. In addition, a feed ramp and a program to shut down pumps in the same sequence and timing was used. A process historian validated the timing of the results entered by the trainer.

Tables 4.1 and 4.2 show the descriptive statistics. Use of safe park significantly impacted operator perceived workload and the amount of material flared. Specifically, operators indicated lower perceived workload and the system released less material to the flare when using the safe park automation compared to without it. In terms of alarm design, operators reported substantially lower perceived workload when using smart alarming versus the non-rationalized alarm schemas. However, about the same amount of flare was released with non-rationalized and smart alarming. These statistically significant reductions in the performance variables measured in this experiment present valuable evidence that implementing alarm rationalization schemas and automation techniques to assist the human operators does improve system performance.

4.4.4 Implications

The risk of a loss of containment event increases during outages because of the sudden changes in pressure and temperature. Freeing up the panel operator from the bulk of the tasks and preventing emergency alarms from being buried in an alarm flood can allow operations to recognize these events early and respond quickly to minimize their severity.

Using alarm automation to reduce the number of alarms to five for a unit outage was new; this was termed a "no brainer" by some of the operators involved in the test.

Table 4.1   Alarm design and automation affected perceived workload.[a]

	Not rationalized mean (SE), 1–100 scale	Smart alarm mean (SE), 1–100 scale	Average
With safe park	58.4 (5.8)	34.5 (6.8)	46.4 (5.5)
Without safe park	73.0 (4.5)	57.1 (5.9)	65.0 (4.5)
Average	65.7 (4.2)	45.7 (5.8)	

[a] SE is standard error of the mean.

Table 4.2   Alarm design and automation also impacted the amount of flaring.[a]

	Not rationalized mean (SE), megagrams/h	Smart alarm mean (SE), megagrams/h	Average
With safe park	19.8 (5.2)	10.5 (3.2)	15.1 (2.5)
Without safe park	31.9 (8.6)	36.6 (6.5)	34.3 (5.6)
Average	25.8 (4.8)	23.6 (4.6)	

[a] SE is standard error of the mean.

The operators were familiar with the upset automation, as it was implemented at the site shortly after the start of the simulator training program 20 years ago and has been very helpful in unit upsets. We now have data that support the benefits. One of the keys to developing upset automation was testing the software on the simulator.

Overall, results showed dramatic improvements for use of advanced alarming techniques (state-based) and upset automation. Operator response time, mental workload and flaring were reduced by 35–70%.

4.5 AI INTEGRATION

4.5.1 Pattern Recognition

In "How to Create a Mind" Ray Kurzweil discusses how humans have a weak ability to process logic, but a very deep core capability of recognizing patterns. Humans use our neocortex which is a powerful pattern recognizer and is not ideal for processing logic (Kurzweil 2012).

Gary Klein's RPD (Recognition Primed Decision) model explains how humans pick up cues, which are then compared to existing patterns. Possible responses are then generated, and a mental simulation is used to validate the chosen response. This is seen as a continual activity that is checking how the situation responds to the actions (Klein 1998).

AI can accomplish amazing things and so can the human brain. Why not design systems that can utilize the best of both. Let us call this **II with AI** (intuitive intelligence with artificial intelligence)

Refer to the *diagram* below which is based on a model Dr. Gary Klein used in his book "The Power of Intuition: How to Use Your Gut Feelings to Make Better Decisions at Work" (Klein 2003):

- **Mental models** – Everything we perceive, and our interpretation of this information is filtered through our mental models (our beliefs and biases).
- **Biases** – Everyone has different mental models depending on their experiences that will influence their decision-making.
- **Situation** – Malfunction, process deviation, anomaly, and contradiction.
- **Physiological responses** – This is how stress can affect us. The Window of Clarity in Chapter 4 covers this in more detail.
- Cues
  - **Panel**, visual: information from process graphics (alarms, process variables, trends, and video), audible alarms.

- ○ **Field,** visual: leaks, gauges, tactile: temperature, vibration: audible: leaks, flow changes, PSVs lifting, and smell.
- **Patterns** – This is where we notice the relationships of all the information we have observed. We use our mental models that we form from past experiences and training.
- **Compare to normal** – Because we work in a man-made environment it is more difficult to recognize patterns unless we are familiar with what the normal conditions are. Trends are an excellent tool for this.
- **Action scripts** – These are the possible responses to the situation.
- **Mental simulation** – Based on the chosen action, imagine what would happen if I _____.

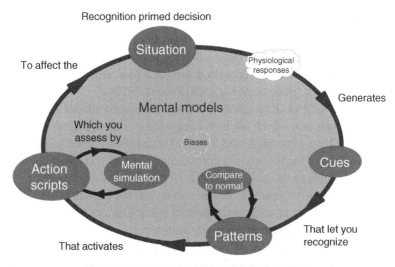

Adapted from "The Power of Intuition" pg. 26, by Dr. Gary A Klein

AI can quickly compare large volumes of data of past process history to detect anomalies. We are unable to fully measure how much data the human mind takes in and processes that helps us detect patterns. They are both very impressive. How do we get them to function optimally together? We know an alarm system can easily be filled with information. However, it is not in the form from which the human mind can easily create and identify patterns.

There has been a reliance on alarming to notify operators of abnormal situations. Alarm summaries usefulness degrade rapidly when more than one page is full. Smart alarming can be very beneficial, however not every situation can be predicted. Notifications could be built directly into the graphics to demonstrate the amount of error visually. For example, it could depict that the alarm is in but also how much deviation there is. This would help operators prioritize right from the same graphic or overview, so they do not have to lose the big picture of the scenario.

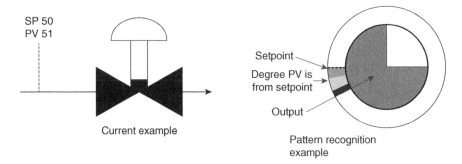

Current example

Pattern recognition example

Objects like the one above on the right could be used on the graphics. The center circle indicates the output of the controller or the amount the valve is open. The color below indicates the degree the process variable is from the setpoint. Progressing from blue to red to indicate severity.

I believe objects like these would increase the operator's ability to assess situations based on pattern recognition. If they were built into an overview of the whole process secondary responses could easily be determined.

4.5.2 Training

Traditional training involves memorizing information, restating concepts in a linear manner, and following procedures. As our control systems, safety systems, and processes become more complex the style of training has not evolved. Dr. Gary Klein's RDP model demonstrates the importance of developing pattern recognition skills.

An effective way to develop pattern recognition skills is scenario-based training. Scenario-based training is a form of experiential learning. It helps the learners to look at the process in new ways and to see the interconnectedness of the different parts of the process. It helps the learner develop more sophisticated mental models, build tacit knowledge, and develop pattern recognition. Rather than asking the trainee how to respond to a known problem, a set of circumstances is given, and the trainee must first identify the problem and then provide the response.

Benefits of scenario-based training:

- Increased retention of information—Applying knowledge in a situation helps promote pattern development and integration with existing patterns.

- Development of tacit knowledge—This is the knowledge based on experience and allows us to respond to situations when there is time pressure.
- More engagement—Interactive training draws in the trainee versus passive training where the trainee is only listening or reviewing information.
- The development of the scenarios will draw out the experience of the experts in the field.

Below are two examples of scenario-based training styles.

4.5.2.1 DMX (Decision Making Exercises)   These were developed for the COP (Center for Operator Performance)[1] by Macro Cognition (Dr. Gary A. Klein). This training was adapted from the Tactical Decision Games that John Schmidt had developed for the US marines.[2] The trainee is given a set of process conditions and first has to detect the problem and then explain how to fix it.

A manual was created on how to develop this style of training which includes the cognitive elements of decision-making. DMX is licensed by Macro Cognition and is available for use by COP members.

4.5.2.2 ShadowBox   A benefit of this training is it breaks down the common decision points used to work through problems. For trainees, this can be helpful as they may not know how to begin to solve a problem. ShadowBox training comprised of three main components: scenario, decision points, and expert feedback. The training presents challenging scenarios through text, images, and/or video. At various points throughout the exercise, the scenario is paused, and the trainee is required to answer a decision. These decision points often ask the participant to reflect on the situation up to that point. This might include prioritizing a list of options (e.g., what is most important to you), or diagnosing what they think is wrong, or anticipating what could happen next. Each decision point taps into the cognitive dimension of the problem, and makes the trainee reflect on their knowledge, goals, and expectations. As part of the decision point,

---

[1] www.operatorperformance.org.
[2] https://mca-marines.org/blog/gazette/designing-good-tdgs/.

the trainees must write down the rationale for their decisions. Finally, after each decision point the trainee gets to see how a group of experts responded to the same prompt. At this point, they get to compare their own responses with the experts' responses and capture insights they might have missed.

ShadowBox training bypasses the expertise bottleneck that many organizations experience. By capturing the expert feedback upfront and packaging it within the scenarios, the experts are not required to sit in and/or facilitate the training. Rather, trainees can complete the scenarios independently or as teams on their own time. Alternatively, trainers can lead ShadowBox training exercises with individuals or groups to enhance engagement and learning.

ShadowBox is also very good at capturing the knowledge of experts. Experts take part in the scenario creation process. As previously mentioned, an expert may not know how they know something or how they arrived at a decision. Developing these scenarios allows them to reflect on how they have made past decisions and break them down into decision points.

There is a Web App available from ShadowBox LLC. A training manual was developed for the COP and is available to COP members or through ShadowBox LLC at www.shadowboxtraining.com.

The original concept was developed by Neil Hintze who was a Battalion Chief with the New York City fire department. He developed the training to capture the knowledge of expert firefighters after the 911 disaster in New York. Dr. Gary Klein and his associates have continued to develop this style of scenario-based training and it is now licensed through ShadowBox LLC.

4.6   CASE STUDY: TURBO EXPANDERS OVER-SPEED

This is example was taken from the authors book "Troubleshooting Tactics: How Process Operators Make Critical Decisions" (Besuijen 2022). In this case, the automation failed, and the operator had to intervene. It involves a Turbo Expander in an ethylene manufacturing facility. The emergency block valve located on the inlet to the expander failed to close during a safety system shutdown. A high level in the expander suction drum initiated the shutdown. Refer to the diagram on the following page.

The shutdown logic included the following steps.

- Close the shutdown valve on the expander suction.
- After 30 seconds close the compressor shutdown valve and the compressor false load valve (this step is to allow some time for the compressor to slow down the machine).
- Close the suction pressure control valve if it is in auto (this was to prevent the valve from winding open when the machine was tripped. It was not part of the safety logic but initiated by it).

This logic looks like an effective way to safely shutdown a rotating machine.

Let us look at the events that led to this machine exceeding the high-speed trip by 12,000 rpm. Refer to the diagram below.

- A new shutdown valve was installed on the expander suction (the step to heat the internals of the valve was missed before the installation. This would have reduced the porosity of the metal and its ability to hold moisture).
- Three days before the event the machine was unstable, and the decision was made to not exceed 19,000 rpm. This was managed by placing the expander suction pressure controller in manual at 70%.
- A large rate reduction was initiated to minimize flaring. This was required as the facility was being shut down due to an internal exchanger leak that was contaminating the product.
- Since the expander suction pressure controller was in manual the pressure in the DeMethanizer and Methane stripper dropped by 250 kPa (35 psi). This led to a large amount of vapor leaving the DeMethanizer overheads and filling up the reflux drum. The reflux drum then overflowed and filled the expander suction drum that initiated a high-level trip to the machine.
- Unfortunately, the expander suction shutdown valve did not close and the machined exceeded the high-speed shutdown.
- The panel operator closed the expander suction pressure controller after 50 seconds and stopped the machine.

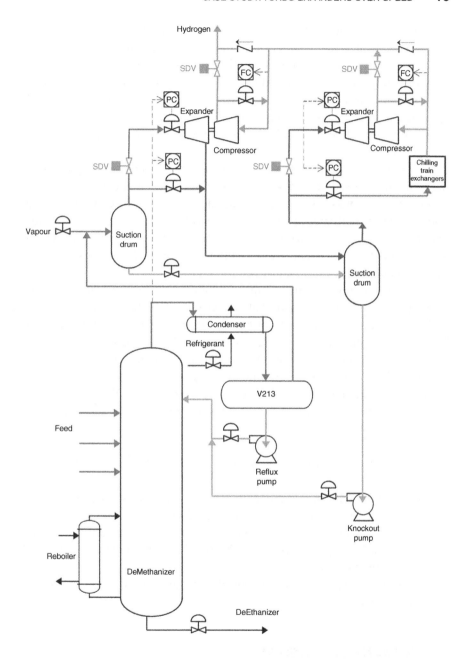

This is an example of an event where the quick response of the panel operator prevented the destruction of the machine. In hindsight we can see this was an obvious solution to solve the problem. It would be easy to overlook that they were also responding to a 50% reduction in feed that would have resulted in several pages of alarms and managing the impact of off specification product in the pipeline. There are many important process moves that would be required by the panel operator that will not be mentioned here.

Somewhere in the middle of all that the panel operator recognized the problem and closed the valve, within 50 seconds of the machine overspeeding. There was no procedure that addressed this issue. There was no documentation that suggested it as a possibility. There was a risk review about 25 years ago prior to the machine being put into service that would have included several experts determining what safety logic to use, with no time pressure.

The intent is not to criticize the individuals' decisions leading to this event, the intent is to recognize that our operators will have to interpret data and make decisions that have not been proceduralized or even thought of. There will be times when not every sequence of events can be anticipated. Which is why operations must be trained on how to sort through data, understand the critical data and how to move to a decision.

As a side note these are the changes that were made to the Turbo to prevent re-occurrence. The more experiences like these we share the safer our industry will be.

- Heat treated the expander shutdown valve internals to prevent ice formation.
- Review the type of valve packing on the expander shutdown valve to ensure it is not impacted by the very cold service.
- Delay the closing of the compressor shutdown valve and false load valve until the speed is below 5000 rpm. Flow through the compressor slows the machine down.
- Close the expander pressure control valve during a shutdown if it is in auto or manual.

4.7 HUMAN-CENTERED AI

In his book "Human-Centered AI" Ben Shneiderman (2022) describes the difference between one-dimensional thinking of automation and two-dimensional HCAI. The diagram below shows the concept of one-dimensional thinking of automation with the ability of humans to interact with the system dropping off as the automation increases.

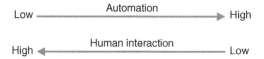

Ben Shneiderman believes to design systems that are safe, reliable, and trustworthy, a two-dimensional approach should be used as indicated in the diagram below. Automated systems are being used more frequently to increase process safety by automating responses to known hazards. The challenge is not only whether you can place the process in a safe place, but also can you prevent events that unnecessarily shutdown production which can potentially cost millions of dollars.

These systems are only as reliable as their weakest link. The sensors or transmitters are as important as the automation. If there is only a single input that has failed, it may not indicate when a corrective action is required or result in unnecessary outages. Newer systems are designed with three transmitters and a two out of three-voting strategy. One of the transmitters can be bypassed at a time which ensures the integrity of the safety system and allows for the removal of a faulty transmitter from the system by operations to improve reliability. This places the system in the upper right quadrant of the diagram.

The final element that isolates the system is also a critical component as demonstrated in the Turbo Expander example. The author is also aware of an event where the trip valve to a large steam turbine did not close during a pump swinging event that bounced the oil pressure low enough to activate the machine trip. The oil pressure recovered quickly leaving the trip valve activated but in the open position. The valve had seized because it had not moved for several years. They are now partially closed periodically to ensure movement. Newer isolation systems are using a double block and bleed setup to ensure isolation, especially on fired equipment.

The safety of the system can be maintained by not allowing the automation to be completely bypassed if it is a safety system. Only redundant components can be bypassed. Since redundancy is expensive it is not usually required for automation that is designed for controlling or optimizing the process. In these instances, the ability to turn off the automation is required to allow operations to take control of the system in the event of a problem with any of the components.

The safe park application described earlier allows the operator to run the whole application, run each system segment separately or to disallow any of the steps separately. This is an example of high automation while allowing high human interaction. All the functions are also on one graphic to allow monitoring of the application as it completes the steps and validate when it has completed.

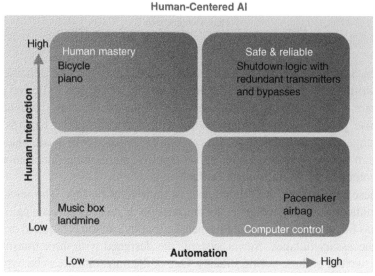

Diagram adapted from "Human-Centered AI" Ben Shneiderman

4.7.1 Case Study: Boeing 737 MAX

In October 2018 and March 2019, Boeing 737 MAX passenger jets crashed minutes after takeoff. These two crashes led to 346 deaths. The 737 MAX was an updated version of the 1960s era 737NG. The crashes were initiated by the failure of one of the two angle of attack sensor. Only one of the sensors was used in the new flight control software, the Maneuvering Characteristics Augmentation System (MCAS). The MCAS software was intended to compensate for changes in the size and placement of the engines on the MAX as compared to prior versions of the 737.

In the first crash the pilots were unaware nor trained in the use of the software. In the second crash the pilots were aware but still not trained in the use of the new software.

4.7.2 AI Mental Models

The Boeing example highlighted the need for pilots to understand how the automated systems work, including their limitations and boundary conditions. The author's experience is the more automation that is added to a system, the more difficult it is to troubleshoot when problems arise. The automation is ensuring the process is kept in a safe state with the added consequence of making the operators job more difficult. Assuming reliability is a consideration.

When one of the components of a complex AI system is malfunctioning, how do we know what the problem is and how to fix it. This could be as simple as a transmitter out of range. Dr. Gary Klein terms this as the Artificial Intelligence Quotient (AIQ 2020).

In order to troubleshoot advanced computer control systems, we will have to have a mental model of how they work. What inputs does the system use and what variables does it use to control the process. There may also be constraints used to prevent known system limitations from being exceeded. How this is displayed graphically will also assist in problem solving.

4.7.2.1 Mental Model Matrix   In 2019 Dr. Gary Klein and Joseph Borders conducted a study with the use of a high-fidelity simulator to understand the mental models of petrochemical operators. This study led to the development of the mental model matrix.

	Positive (+)	Negative (−)
System	How the system works: Parts, connections, causal Relationships, process control logic	How the system fails: Breakdowns and limitations
Person	How to make the system work: Detecting anomalies, appreciating the system's responsiveness, performing workarounds and adaptations	How users get confused: The kinds of errors people are likely to make

An understanding of how the system works is a common explanation of what a mental model is. What is not often considered is an understanding of how a system fails. What are the system's limitations and vulnerabilities to failing? Most of this understanding is what we gain from experience and will fall into the tacit knowledge category. This is not always shared in our manuals and procedures.

A key operating skill is to find a system's weakness. Many engineers have left the control room deflated after an interaction with a highly suspicious panel operator after requesting they try something new with the process. A tip for engineers: before you go to the control room, think of all the problems that could happen and develop mitigations to manage them. Be clear of what you would like to achieve, and operations may have some suggestions. Some operators will insist that this be well documented.

The next step after a failure is developing workarounds to stabilize the process, keeping on specification, and staying online when possible. Do not rule out the possibility that the best decision to make is to shutdown and isolate the facility (safe park).

One last aspect to consider about mental models is considering confusion. This could include the information presented in the graphics and when communicating with others. Being very concise and precise with requests of the field operator can accelerate problem-solving.

REFERENCES

AIQ (2020). Artificial intelligence quotient, helping people get smarter about smart machines. In: *Psychology Today*. Posted July 1, 2020, https://www.psychology today.com/us/blog/seeing-what-others-dont/202007/aiq-artificial-intelligence-quotient.

Besuijen, R. (2022). *Troubleshooting Tactics: How Process Operators Make Critical Decisions*. KDP.

Borders, J., Klein, G., & Besuijen, R. (2019). An operational account of mental models: A pilot study. International Conference on Naturalistic Decision Making, San Francisco, CA.

Klein, G.A. (1998). *Sources of Power: How People Make Decisions*. Boston, MA: The MIT Press.

Klein, G.A. (2003). *The Power of Intuition*. New York: Doubleday.

Kurzweil, R. (2012). *How to Create A Mind, The Secret of Human Thought Revealed*, 41–53. New York: Penguin Books.

Loft, S., Sadler, A., Braithwaite, J., and Huf, S. (2015). The chronic detrimental impact of interruptions in a simulated submarine track management task. *Human Factors* 57 (8): 1417–1426.

Shneiderman, B. (2022). *Human-Centered AI*. Oxford University Press (February 10, 2022).

Simonson, Richard J., Joseph R. Keebler, Elizabeth L. Blickensderfer, and Ron Besuijen, "Impact of alarm management and automation on abnormal operations: a human-in-the-loop simulation study", Applied Ergonomics, April 2022.

Wickens, C.D. (1992). *Engineering Psychology and Human Performance*, 345. HarperCollins.

5

AI AND ALARM ANALYTICS FOR FAILURE ANALYSIS AND PREVENTION

Steven Apple

Global Director of Operator Performance Services, Sneider Electric USA, Belton, TX, USA

OVERVIEW

Alarms are what might be considered the "garbage bin" of plant automation. Alarms are where you stuff those things not accomplished by automation or human machine interfaces. It is where you default to asking the operator to intervene in the running of the operation to prevent failure, improve product quality or maintain production capacity. As a result of this, the data in the alarm system is a ripe target for investigation of potential plant improvement and failure reduction.

Alarms can be either a help or hindrance. Bad alarms basically are thieves within the control system. They steal both an operator's time and attention during normal operations, and they steal an engineer's time figuring out what went wrong when they are at the center of an incident. The best alarms systems have a perfect alarm balance. Not too few and not too many.

Artificial Intelligence in Process Fault Diagnosis: Methods for Plant Surveillance,
First Edition. Edited by Richard J. Fickelscherer.
© 2024 John Wiley & Sons, Inc. Published 2024 by John Wiley & Sons, Inc.

In alarm rationalization projects, it is normally the goal to reduce the total number of alarms to meet an established metrics (as recommended in EEMAU 191 for instance). And yet, the goal of an engineer is to be sure that everything that can go wrong has an associated warning. For a long time, the latter of these was an accepted practice. With the passage of time, alarms started being seen as being contributory as opposed to assistive to industrial incidents. An overabundance of alarms led to missed important alarms while handling unimportant alarms. (See HSE information sheet 6—Better Alarm Handling for one example.) With this as a driving force, then the former took precedent, and people started reducing alarms—sometimes with almost reckless abandon, and that also led to incidents. A good balance is now understood to be the correct path.

5.1 INTRODUCTION

Previous chapters have already emphasized the use of alarms, how they are managed, and how they are designed for use in an automation system. Here the potential methods for improving alarm systems using artificial intelligence and machine learning will be discussed. This can be done in two fashions:

1. Post-alarm assessment for future prevention or improvement (i.e., using AI/machine learning to troubleshoot and engineer solutions).
2. Pre-alarm (real-time) assessments that either reduce or improve the quality of alarms (i.e., deployment of AI or machine learning for real-time benefit).

A warning is appropriate at this point. Always fix the basics of the alarm system before trying to complicate the alarm system with complex analytics and/or enhanced/advanced alarming. Experience indicates that a high percentage of alarm system issues can be remedied with simple fixes. Perhaps this might even negate the desire for a more complicated analysis. So, adhere to the KISS principle as in all engineering endeavors.

In the case of post-alarm assessment, this means automation engineers should not attempt to work with an overloaded alarm system. Much time would be wasted assessing alarms that are either unimportant or lead to arbitrary conclusions.

And in the case of pre-alarm analysis—it is entirely possible to complicate a system that may already be confusing in its application and practical use.

I have yet *to* see *any* problem, however complicated, *which, when you looked at it in the right way, did not become* still *more* complicated.- *Poul Anderson. American Science Fiction Author 1926–2001 (Anderson's law)*

Also, as a further warning—try to make any applied fixes become a part of the Basic Process Control System (BPCS)—the control system in use. Some control systems have advanced processing capabilities that allow for direct integration of complex solutions. Others may require more roundabout ways of applying a discovered improvement. Always balance the risk of non-use against the value of the fix to the operation. And be sure to add clear guidelines into the background, development, and application of any fix for the benefit of future maintenance. If implementation on an external system is required, consider the following prior to installation:

- Will the operators use it? Many systems have been installed that lay unused less than a year after installation due to different user experience, operating system, accessibility, etc.
- Will corporate cybersecurity rules allow for the necessary interchange of data?
- Will the control and external systems integrate and communicate effectively together?
- Does the cost of proper installation and maintenance outweigh the value?
- Who will maintain and proliferate upkeep?
- Is there another way to get to the same solution?

One other note is that in the past, some BPCS vendors have insisted that anything NOT deployed on their system may have warranty implications. This may need investigation before proceeding with any non-BPCS deployment.

It may seem ironic to be issuing such warnings in a book that explains their use. A control engineer should always be aware of the limitations and drawbacks.

5.2   POST-ALARM ASSESSMENT AND ANALYSIS

Post-alarm assessment and analysis is performed on an existing alarm system and can be performed on two unique databases normally held within the databasing structure of the control system. These are the Alarm Configuration Database and the Real-time Alarm Database. These can be used for analysis separately or combined, depending on the type of analysis being performed. Note: The author does not know of a control system that does not allow these databases to be established and accessed. However, data management policies users sometimes do not enable the collection and retention of these data for various reasons.

5.2.1 Alarm Configuration Database

The alarm configuration database is a static database. It is a subset of the complete control system database. It contains settings for each I/O point in the control system to which an alarm can be assigned. The fields, and the richness of these fields vary from system to system and contain information related to the alarm assignments envisioned by the system manufacturer. Some systems have capabilities to add user fields if desired (e.g., lot number for batch processes). This alarm configuration database consists of fields for the tag name, definition, and settings (HI/LO/etc.) for each point to be alarmed.

This static database may often be represented as what is referred to as a Master Alarm Database (MADB). ISA 88 is also establishing the need for a similar database for batch processing. Many activities within the alarm management life cycle, as defined in ISA 18.2, make use of the MADB. It becomes a certified MAD when it has undergone a process of rationalization in which the rationale for each alarm has been justified and qualifying information has also been included as part of the database. Such information includes (but is not limited to) The Cause(s), Consequence, Corrective Action, and Criticality (the four Cs) of each alarm. These are included in addition to the information already contained in the configuration database. Various control systems contain differing configuration parameters. Be certain to educate yourself of the specific parameters available for any given control system.

Completion of a rationalization exercise is covered in other sections of this book. However, it may be recognized that the existence of this database yields the simplest layer of AI application in an alarm system. For example, this database can be utilized in real-time to allow operational access to alarm information in the form of a KnowledgeBase (KB) associated with each alarm. Some control systems allow right-click context-sensitive access to this KB.

The Configuration/MADB database itself can also be statistically mined post-development for many important pieces of information relating to the alarm system. First, it should be statistically analyzed to establish the mixture of priority assignments with the alarm system. According to EEMUA, the best systems use four tiers of alarms. The highest tier should be assigned to less than 5% of the alarm total. And the third tier to 80% or greater. The fourth tier being used for informational, or alternatively directed alarms (e.g., not to the operational console). That fourth tier is not considered in comparing total alarm priority distribution, as it is not limited in scope or size, and has little effect on alarm activation or operator loading. ISA 18.2 also has content on recommended distribution of configured alarm priorities. Refer to ISA 18 TR8 (not yet released at the time of this writing) for proper use of operator ALERTS.

Not yet discussed here is the real-time alarm database (next section). However, it is also possible to begin to step through various layers of this priority distribution utilizing actual alarm activation and discover such things as:

- What is the priority distribution of alarms actually activated (i.e., not just assigned).
- What is the priority distribution of alarms during specific events or incidents.
- How does the priority of alarms affect the operator actions (utilizing the operator action journal).

Other layers of analysis might be imagined. Probably some with diminishing returns. The key to this analysis is in the prevention of failures by proper prompting of operator actions, and proper prioritization to be certain the human interaction—when called for—is focused on speedy resolution. Incidents stemming from improper focus of such interaction should be a flag to the engineer/scientist to investigate better methodologies to prevent repeats of similar incidents.

Recently, a special priority of alarms has been defined. These are referred to as Independent Protection Layer (IPL) alarms. They are included as part of a safety program's Safety Integrity Layer (SIL) as outlined in ISA 84. It should be obvious that these are highly important alarms requiring special attention because the human is now being utilized as a recognized safety layer in the plant. In other words—these alarms require immediate action above and beyond even priority one alarms. For statistical treatment, these alarms should be included with priority one alarms in any analysis.

5.3   REAL-TIME ALARM ACTIVITY DATABASE AND OPERATOR ACTION JOURNAL

These databases (often combined as one—The Alarm and Event or A&E database) contain the historical archiving of alarms and operator actions as they occur. A typical alarm activation entry will contain a time stamp, the

I/O that was activated, the label for that alarm, the type of alarm (Hi alarm for instance-batch number if a batch operation) and perhaps a bit more information depending on the configuration setup. A typical operator action contains similar information, with the exception being that the type being an operator action (ACK or CLEAR for instance). It may also indicate a setpoint change or similar activity.

A quick warning is in order before utilizing data from such a database. Be extremely careful when analyzing events to be certain that the internal clocks on any databasing operations are properly aligned. Analysis of abnormal situations often involves data originating from a variety of sources such as LIMS systems, plant historians, material inventory systems, etc. Each of these systems typically provides their own unique time stamp to the data they store. Many data scientists have chased ghosts simply because of poor clock synchronization between separate sources.

EEMUA 191 and ISA 18.2 call for some very specific statistical analytics to be done with these databases, and most commercial packages for alarm analysis do that and even a few more. These assessments are discussed in other chapters or can be found in EEMUA and ISA documents.

Following are some areas of investigation that might prove valuable. Those analytics fall into two basic arenas:

1. Post-event analysis
2. Real-time analysis

Post-event analysis refers to the examination of data using the historical database to establish patterns of behavior that can be assessed and identified as being causal in their effects on avoidable events or incidents.

Post-event analysis is very useful in assessing process faults, and their taxonomy. Once the process of an event has been established, several things can be done:

1. Identify changes necessary to either the design or procedures within the process operation that will prevent or alleviate damages.
2. Set filters in place to identify any repeat of this pattern and either automate prevention or allow awareness of the potential for occurrence. These filters are intended to trap or isolate known conditions that lead to failure.
3. Set operational regions/envelopes that alarm when an abnormal operating state is seen based on previous data patterns within previously reviewed operating data. Sometimes these sorts of methods can interact with first principles models to identify potentially hazardous states.

4. Assessment of operator actions can yield similar results and can sometimes reveal non-acceptable behavior of operators during either normal or abnormal operations. Such behavior is usually remedied by either training or enforcement of approved procedures. Alternatively, studies of repetitive operator actions may reveal automation of the actions as a consideration. ISA 106 has a standard that discusses methods of procedure automation and control.

5.4 PRE-ALARM ASSESSMENT AND ANALYSIS

This practice encompasses what might also be termed "smart alarming" or Expert Systems. Caution is advised when invoking such methods because there may be a tendency by some engineers to become entranced by these methods and overuse. It may be prudent to establish utility and effectiveness of such systems in pilot plant environments before considering their use in manufacturing.

You have not seen an alarm flood until you have seen a "smart alarm" flood. Steve Apple (author)

The following practices should be followed:

1. Do not invoke expert rules or smart alarming on a separate system other than the BPCS unless it is either a) the only way to invoke the necessary alarms or b) the value of the alarms is so great that operators can be expected to actually utilize that separate system.
2. Utilize as much of the internal calculation, importing and reporting tools available on the BPCS as possible. Modern BPCS vendors are including much more of this sort of capability contained within their systems. Experience shows that an operator will learn to use and devote all their attention to a singular system.
3. Cyber security protections and sign in procedures have lately made this an even bigger issue. This is especially true in a system where the values are sent from the BPCS to another system and annunciated there or sent back to the BPCS for alarming.
4. Do not invoke smart alarms mildly—that is, alarms that can be easily discerned by other methods, or directly on the BPCS unless the BPCS alarming system is being replaced with an external system.
5. Be careful to NOT alarm on potential events that may NOT happen. Users show little patience for a system that produces too many "false positives.". This is somewhat akin to the little boy who cried wolf and instills immediate distrust in a system.

6. Other than Independent Protection Layer (IPL) alarms (discussed later) do not include safety system alarms in the mix. When the safety system takes over, the operator is out of the loop.

While these guidelines act as a warning to those undertaking AI activities, it should be clarified that third-party systems may offer many advantages. After-market add on systems often offer ease of use, and implementation advantages that have not yet been implemented in BPCS vendors' products. Carefully review the advantages of both paths when making a decision about implementation of AI.

See articles in InTech April 1999/July 1999 referring to manufacturing use of AI (i.e., real-time expert systems) in pharmaceutical processes. These were authored by Joe Alford and others at Eli Lilly.

5.5   UTILIZING ALARM ASSESSMENT INFORMATION

Good alarming is intended to prevent failure, and potential incidents in a processing plant. There are facets of the alarm information that can be utilized to resolve failure issues. Alarms are intended to point to PROCESS upsets or abnormal situations. The first rule of a good alarm is that it must be associated with an intended operator action. Any other uses of the alarm system, begin to fall in the category of notifications (e.g., alerts, prompts and notices). Also, whether an alarm is simply journaled for historic purposes or made available for operations use is an important consideration.

Alarm assessment has two potential uses for the operator.

1. Presentation of additional assistive information for operator actions
2. Design of a better alarm (discussed in more detail in following sections)

One example of utilization of these analytics is to make the results available to the operator in the form of notifications/decision support systems. ISA 18.2 is in the process of releasing a Technical Report (TR-8) addressing the use of operator notifications. This TR addresses how to utilize additional warning/notification type. This TR can be obtained from ISA for guidance on presentation of assistive notifications.

A technique already in use by many BPCS vendors is to add "right-click" access to additional alarm information. In this manner, operators who see an alarm on a control system display may request additional information to resolve the alarm. This information may be something simple- such as the information from the MADB. This includes causes, consequence, corrective action, and criticality. It also may be something more complex such as complete procedures to be followed or access to calculation engines to analyze what to do in a complex situation.

Such information should be directed to an Human Machine Interface (HMI) screen separate from the current display being used for operations monitoring. Alternatively, sidebars, or lower screen areas may be used to display such information. Always keep the main HMI screens clear for operational use.

5.6 EXAMINING THE ALARM SYSTEM TO RESOLVE FAILURES ON A WIDER SCALE

As previously stated, much of what the alarm system produces is to cover for the shortcomings of the existing control system. However, alarms are also a very important part of the operation of a process. Alarms cover failures that are either mechanical or systemic in nature for which automation does not suffice.

Traditional/historically successful methods focus on:

- Sequence of events module;
- Use of first principles to determine likely root causes;
- Use of simple data analytics to identify redundant, symptomatic, consequential, or correlated alarms with heuristic thresholds;
- Use of data analytics to identify problem areas with upsets related to transitions, out of service and out of suppression states;
- Use of data analytics to identify problem areas with chronic alarm shelving.

These are discussed below.

5.6.1 Sequence of Events (SOE) Module

Most DCS vendors make available a SOE module. That module extracts— usually on a trigger of a specific event—a chronological record of the data and events that led to that event. It is often capturing events on a sub-second or even microsecond level. Without use of an SOE module, SOE may also be established after the fact utilizing a time-stamped list of alarms and operator actions. SOE information may then be cross-referenced with other data in the control system at the time of any given incident or upset. This may require accessing data from more than one data storage location. Thus importance of synchronization of time clocks used in examining data of such a high-fidelity nature.

SOE is often useful in post-event analysis of failure patterns to gather sufficient data to examine the taxonomy of a given event. In some regulated industries, post failure event analysis is mandatory whenever a failure event includes parameters that may affect product quality.

Sometimes, multiple events must be captured to begin pattern recognition signature analysis. However, a singular examination may sometimes yield results significant enough to further failure mode assessments or root cause analysis. Software packages can be purchased with a variety of tools for such analysis.

5.6.2   Use of First Principles to Determine Likely Root Causes

First principles models are useful in defining the thermodynamic, material balance, and physical events that can lead to failures in the system. These models may be used in the form of a singular unit process model representing the unit under consideration, or a more complex Operator Training System (OTS) application. The OTS may prove to be more useful for the interactions that may occur from multiple processes external to that under consideration. OTS models are usually broader in nature. Additionally, OTS systems often facilitate the investigation of different "what-if" scenarios. Though, these can often be replicated in a simple model by running several case studies of the process in question and altering the input parameters to simulate the desired effects.

For one-off studies of odd behavior, the simple model may prove desirable because full OTS systems are costly to purchase and build. A full OTS model is justified when undertaking multiple studies, or where multiple units are to be considered with their interactions.

These types of tools have been widely in use in the process industries both for investigation and training since the mid-1990's and have become much more viable as the market has grown.

5.6.3   Use of Simple Data Analytics to Identify Redundant/Repetitive Alarms

As the use of digital control systems became more prevalent and alarms became easier to introduce to these systems, the overloading of these alarms became more noticeable. A few early events drove studies that began to examine the depth and breadth of this issue. Many incidents have been caused by overloaded alarm systems. And thus, the general field termed Alarm Management.

In the early 1990s EEMUA (The Engineering Equipment and Material Users Association) in the UK published their EEMUA 191 guideline on the effects and maximum tolerance limits of alarms. This publication established (specifically for DCS-controlled industries) the types of nuisance alarm issues seen by operations. In addition, they established key performance indicators (KPIs) for various types of alarm metrics. In addition to that, they recommended a balance of alarm priorities to properly direct operators' attention to the most important alarms.

These KPIs can be quickly calculated utilizing simple statistical tools. They require a sufficient (e.g., 30–90 days worth) amount of data from the alarm system and a statistical analysis toolkit. The ISA 18.2 Committee also points to EMMUA as the official source for their KPI numbers in their published standard. Please look to these publications to find suggestions for limits and methods for remediation. Examples shown below are from the ISA 18.2.

Alarm Priority Matrix			No Impact	Low	Medium	High
M **T** **T** **R**	Immediately	0 mins - 5 mins	4	3	1	1
	Soon	5 mins - 10 mins	4	3	2	1
	Later	Greater 10 mins	4	3	3	3

Alarm Performance Metrics		
Alarm Performance metrics Based upon at least 30 days data		
Metric	Target Value	
Annunciated Alarms per Time:	Target Value: Very Likely to be Acceptable	Target Value: Maximum Manageable
Annunciated Alarms Per Day per Operating Position	-150 alarms per day	-300 alarms per day
Annunciated Alarms Per Hour per Operating Position	-6 (average)	-12 (average)
Annunciated Alarms Per 10 Minutes per Operating Position	-1 (average)	-2 (average)
Metric	Target Value	
Percentage of hours containing more than 30 alarms	<1%	
Percentage of 10-minute periods containing more than 5 alarms	<1%	
Maximum number of alarms in a 10 minute period	10 or less	
Percentage of time the alarm system is in a flood condition	<1%	
Percentage contribution of the top 10 most frequent alarms to the overall alarmload	1% to 5% maximum, with action plans to address	
Quantity of chattering and fleeting alarms	Zero, action plans to correct any that occur.	
Stale Alarms	Less than 5 present on any day, with action plans to address	
Annunciated Priority Distribution	3 priorities: -80% Low, -15% Medium, -5% High or 4 priorities: -80% Low, -15% Medium, -5% High, <1% "highest" Other special-purpose priorities excluded from the calculation	
Unauthorized Alarm Suppression	Zero alarms suppressed outside of controlled or approved methodologies	
Improper Alarm Attribute Change	Zero alarm attribute changes outside of approved methodologies or MOC	

Be warned that 90 days' worth of alarm data from some systems can be overwhelming, and can potentially overload the capacity limits of some statistical tools.

Some examples that are useful for failure analyses:

- *Redundant alarms*—these are alarms that yield redundant information for an event for which another alarm already reports the occurrence. The correct alarm should be identified, and the redundant alarm(s) eliminated. Alternatively, if reliability of the measurement is in question, voting methods (all occurring or one of many occurring) should be incorporated.
- *Symptomatic alarms*—this is generally referred to as a first-out alarm in a sequence that occurs together. Various methods- such as SOE mentioned above may be utilized to identify the important alarm and eliminating the now superfluous alarm(s).
- *Consequential alarms*—this type of alarm is one that occurs regularly— with some level of percentage certainty attached and within a given time interval—after a given alarm.

  The inclusion of the certainty (percentage occurrence) and time interval require a bit more complex statistical analysis and therefore more computational effort. And longer time intervals or looser percentage requirements might lead to a bigger list. That bigger list also takes more time to calculate, so results may be slow in coming on the average laptop computer. This is due to the fact that the entire alarm occurrence list must be examined to establish the percentage occurrence.
- *Correlated alarms*—correlated alarms are very similar in nature to consequential alarms and contain the same sorts of time indices. However, to be "correlated," such alarm should occur with 100% certainty of the initial alarm.
- *Chattering alarms*—this type of alarm is one that occurs when an alarm goes continuously in and out of alarm with no action by the operator. It may be an indication of an alarm set too close to the desired operating limit. Such alarms are an indication of need for examination of limits and/or on/off hysteresis settings.

All of these measures are particularly useful in identifying grouped alarms for which only one indicative alarm may be necessary depending on the reliability of the instrumentation from which they are produced. It is easy to see that many alarms can be eliminated in a system that has such characteristics.

It should also be noted that consequential and correlated alarms may lead to identification of specific failure patterns if their connection is nonsensical or if they only occur during specific events.

5.6.4 Use of Data Analytics to Identify Problem Areas with Upsets Related to Transitions, out of Service, and out of Suppression States

When a process is transitioned—by design or by upset—between modes of operation, the alarms during that transition may not be useful. Therefore, the alarms set for one mode of operation may not be conducive to the new mode of operation. The word "mode" may be interchangeable with "state." Batch processes are particularly notorious for including multiple transitions, as such processes consist of a sequence of steps with a transition usually existing between steps. For example, it is common that in advancing from one step to the next, alarm attributes such as setpoint and priority may change, or an alarm's suppression state may change, depending on whether it is relevant to a particular step.

Analysis of events for these sorts of process modes requires two elements:

1. A trigger initial event that can be captured or measured.
2. The list of alarms that annunciate within a defined interval following the event. OR
3. The list of alarms that needs to be changed at the new mode of operation.

The initiating event can be identified as either an alarm, a process measurement, a process calculated parameter, or a process step relative time. For instance, when a pump shuts down, an alarm identifying the loss of the pump (a flow measurement for instance) may be the initiating event. Alternatively, in a process that is changing to a new operating mode may require a change in a flow rate of a critical operating variable. One example might be a polymer operation that is changing to a new melt-index (the quality parameter for that process). And the hydrogen feed to the reactor is changed to alter that melt-index. In both cases, the alarms following the change and down-stream from the initial element need to be altered to reflect the new reality. Another example using time may be that of a sterilization operation where cool-down is the next phase of operation at some identified time after the process has remained at sterilization temperature. (Tip of the hat to Joe Alford for this specific example.)

In the case of equipment outages (like a pump shutting down), downstream alarms need to be silenced in some fashion to prevent spurious alarms from the equipment that will naturally want to report issues that have nothing to do with their operation. Some equipment may need to be shut off, or some valves closed and some balancing take place in the unit to protect the parts of the process that remain viable during this upset condition. Alarms for "still active" equipment should not be shelved. ISA 18.2 refers to a term "suppression by design" which discusses methods for automatically suppressing alarms when appropriate.

In the case of a mode change by design in the unit, a temporary silencing of alarms may be called for or a resetting of the new operating limits and therefore the resultant alarms may need to be addressed. Some control systems have automated ways of taking care of this. Some third-party add-on systems may also take care of this but may not satisfy cybersecurity policies. Always investigate which is the best route. Most modern BPCS vendors offer shelving capabilities that allow a myriad of ways to resolve this. And some even offer state-based alarming capabilities that can resolve such issues.

In either case, the use of Boolean logic solvers is usually called for in such cases. State tables are set up and applied as per pre-identified logic states are satisfied. In order to set these states, a thorough review of the process P&IDs are used in conjunction with process knowledge to be sure a full understanding of the states are appreciated. In some cases, some downstream alarms may need to be left in place if pressures or flows still exist that may lead to a fault state.

5.6.5   Use of Data Analytics to Identify Problem Areas with Chronic Alarm Shelving

Alarm shelving is a very useful tool if available to the BPCS. Alarm shelving does not necessarily disable the alarm. In most cases, it simply silences the alarm and halts its notification to both the operator and the process screens they use to operate the plant. Such alarms are still logged in the system for later use and review—handy in cases of auditing and/or checking the shelving logic. This should also allow a list of all currently shelved alarms.

Shelving can be used to both shelve a current alarm in an ad-hoc nature (e.g., a bothersome alarm) or to automate shelving of groups as highlighted in the section above. All shelving of an ad-hoc nature should have some sort of duration attached to the system so that the alarms are re-enabled after some period has elapsed, except in the case of those that are shelved due to a logic state as outlined above. This duration is usually the length of a shift. In this way, a new shift operator knows which alarms they have personally shelved. All shelved alarms should be reviewed at shift handover to be certain a new operator is not surprised by a sudden un-shelving of a handful of alarms.

In cases where alarms are being regularly shelved, review of that alarm should be undertaken to understand why it has been shelved. And any correlations of those alarms to process operations should be reviewed as well, possibly utilizing some of the methods outlined in this chapter. This helps to avoid "chronic alarm shelving."

Also note that shelving should never be looked at to fix a broken alarm system. Achieving a clean alarm base should always precede any alarm shelving. In the early days of the availability of state-based alarming, smart alarming, and alarm shelving many looked to these tools to fix their alarm problem (I'll just make the bad ones go away). This added a layer of complexity on top of a problem that was already bad and made matters more difficult to resolve when post-event investigations were performed.

5.7 EMERGING METHODS OF ALARM ANALYSIS

In considering a BPCS system, and all of the interlaced instrumentation, communication networks and HMI, a challenge exists in how best to exhibit information to represent the process. So, temperatures and pressures are often used as surrogates to help represent either product quality or thermodynamic state. And flows rates are combined to complete the mass balance necessary to get it right. Despite all those measurements, they are only indicators of what is truly happening on a thermodynamic or mass balance level in the unit itself.

With those parameters available—valves, valve positioners, pumps, compressors, heaters, heat exchangers, feed belts, grinders, and other such equipment are then utilized to condition the material into the form or state necessary for the desired treatment.

Obviously, these parameters are only approximating what is necessary to manufacture the desired end product. And engineers use analyzers, lab checks and other instruments along the way to guarantee that a measure of quality is being reached. The commercialization of instrumentation is more refined or more readily available when higher quality is necessary. And then there is failure of equipment to deal with. All these are the responsibility of the design, process, control, and maintenance teams to work together to make a processing plant profitable. As a result, analytics has a lot of value when properly applied to point the efforts of these teams in the right direction. It may also yield results not directly available from any given sensor in the process.

5.7.1   Use of Advanced Modeling Methods to Determine Remediation

Advanced modeling in this case refers to use of first principles or high-fidelity empirical models. And normally, these fall into the category of the types of models used in OTS. OTS systems are available from most BPCS vendors. We will discuss how to use such models with alarm systems.

High fidelity empirical modeling refers to models using highly accurate and sufficient frequency data to accomplish the level of accuracy necessary to define both the dynamic changes and precise values in the data to be modeled. It may take the form of statistical models, Neural Network models, or even hybrid models combining one or more types of method.

Such highly accurate models allow the user to create what is referred to as a "digital twin" of the process. It is highly important that this digital twin accurately represents the quantitative and qualitative changes both in directionality and magnitude of change. As a result, such modeling systems are quite costly, and require extensive maintenance and updating to be certain they are matched to the operation.

Such a model is highly useful to create a simulated world in which no product can be wasted, and only simulated incidents take place. When available, alarm management solutions can first be applied to this simulated environment and tested with operations staff to verify that the intended solution achieves the desired result. This is especially valuable when the abnormal situation to be alarmed represents a safety or other hazardous condition.

As regulatory bodies make stricter requirements on operator training, these models are becoming much more in vogue. If your company has a simulator, gaining access to that simulator may prove valuable. Note also that once the change is tested, and then deployed, it means the simulator is also brought to current plant state so long as the alarm scenarios are retained.

5.7.2   Use of Automated Machine Learning to Determine Causes and Assess Interventions

Machine learning has been with us for a while. Machine learning algorithms build a model based on sample data (called training data) from multiple variables in order to make predictions or decisions not obviated by any of those individual pieces of data alone. Machine learning algorithms are utilized where it is difficult or unfeasible to directly infer the information needed to perform a task or make a decision. Machine learning is often used for other areas of process control and decision making. The following section attempts to narrow the use to areas where alarms come into the picture. And, as might be expected, there are some areas of overlap of the functionality.

Some machine learning methods are more applicable than others and have various toolkits to utilize when addressing specific classes of problems. For instance, some are better with discrete data such as switching, and some are more applicable to continuous data such as empirical process data.

One of the uses of machine learning is to make "soft sensors" to replace or replicate process readings that are sometimes only available from laboratory analysis. A soft sensor (indicating using software analysis to replicate an imaginary sensor) means other readings are used that either react or are correlated to the needed value. The extremely valuable part of this is that a lab-evaluated sample of that parameter may not be available for several minutes or even hours and the parameter may require immediate adjustment to keep the process viable.

The method for doing this is to take readings from several process parameters (usually three or more) and take lab samples at a time when those process parameters line up given process delay/transit times. To establish this delay, the amount of delay time it takes for the measured variable to transfer (via pumping of normal flow) whatever process material to the point where that lab sample is taken must be established. Then the process parameter measurements must be aligned with the lab sample analysis at a time interval that removes that process delay for analysis purposes. So, this is accomplished as an offset in the time stamps of the associated data parameters. This essentially makes the "soft sensor" as if the reading is being delivered immediately (i.e., with no lag) for control purposes.

Some industry examples:

This first example is in the polymer industry where samples to determine product quality parameters are taken almost 2 h after the actual process (the reactor) that produces the quality such as melt index or density. And then the lab takes a couple of hours to process the sample. So, in the case of this process, an instantaneous estimation of the expected lab parameter can be rendered to avoid off-spec product or equipment failures that may have not been known for another 4 h. These readings can be added to the control system for automation use and alarm use.

The second example is in the case of emission stack gases. Again, several parameters are taken at the same time a stack gas sample is taken (or aligned with lags as per above) and instantaneous stack gas analysis can be had rather than the sporadic readings from a stack gas analysis at the lab. This also helps avoid outages, and regulatory fines. Such readings are added to the automation of the system for control and alarm use.

A third example is one of predictive analysis of failure modes in water systems. Consider water treatment facilities for large municipal water companies. Unlike process systems water systems are not as highly instrumented. Consequently, expertise on the topic is usually low. Predictive analytics for "soft sensors" to predict upcoming issues is therefore quite valuable. Usually, these models are simple statistical models yielding predicted future results. In other words, they are watching trends and dynamic activity of those trends to offer the operations personnel a pre-warning that things are headed in a direction that is not desirable. It's simply a future look at the potential result of the slope of the curve. Such systems are often backed by redundant measurements to be certain false predictions do not slow or stall the process at high demand times.

A fourth example is rather unique in that an expert system was utilized to establish an entirely different method of viewing a batch cycle in a process. The uniqueness lies in that the data is viewed as a beginning to end of batch and an "operating envelope" established. Within the envelope, a proper fermentation takes place. Should the parameters exceed the bounds of the envelope, then an exception to a proper fermentation requires a deeper investigation to see if the batch is viable. A picture is worth a thousand words—here's the example. Reference provided by Joe Alford (retired pharmaceutical control engineer).

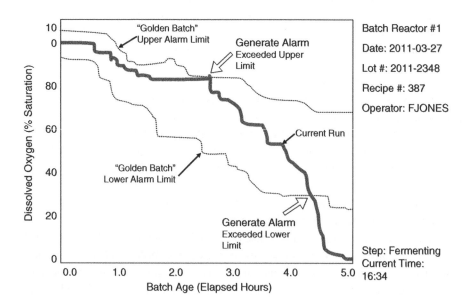

Ref ISA 18.2: Used with permission of ISA, www.isa.org

The last example is one that has been in our industry for a while and offered by an equipment vendor. A rotating equipment monitoring vendor had gathered volumes of real-time data on how rotating equipment failed over time as witnessed by their monitoring systems. So, rather than leave the interpretation to an operator, they built statistical models (and several sub classes of each) to yield early predictions of failures in bearings, motors, pumps, compressors, and any other rotating equipment for which their equipment had been used to monitor. I salute such a company for undertaking the effort, and for continuing to collect data to improve the quality and depth of the models they continue to offer. Others have followed suit.

5.8 DEEP REINFORCEMENT LEARNING FOR ALARMING AND FAILURE ASSESSMENT

Deep reinforcement learning is often used to narrow patterns to reasonably accepted norms or operating ranges. The deep part comes to play when those learned methods are applied to similar equipment or processes for which the data is not yet available but is of similar nature.

These methods can often be valuable for processes where the actual mechanisms—whether thermodynamic, material balance or mechanical are not readily apparent or not easily extracted using the instrumentation provided. However, control systems can be set to maintain a statistically narrow standard deviation of result so long as the control parameters are not allowed too much variance from the envelope of operation.

The drawback of this sort of training/learning is that it involves trial and error. And some processes pay dearly for quality or yield losses when experiments are performed.

There are relatively few examples (in fact an extensive search yields only passing references) of this sort of learning in the process industries. However, the methods have been implemented to some extent in some modern tools. More in-depth lookup of the methodologies can be undertaken if interested.

5.9 SOME TYPICAL AI AND MACHINE LEARNING EXAMPLES FOR FURTHER STUDY

A person would rather live with a problem they cannot solve than a solution they do not understand. Benzi Livneh—Owner/President Knowledge Based Engineering (KBE) South Africa

5.9.1 Boolean Logic Tables

Boolean Logic is probably the most useable AI tool for alarm management work. It is used in a large variety of applications and methods. Boolean logic is most useable when coupled with some sort of expert system or programmatic triggering mechanisms via which the rules can be engaged upon identified need. Boolean logic precedes AI by perhaps hundreds of years. It has been a mainstay of the basis for if/then/else programming since its inception. It also became the basis of many early expert system programs, and is at the heart of expert system design (discussed below). The idea being that an expert can tell you the decision that should be made given the parameters and a series of potential decisions.

Boolean logic allows methods of using combinatorial logic for complex alarming situations. For instance, when the pressure is one value and the temperature another, it may call for an alarm to prevent failure or runaway process issues. Boolean logic is also used often in what is termed "smart alarming" or "state-based alarming." In these cases, a process state is established either through the interaction of other alarms or process variables which then identify a specific process state. In the most complex uses of these sorts of Boolean tables, entire alarming profiles might be changed to match newly identified (and pre-planned) operating regions of the process. In terms of thermodynamically oriented processes, one type of feed stock may entail entirely different pressure and temperature profiles to produce a profitable product and/or maintain product qualities.

Boolean logic operations can also be combined with most of the other assessment/calculation techniques shown in this section. For example, a calculation of some sort or an identification of some complex situation may be necessary either to kick off or to be included in the midst of a Boolean set of instructions. For example, mass balances may have great value within Boolean instructions to establish next steps. Or a statistical or other complex sort of calculation may be required as part of the instruction set.

Some good examples of useful Boolean alarm techniques can be found in start-up/shut down routines. Product transitions may warrant their own set of Boolean rules to avoid unnecessary an unwarranted alarm activity. And of course, processes that see wide variation in final product qualities or initial feed stocks are certainly candidates. There are many examples of application in all these areas. Boolean logic is most often successful when the experts can establish or identify the expected results—either directly or through calculation as mentioned. They may be also used to kick the decision out to a knowledgeable human for further analysis. And they may also be used to trap unforeseen activity to an investigatory database for off-line analysis after the fact.

5.9.2   Statistical Regression and Variance

Statistical regression is quite useful in continuous data analysis. It can also be useful for event analysis that can be aligned in a logical comparative fashion for changes in a parametric system.

Variance is a basic part of regression and proves quite useful in assessing the fit, or applicability of any regression to new or unseen data. Regression should always account for this goodness of fit—usually reflected as an R squared value. The closer R squared is to one, the better the fit. Variance is used to make regression fit better to empirical data. In addition, those more skilled in the techniques may employ principal component analysis techniques to reduce the dimensionality of datasets that become overly cumbersome when involving multi-variate analysis.

There are many forms of regression. Linear regression, non-liner regression (sometimes called least squares regression), and partial Least squares regression. These have always been considered to be some of the main tools of statistical analysis, and they should be in the toolkit for any analytics process. Further study should be undertaken to fully use the capabilities, though some toolkits allow good ease of use without a lot of deep understanding.

5.9.3   Artificial Neural Networks (ANNs)

ANN's were popularized in the early 1990s when the power of computer systems became capable of handling the massive amount of computing necessary to resolve their inner workings. They were called neural networks because they were thought to mimic the way neurons work in the brain. One node passes results to another node until the nodes balance to a result to fit the data to which they are applied—therefore like neurons. They are capable of Multiple-input, multiple-output models and have several areas of applicability in the industry. There are several commercial industrial packages available for ANN use on data sets. The first two examples shown in industry examples above are quite amenable to ANN-type models. More reading can be undertaken to see the potential value of ANNs.

A valuable use of ANNs is to determine (without having to postulate a model structure such as is required for first principle and statistical models) whether or not a relationship (i.e., significant correlation) exists between a set of inputs and outputs. If a relationship is shown to exist, various modeling techniques (including keeping the model as an ANN) can then be pursued to determine which form of model is most appropriate. The application of ANN's often leads to discovery of relationships that may not be obviated by other means.

ANNs have shown to be fast in their predictions once trained. And when trained to a first-principles model, they are quite accurate in their renderings. They suffer from good fit in interpolation for areas where sparse data is available or in extrapolation beyond the data used to train the models. Some work by Kramer in the 1990s showed that by combining the results of ANNs with linear and non-linear regression techniques some of those shortcomings could be overcome. Later work by others showed that they could be combined with first principles models also to overcome interpolation and extrapolation issues. Some data reconciliation was necessary to accomplish empirical/ideal data combination. More can be found on ANNs through additional study.

5.9.4 Expert Systems

Expert systems are quite useful and rely on a KB to feed their use. The basics of any expert system is that they contain the knowledge of an expert to yield advice when specific conditions or parameters are met. Their use is often in an advisory mode, though they can also be triggered by invoking rule sets that guide or force operational states.

One commonly used installment of expert system is logic trees (or Boolean tables as identified above). Think of it as a (potentially endless) set of if-then-else statements. And think of any exception to any rule having a series of alternative solutions. If one knows enough, one could make decision trees based on these rules. And that is likely their biggest drawbacks. Early installations of these sorts of systems soon became overloaded with endless possibilities of alternatives as each caveat to each situation was examined and expanded. However, these sorts of systems were immensely useful in solving problems where the answers were known by some expert somewhere, but access to that expert was not always readily available. When combining proper alarming and advisory knowledge with human assessment, they have proven quite useful. They were used to improve production, guide operations, and avoid failures in a myriad of ways.

One example:

The largest and most successful installations of expert systems known to the author was on the old age film machines at a photographic materials vendor. The process for those materials was a "film machine" a block long. Feedstocks went in one end, and film came out the other. Start-up of such machines was 24-hour process, and so stretched across potentially three shifts. The entire process was guided by expert systems to maintain the continuity of the startup, and then remained as a guidance mechanism for ongoing operation.

The startup and operation of this unit made extensive use of many of the methods discussed in this section—including wide use of Boolean operators to guide decisions combined with various calculation techniques to decide the direction of following steps. So, they are VERY useful when the problem cannot be solved any other way.

Expert systems are widespread in their use. They are used to schedule ships to port. They are used to adjust guidance systems in spaceships. They monitor and maintain the proper arrangements of satellites in the skies. They guide the production and packaging of pharmaceuticals to a high degree of quality. Many more examples exist.

For use in alarm systems, expert systems have proven quite useful as the deployment engine for any of the aforementioned analysis techniques. Once alarm patterns and/or data patterns are identified that lead to events or incidents, Expert Systems can give either automated guidance to an operator that is above and beyond that provided by the typical basic alarm system. Alternatively, they can be used to shift control scenarios in automated fashion. There is much information available to further study expert systems.

5.9.5 Sensitivity Analysis

Sensitivity analysis is a topic whose value is not fully appreciated. However, it is easily understood. Sensitivity analysis means measuring the sensitivity of any one operating variable to another over which it has a controlling effect. By studying the sensitivity of a group of variables surrounding any piece of equipment in a process, their relative immediate and long-range contribution to control of that process can be better understood. It has been used in an advisory manner to alert operators to upcoming failures or process issues. For instance, imagine the operator knowing that he can turn up the flow on a certain incoming line, and it will only alter the operation of a compressor by a little, or if he turns up the flow in another incoming line, his compressor will very quickly head to failure. This can allow the operator to use learned judgment to select the appropriate settings in the appropriate situation.

Alternatively, if the process is off spec—is there any one variable under his control that can more quickly recover it? Or when running near limits is there any one variable that has little effect but will allow them to inch up on the limits without fear of failure? This also means you can know which parameters are best left alone when operating near the limit due to their highly sensitive nature.

In the case of sensitivity of alarm assessments—when do too many of one alarm over a period become indicative of a pending failure? When do certain alarms occurring together lead to certain process, system or equipment faults? Sensitivity studies can yield such answers.

5.9.6 Fuzzy Logic

Fuzzy logic has been around since the mid-1960s, but is "fuzzy" to understand with relatively few practical applications known. The paradigm essentially deals with "partial truths." Fuzzy logic was a hot topic for a few years and much was invested. People were making fuzzy toasters, cameras and such.

The concept of fuzzy logic is that values in analytics are never exactly as they should be, but they are more likely to be one than the other. So, you can assume certain things about the existence of variables based on their likeliness or closeness to a condition than their actual value. Their value to alarm systems is that where an alarm might have been used, and perhaps in redundant or nuisance fashion, a fuzzy logic controller can intervene to enhance control away from potential incidents.

Any study of fuzzy logic will lead quickly to better understanding of membership functions, and how to treat them as they vary between zero and one. It is most useful for Boolean logic operations where the absolute value of something cannot necessarily be established, but its closeness or likelihood of being one thing as opposed to another can be established. And conclusions can be reached—often with very incomplete data. More can be read on the topic by interested parties.

5.9.7 Bayesian Networks

Bayesian networks are a bit obscure in their use—not widely spread. Their value comes when an expert may have an opinion that a relationship exists that is not normally revealed by other data analytics. Bayesian models are often able to show the inter-dependencies that other methods might not reveal. Therefore, the warning is that they should not necessarily be a go-to solution but may show to be valuable when suspicions tell you unseen relationships exist.

As a field of sparse research, there are not many instances of Bayesian network usage within the process manufacturing industries. More can be read on Bayesian Networks by interested parties.

5.9.8 Genetic Algorithms

Also a bit obscure in use, genetic algorithms are applicable when finding matching sets of information dependent on one another. For instance, comparing parts analysis for best matches of subcomponents when analyzing failure rates. Available literature on GAs will quickly get you into parent/child relationships. Like Bayesian nets, they are not something that should be a go-to solution except in cases where other systems yield little result. And perhaps as an expert you are convinced there is more to the story that may be due to inter-related parts or processes. In terms of use for alarming systems there has been some work done to find correlations in recurring alarms sequences. More can be studied on these by the interested reader.

5.9.9 SmartSignal, PRiSM (AVEVA), and PPCL

These products have their own category mostly because they do not fit the other categories discussed so far. Perhaps PRiSM and Smart Signal can be considered one category because they have both been designed for the same sort of target. A user frustration is that the developers of these technologies have refused to reveal the underlying algorithmic engines that drive their results. They all seemed to take the same approach of "This information is on a need-to-know basis only." Both have shown promise in the area of asset maintenance monitoring. The following is a summary of each.

5.9.9.1 *SmartSignal/PRiSM*   These were both developed as analytics engines to examine failure mechanisms in the energy industry. SmartSignal came out of Argonne National Labs and was used to predict failures in energy processes. It was then applied to such things as failure prediction in locomotives and other systems. PRiSM similarly was developed in the power industry and used for power generation and distribution systems to predict equipment failures and reliability in such complex systems.

Both of these systems have also been adapted with some degree of success in other mechanical and thermodynamic processes. The author has not personally had a lot of experience with these packages, but they suffer from adoption because their inner workings are not well understood. However, they solve a class of problem and show promise to continue solving the problem when applied to similar problems, so they may be the right answer in given situations.

5.9.9.2 *PPCL's CVE and CPM*   This is a totally different approach to analysis that is highly visual in nature. One must be trained in its use to fully understand the techniques. PPCL (a UK Company) was originally offered as Curvaceous software out of the UK. It proposed the situation that if you could just visualize the interactions between various variables, then their inter-operation with each other would be more apparent. CVE stands for C Visual Explorer, and CPM stands for C Process Modeler.

Therefore, it does just that. Operating parameters themselves are lined up on the *x*-axis, and then their operating ranges are aligned on the *y*-axis. When a single point of operation is established, the condition of each point gets a line through the entire set of variables at one given point in time. Once established for an entire time series (months of data for example), the "operating envelope" becomes visually apparent. The user can then select any given operating parameter and examine where every other variable can operate when that parameter is at a selected value, and all ranges of other variables becomes apparent.

The product has proven useful when establishing alarm limits at acceptable values, given a true understanding of the operating ranges of the parameters to be alarmed. They offer both a past-look and forward "what-if" look capability. This is identified as a separate category because it is vitally different than other analytics offerings. Readers can investigate this further if interested.

5.9.10 Control System Effectiveness Study

One last emerging method is worth a quick introduction. It is patented, but currently undergoing industry assessments. Let us briefly term it Control System Effectiveness for Operations. Essentially it is a way of examining how operators use a control system- the HMI, alarms and all other operator assistance applications to effectively control the process.

For background, the intent of process control is to automate as much as we possibly can. The more that can be automated the less is left to chance, and the associated risks of human decision. What we cannot control we pass to the control system operator. We train them, and hand the system over to them. And their first layer of system operations is the HMI. Those can be either well matched or badly adapted to the system. Much study was done by the ASM Consortium™ and there are books written on effective HMIs. ISA has a standard (ISA 101) regarding proper implementation. The second layer is the alarm system. Whatever automation is not achievable, and whatever alterations are not obviated by the HMI, we then turn into an alarm to prompt a change to be made by the operator.

If we gather sufficient operating data (a few production KPIs) and operator alarms and actions over a long period of operation, we can assess how well individual operators are using the system. This tells a story of not how good or bad the operators are, but how well the system is designed and implemented to help the average human resolve problems (assuming operators are pretty much the same). Once operational gaps can be established and compared with the ongoing interactions, the shortcomings of any system design can be evaluated.

Alarms are the first point of assessment. Why does one operator see an alarm and change the heat input, and another see the same alarm and change the feed rate. Assuming there is likely one correct answer from a thermodynamic or physical sense, why would they reach different solution paths? In this way we can identify better alarming, better HMI, better training or whatever is the assessed shortcomings of the system involved.

These comprise the main AI and machine-learning techniques we see in use or being tested for alarming techniques to date.

5.10 WRAP-UP

At this point I would like to nominate Greatest Engineer award to Bridget Fitzpatrick, in whose honor I write this chapter. One of my greatest process analytics experiences was with Bridget using sensitivity analysis. We were evaluating a model in a specific part of a plant. Bridget had been working with a process in her plant that had suffered from non-stop and inconsistent alarms that made little sense. They were in dire straits because the upsets were causing anomalies in the effluent wastewater from the system. These anomalies were causing regulatory issues that could lead to the need for a multi-million dollar deep-well treatment facility.

Bridget decided to throw in the kitchen sink and perform a sensitivity analysis on ALL variables surrounding this specific process. In doing so, she discovered that a very specific variable had a high sensitivity to the process that should NOT have been affecting the process at all. And then she investigated WHY. And she discovered that there was a bleed over between two co-located lines that had been dragging down the quality and producing regulatory unacceptable effluents of the product for months. Bridget fixed two lines in the plant, and by doing so avoided several million dollars of down-hole water treatment that was the next alternative.

Bridget also worked with me on First principles modeling for similar issues and worked years later with expert systems for her company's pharmaceutical grade products. I do not think I have ever met an engineer quite like her, and I do not think I ever will. Bridget looked past the models to the real world that they represented. Let us all try to be a little more like Bridget. Data can tell you nothing until you put it to practical use.

This chapter closes with a quote from Earl C Kelley (American author and educationalist 1895–1970):

We have not succeeded in answering all our problems. The answers we have found only serve to raise a whole set of new questions. In some ways we feel we are as confused as ever, but we believe we are confused on a higher level and about more important things.

5.A

PROCESS STATE TRANSITION LOGIC EMPLOYED BY THE ORIGINAL FMC FALCONEER KBS

Richard J. Fickelscherer[1] and Daniel L. Chester[2]

[1]Department of Chemical and Biological Engineering, State University of New York at Buffalo, Buffalo, NY, USA
[2]Department of Computer and Information Sciences (Retired), University of Delaware, Newark, DE, USA

5.A.1 INTRODUCTION

The following describes in great detail the underlying logic of how the original (i.e., hand compiled) version of the FMC Tonawanda ESP knowledge-based system (KBS) (also known as *FALCONEER*) (Skotte et al. 2001, Fickelscherer et al. 2003) determined the current operating state of FMC's electrolytic sodium persulfate (ESP) process. For the same reasons doing so was required in the original FALCON System (Fickelscherer 1990), this was again essential in order for the KBS to perform the proper analysis on the collected process sensor data. The program

Artificial Intelligence in Process Fault Diagnosis: Methods for Plant Surveillance,
First Edition. Edited by Richard J. Fickelscherer.
© 2024 John Wiley & Sons, Inc. Published 2024 by John Wiley & Sons, Inc.

was able to be turned on at any time and then would automatically determine its current process state and monitor for all state transitions possible from that state. This allowed it to minimize **"nuisance alarms"** while doing its **"intelligent supervision"** of that process' operation.

5.A.2   POSSIBLE PROCESS OPERATING STATES

There are seven possible process operating states in FMC's ESP Process:

0)   The process has been previously shutdown and is awaiting the next startup.

1)   The process is being started up—current to cells is on but cell products are being recycled to the neutral tank rather than being sent to the crystallizer. Once this switch to the crystallizer is accomplished, the process is considered to be in production mode.

2)   The process is in production mode and is running within all standard operating conditions (SOCs) of its key sensor points (i.e., relevant state process variables). These points are those with associated high and/or low interlock activation limits.

3)   The process is in production mode but is not running within all SOCs of its key sensor points.

4)   The process is in production mode but is rapidly approaching one or more interlock regions of operation.

5)   One or more process interlocks should have occurred but the process is still in production mode.

6)   The process is just shutting down. (Current to the Cells has just been shutoff).

Possible ESP Process State Transitions

It is possible to go from state 0 to states 1, 2, 3, 4 or 5.
It is possible to go from state 1 to states 2, 3, 4, 5 or 6.
It is possible to go from state 2 to states 1, 3, 4, 5 or 6.
It is possible to go from state 3 to states 1, 2, 4, 5 or 6
It is possible to go from state 4 to states 1, 2, 3, 5 or 6
It is possible to go from state 5 to states 1, 2, 3, 4 or 6
It is possible to go from state 6 to states 0, 1, 2, 3, 4 or 5

The above possible ESP process state transitions indicate that almost any other state can be reached from any starting state. This is possible because these transitions can happen faster than the monitoring data sampling rate of the FALCONEER KBS (currently the analysis sampling interval is 1 minute). It can consequently miss the commencement of significant operating events leading to abnormal operating behavior. Thus, the KBS must be prepared to encounter any state at any time. By definition, transversal of the possible process states from state 1 to state 6 will be defined as a completed production campaign.

5.A.3 SIGNIFICANCE OF PROCESS STATE IDENTIFICATION AND TRANSITION DETECTION

The program must determine which state the process is in so that only meaningful data is analyzed and validated. The fault analyzer will be active during process states 2, 3, and 4 for determining all possible assumption variable deviations and will perform interlock malfunction analysis during process state 5. The KBS is in waiting mode (state 0) after shutdown (state 6) occurs and will passively monitor the process to determine when the next startup (state 1) completes. It will then automatically become active again.

5.A.4 METHODOLOGY FOR DETERMINING PROCESS STATE IDENTIFICATION

The process state that the process is currently in when the program is turned on is inferred from the present value states and predicted next value states of the various key sensor variables. Again, a process sensor variable is considered key if it can cause an interlock shutdown. The present value states and predicted next value states of these measurements are determined in the following manner.

5.A.4.1 Present Value States of All Key Sensor Data

Each key sensor point will be used to populate the following sensor record. Each field in this record will be updated if required whenever a new vector of data is obtained by the Process Historical Data (PHD) Server connected to the Distributed Control System (DCS).

(1) PV = Present Value
(2) PVS = Present Value State
(3) PPV = Previous Present Value
(4) PNV = Predicted Next Value
(5) PNVS = Predicted Next Value State

(6) MaxDCS = Maximum DCS Present Value Limit

(7) HIntlk = High Interlock Limit

(8) MaxSOC = Maximum Standard Operating Condition

(9) MinSOC = Minimum Standard Operating Condition

(10) LIntlk = Low Interlock Limit

(11) MinDCS = Minimum DCS Present Value Limit

At any given time, each key sensor can be in one and only one state of the seven possible states for its present value state (PVS). These states are determined as follows:

Possible present values (PV)			PV state (PVS)	PVS ID
MaxDCS	< PV		Bad value high	3
HIntlk	≤ PV	≤ MaxDCS	High interlock expected	2
MaxSOC	< PV	< HIntlk	Outside max SOC	1
MinSOC	≤ PV	≤ Max SOC	Within SOC	0
LIntlk	< PV	< MinSOC	Outside min SOC	−1
MinDCS	≤ PV	≤ LIntlk	Low interlock expected	−2
	PV	< MinDCS	Bad value low	−3

5.A.4.2   Predicted Next Value States of All Key Sensor Data

The predicted next value (PNV) for a given key sensor PV is calculated by a simple linear extrapolation of that PV and the last previously measured PV, referred to as the previous present value (PPV). The formula for calculating the PNV is:

$$PNV = 2*PV - PPV \qquad (C.1)$$

The PNV is useful for anticipating process state transitions to state 4. This is determined as follows:

Possible	PNV		PNV state (PNVS)	PNVS ID
HInlk	≤PNV		Expect high interlock soon	1
LInlk	< PNV	< HInlk	Interlock not expected soon	0
	PNV	≤ LInlk	Expect low interlock soon	−1

Anticipating interlock activations before they happen will help eliminate spurious or incorrect diagnoses by the KBS from occurring. After an interlock trip occurs, the various quantitative model residuals required in the fault analysis are not evaluated in order to avoid a GIGO (i.e., Garbage In, Garbage Out) situation with those residuals. In these situations, the KBS would passively monitor the process until it could determine that state 1 (Startup) is completed for the next production campaign and the KBS data buffers are flushed. It would then automatically begin to analyze process data once again.

5.A.5 PROCESS STATE IDENTIFICATION AND TRANSITION LOGIC PSEUDO-CODE

The following describes the logic for determining the current process state and state transitions for the FMC ESP process. This logic is written in object oriented programming pseudo-code, where objects have both specific attributes and methods for determining values for those attributes. The logic described assumes that all process variable measurements used by the KBS arrive periodically as a complete time-stamped vector. The KBS then uses those values and any past results it requires from previous KBS analyses to analyze this current vector and display its results. This analysis must be completed before the next vector of process variable measurements arrives to allow the KBS to run in real-time. Currently, the FALCONEER KBS uses a period of 1 minute as the adjustable update frequency for data analyzed by the KBS.

5.A.5.1 Attributes of the Current Data Vector

Attributes	Data type	Description
1) first_data_ received	Boolean	This is the first vector of process sensor data to be analyzed by the KBS for this production campaign
2) KBS_initialized	Boolean	KBS data buffers have been flushed with legitimate data (i.e., that collected during actual operating conditions)
3) FALCONEER_ monitoring_ mode	Integer	**0** – KBS is passively monitoring process until next startup is complete
		1 – KBS is actively performing Sensor Validation & Proactive Fault Analysis (**SV&PFA**) of all possible modeling assumption deviations being directly monitored for.
		2 – KBS is actively performing interlock failure analysis for all interlocks within its realm of process system operation. (**This analysis is currently outside the scope of the FALCONEER KBS but the KBS's current design allows for it to be directly added**).
4) Current_to_ Cells_is_on	Boolean	Current is now on to Cells

Attributes	Data type	Description
5) Product_being_ sent_to_ Crystallizer	Boolean	Cell product is not being recycled but actually sent to Crystallizer.
6) all_process_ variables_ within_SOC	Boolean	All key process sensors are within their Standard Operating Condition limits.
7) Sensor_ PNVS(I)	Integer array	The Predicted Next Value State for each key process sensor being monitored; it is assigned one of three possible state values, two which indicate that an interlock low (-1) or high (1) is expected from it soon.
8) interlock_ expected_soon	Boolean	Interlock trip anticipated within the next time step interval.
9) Sensor_PVS(I)	Integer array	The present value state for each key process sensor being monitored; it is assigned one of seven possible state values, four which indicate that an interlock low (-2 or -3) or high (2 or 3) is expected now.
10) interlock_ expected_now	Boolean	Interlock trip anticipated now.
11) startup_ commenced	Boolean	Current to Cells is on but product is still being recycled to Neutral Tank.
12) startup_ completed	Boolean	Process is in production mode.
13) shutdown_ commenced	Boolean	Current to Cells has been shut off.
14) current_ process_state	Integer	**0** – Process state is unknown. (KBS is idle awaiting next process startup). **1** – Process startup is occurring (Current to Cells is on but the product is being recycled). **2** – Process is in production and is within all SOCs. **3** – Process is in production but is not within all SOCs. **4** – Process is in production but is rapidly approaching one or more interlock(s). **5** – Process interlock(s) should be occurring. **6** – Process has just shutdown.

5.A.5.2 Method that is Applied to Each Updated Data Vector

Before any data vectors have been received, attribute `first_data_ received` is set to FALSE. Then, as each data vector is received, the following sequence of statements are executed.

determine_if_first_data_received Initialize all the various Booleans after the KBS receives its first vector of sensor data collected for either the current or next (if the process is in state 6) production campaign.

```
IF (first_data_received EQ FALSE) THEN
            startup_commenced        = FALSE
            startup_completed        = FALSE
            shutdown_commenced       = FALSE
            KBS_initialized          = FALSE
            first_data_received      = TRUE
END IF
```

determine_if_startup_commenced Determine if current is on to cells.

```
IF ((IIC8000R.PVS>-2) AND (IIC8000R.PVS<3)) THEN
        Current_to_Cells_is_on = TRUE
ELSE
        Current_to_Cells_is_on = FALSE
END IF
IF ((Current_to_Cells_is_on EQ TRUE).AND.
        (startup_commenced EQ FALSE)) THEN
        startup_commenced = TRUE
END IF
```

determine_if_startup_completed Determine if process is in production mode.

```
IF (XCV6272.PV EQ NORMAL) THEN
        Product_being_sent_to_Crystallizer = TRUE
ELSE
        Product_being_sent_to_Crystallizer = FALSE
END IF
IF ((startup_commenced EQ TRUE) AND
        (Product_being_sent_to_Crystallizer EQ TRUE) AND
        (startup_completed EQ FALSE)) THEN
        startup_completed = TRUE
ENDIF
IF ((startup_completed EQ TRUE) AND
        (Product_being_sent_to_Crystallizer EQ FALSE))
THEN
        startup_completed = FALSE
ENDIF
```

determine_if_shutdown_commenced   Determine if the process is being shutdown.

```
IF ((Current_to_Cells_is_on EQ FALSE) AND
        (startup_commenced EQ TRUE) AND
        (shutdown_commenced EQ FALSE)) THEN
        shutdown_commenced = TRUE
ENDIF
```

determine_current_process_state   Assigns one of seven possible values to the current_process_state.

Determine all process variables within SOC

```
all_process_variables_within_SOC = TRUE
FOR EACH I FROM 1 TO Number_of_Key_Sensor_Variables DO
        IF (Sensor.PVS(I) NE 0) THEN
                all_process_variables_within_SOC = FALSE
        END IF
```

Determine interlock expected soon

```
interlock_expected_soon = FALSE
FOR EACH I FROM 1 TO Number_of_Key_Sensor_Variables DO
        IF (Sensor.PNVS(I) EQ (-1 OR 1)) THEN
                interlock_expected_soon = TRUE
        END IF
```

Determine interlock expected now

```
interlock_expected_now = FALSE
FOR EACH I FROM 1 TO Number_of_Key_Sensor_Variables DO
        IF (Sensor.PVS(I) EQ (-2 OR -3 OR 2 OR 3)) THEN
                interlock_expected_now = TRUE
        END IF
```

Assign the Current Process State.   If the current operating state does not match any of the first six possible states (1, 2, 3, 4, 5, or 6), it is assigned the unknown state (0):

State (1): Process is being started up (KBS will not allow Auto-pilot mode and it is not performing fault analysis).

```
IF ((startup_commenced EQ TRUE). AND.
        (startup_completed EQ FALSE)) THEN
        current_process_state = 1
```

State (2): Process is running and all key sensor variables are within SOCs (Auto-pilot mode allowed once KBS is initialized).

```
ELSE IF ((startup_completed EQ TRUE) AND
        (all_process_variables_within_SOC EQ TRUE) AND
        (interlock_expected_soon EQ FALSE)) THEN
        current_process_state = 2
```

State (3): Process is running but not all key sensor variables are within SOCs (ask the operators if they still want Auto-pilot mode after KBS is initialized).

```
ELSE IF ((startup_completed EQ TRUE) AND
        (all_process_variables_within_SOC EQ FALSE) AND
        (interlock_expected_soon EQ FALSE) AND
        (interlock_expected_now EQ FALSE)) THEN
        current_process_state = 3
```

State (4): Process is running and is quickly approaching interlock shutdown (drop out of Auto-pilot mode and alert the operator that immediate action is necessary).

```
ELSE IF ((startup_completed EQ TRUE) AND
        (interlock_expected_soon EQ TRUE) AND
        (interlock_expected_now EQ FALSE)) THEN
        current_process_state = 4
```

State (5): Process is just about to interlock or there has been an interlock activation failure—actively perform interlock failure analysis (not currently within the original FALCONEER KBS's intended scope).

```
ELSE IF ((startup_completed EQ TRUE) AND
        (interlock_expected_now EQ TRUE) AND
        (Current_to_Cells_is_on EQ TRUE)) THEN
        current_process_state = 5
```

State (6): Process has just been shutdown; reset all the state transition Booleans once process shutdown occurs so that the KBS will properly monitor for the next process startup.

```
ELSE IF (shutdown_commenced EQ TRUE) THEN
        current_process_state = 6
        first_data_received = FALSE
```

State (0): Process has previously been shutdown and its restart status is unknown—KBS passively monitoring data until the start of next production campaign occurs.

```
ELSE
        current_process_state = 0
END IF
```

determine_if_KBS_initialized   Sets a Boolean indicating that the KBS is actively performing its analysis.

```
IF ((current_process_state Eq. [2 OR 3 OR 4 OR 5]) AND
        (KBS_initialized EQ FALSE)) THEN
        KBS_initialized = TRUE
END IF
```

determine_FALCONEER_monitoring_mode   Sets the monitoring mode of the KBS on current process sensor data.

```
IF ((KBS_initialized EQ TRUE) AND
        (current_process_state Eq. [2 OR 3 OR 4])) THEN
        FALCONEER_monitoring_mode = 1
ELSE IF ((KBS_initialized EQ TRUE) AND
        (current_process_state Eq. 5)) THEN
        FALCONEER_monitoring_mode = 2
ELSE
        FALCONEER_monitoring_mode = 0
        KBS_initialized = FALSE
END IF
```

5.A.6 SUMMARY

The temporal state transition logic described above as object-oriented program pseudo-code has been demonstrated effective in the original FMC FALCONEER KBS for determining the current KBS context. Doing so allows the KBS to continuously monitor on-line, real-time process operations and perform its Sensor Validation and Proactive Fault Analysis (SV&PFA) on all legitimate process data, shutting off its analysis when it is not appropriate to perform it. This thus automatically eliminates GIGO situations.

APPENDIX 5.A REFERENCES

Fickelscherer, R. J., *Automated Process Fault Analysis*, Ph. D. Thesis, Department of Chemical Engineering, University of Delaware, Newark, DE., 1990.

Fickelscherer, R. J., D. H. Lenz, and D. L. Chester, "Intelligent Process Supervision via Automated Data Validation and Fault Analysis: Results of Actual CPI Applications," Paper 115d, *AIChE Spring National Meeting*, 2003, New Orleans, LA.

Skotte, R., D. Lenz, R. Fickelscherer, W. An, D. Lapham III, C. Lymburner, J. Kaylor, D. Baptiste, M. Pinsky, F. Gani, and S. B. Jørgensen, "Advanced Process Control with Innovation for an Integrated Electrochemical Process," *AIChE Spring National Meeting*, 2001, Houston, TX.

5.B

PROCESS STATE TRANSITION LOGIC AND ITS ROUTINE USE IN FALCONEER™ IV

5.B.1 TEMPORAL REASONING PHILOSOPHY

Temporal reasoning in both the original FALCON and the various FALCONEER™ IV knowledge-based systems (KBSs) is limited to adjacent time intervals. Sampling of all the process data occurs at a predetermined frequency. However, context within the KBSs changes only when it is appropriate. The logic of these KBSs is concerned with the previous state (the one existing one data time sample ago), the current inferred state (the one determined from the current sampled data), and the forecasted next state (predicted from the previous and current inferred state) when determining its context. This is similar to the temporal reasoning employed by the ventilator manager (VM) KBS (Stefik et al. 1983) and greatly simplifies these programs' underlying temporal logic, which otherwise would require more elaborate representations of events and time.[1]

---

[1] See Appendix 5.A. for a detailed pseudo-code example of a more complex temporal logic reasoning routine used by the original FALCONEER KBS at FMC.

Artificial Intelligence in Process Fault Diagnosis: Methods for Plant Surveillance,
First Edition. Edited by Richard J. Fickelscherer.
© 2024 John Wiley & Sons, Inc. Published 2024 by John Wiley & Sons, Inc.

This problem that the process state continuously changes over time requires that the fault analyzer also changes its prior presented plausible fault hypotheses continuously over time. This problem is widely known within the Artificial Intelligence community as the frame problem (Rich 1983). In such situations, new information causes an old conclusion to be withdrawn. Such a conclusion is said to be defeated (Charniak and McDermott 1985).[2] The entire reasoning system is classified as nonmonotonic. Nonmonotonicity is not an unusual property of a reasoning system. On the contrary, most of the inferences done by real programs are defeasible (Charniak and McDermott 1985). For FALCONEER™ IV process fault analyzers, the reasoning employed during each analysis step is monotonic, meaning the program firmly believes its conclusions at that time stamp. On the next analysis cycle, it completely forgets its previous conclusions and draws all new inferences. These new conclusions may or may not, depending entirely on what process operating state is currently occurring, confirm previous conclusions by the fault analyzer. The program's awareness is thus always focused on the current process state.

5.B.2 INTRODUCTION

Once the MOME diagnostic strategy was adopted as the desired model-based diagnostic methodology in the original FALCON Project (see Chapter 11), a comprehensive attempt to formally verify the FALCON System's knowledge base commenced. A major problem that became more apparent during further testing with plant data concerned the primary models being used to detect and diagnosis faults. There was a definite need to limit the conditions under which those various models were considered valid representations of the actual process behavior. It was evident that major process upsets such as pump failures and interlock activations invalidated many of the fundamental assumptions used during the development of those models. Consequently, it became necessary for the fault analyzer to always determine the current operating state of the process before the diagnostic rules were applied. Such determination was required in order to determine which of the specific modeling assumptions were still valid and thus which of the Primary models would be appropriate in the fault analysis. In order to do this, process events such as process startups, shutdowns, and interlock activations needed to be explicitly monitored for by the FALCON System.

---

[2]A conclusion that may later be defeated is defeasible.

This monitoring was accomplished by developing rules directly detecting the occurrence of such events and could subsequently determine the current operating state of the process system. The possible operating states became (1) the process was being started up; (2) the process was in production; (3) the process was in production, but was rapidly approaching an interlock activation; (4) the process had interlocked, but production had ceased for less than 2 minutes; and (5) the process had interlocked and production had ceased for more than 2 minutes. The transition between these various process states was automatically determined by the fault analyzer, which then limited its analysis of the process data accordingly.

Two minutes after an interlock was chosen as the cut off point for fault analysis because it represented the typical amount of time required for the effects of major process transients to subside after emergency shutdowns. After 2 minutes it is no longer possible to analyze current process data in order to determine the fault situation that originally caused the emergency process shutdown. Moreover, after an interlock the majority of process models no longer made reliable predictions of the observed process behavior. It was therefore necessary to halt the diagnosis of all process faults which relied upon those models, which turns out to be most of the target process fault situations monitored for by the fault analyzer. Thus, 2 minutes after a process interlock the FALCON System would go to "sleep" until the process system was started up again. While "sleeping," the fault analyzer would continue to monitor plant data, but it would be incapable of diagnosing process fault situations. When the process restarted, the FALCON System would automatically "re-awaken," that is, automatically commence analyzing for all process fault situations again. Messages informing the process operators of these state transitions were automatically displayed by the human/machine interface.

Adding this capability for automatically determining current process state had several consequences. First, it allowed the FALCON System to be turned on when the process was either operating or shutdown: the fault analyzer could automatically determine which state it was in. It also allowed the FALCON System to be run continuously, regardless of the plant state transitions that could occur. More importantly, adding this capability also logically structured the entire knowledge base according to the patterns of evidence contained within the various diagnostic rules. This reduced the fault analysis of the incoming process data to an ordered, sequential search through the entire set of possible process faults. This search sequence in effect constitutes a priority hierarchy between the various possible process fault situations. The logical structure of the fault priority hierarchy within the FALCON System's knowledge base is illustrated in Fig. 11.4. As discussed in Chapter 11, discovering the rationale behind this hierarchy represented a major development in the MOME paradigm.

The enhancements caused a substantial increase in the size and complexity of the knowledge base. The upgraded version of the FALCON System's diagnostic knowledge base contained approximately 800 diagnostic rules (over 10,000 lines of Common Lisp code). It was capable of detecting and diagnosing all of the 60 target process fault situations plus about 100 additional fault situations, most of which were malfunctions in the interlock shutdown system. It was also capable of detecting and diagnosing a few extremely dangerous fault situations that could occur during process startups.

A similar type of analysis was performed in the original FALCONEER KBS (pre-FALCONEER™ IV availability) developed for the FMC Corporation. The details underlying this analysis are described in pseudocode presented in Appendix 5.A. The following describes how this analysis is currently performed in FALCONEER™ IV.

5.B.3   STATE IDENTIFICATION ANALYSIS CURRENTLY USED IN FALCONEER™ IV

State ID variables are by definition those used for determining the current operating state of the process; that is, starting up, operating, shutting down, shut down, etc. These variables can be any measured system variable, continuous, or discrete. When operating (i.e., actively analyzing actual process data), FALCONEER™ IV always first determines whether the process is operating within acceptable conditions or not. This is accomplished with the State Identification (State ID) Module. This module determines if the process is currently operating in a mode in which the FALCONEER™ IV analysis can derive meaningful results. FALCONEER™ IV is idle if the process is not operating. The program automatically begins its analysis of sensor measurements once process startup is complete. It continues this analysis until the process is shutdown and then is idle again (i.e., passively monitoring process data) until the next process startup completes. This process condition monitoring suite thus runs continuously and appropriately adjusts its analysis to current process operations accordingly.

State ID variables are any sensor measurements used by the program to determine if the process is currently operating in production mode. These variables tend to monitor either key process feeds or process configuration states. They generate GREEN alerts (used to denote normal operation) when the process is operating and each of these variables is either within its associated SOCs or its normal state. If outside their SOCs but within their interlock limits, these variables generate YELLOW alerts. In either case, the program status is "In Production Mode" and "Not Expecting Interlocks." If outside their interlock limits or normal configuration, these variables generate RED alerts and the other FALCONEER™ IV's analysis modules (SV&PFA and Virtual SPC) are preempted from running. In this situation, the program status becomes "Not in Production Mode" and "Interlock Expected." It stays in this mode until **none** of the State ID variables are in RED alert.

The State Identification (State ID) Analysis is always performed first when the process interval time since the last FALCONEER™ IV analysis cycle elapses.

If all State ID variables and all other measured variables enabled in the State ID module are in their acceptable operating ranges (if continuous variables) or configurations (if discrete variables), the following analysis occurs in this order:

(1) Any non-State ID module variables which have "BAD" values are identified and all Primary models which depend upon them are disabled.

(2) The SV&PFA analysis is performed with all still currently enabled primary models.

(3) Any resulting variables which are determined to be either in RED SV&PFA alarm or "BAD" values and are also used in any Performance Equations cause the results of those calculations to be classified "calculation invalid" (i.e., performance equation calculated results are suspect because of possibly invalid values of variables are being used in this calculation [or "BAD" for short]).

(4) The Virtual SPC analysis is performed on all "GOOD" (i.e., not "BAD") measured variables and performance equations for which their individual Virtual SPC Analysis time interval has been surpassed.

(5) The program actively monitors the system time to determine when the next FALCONEER™ IV analysis cycle should commence‧ again (i.e., whenever the analysis time interval defined for the particular process being monitored has been surpassed).

If any one of the State ID variables or other measured variables enabled in the State ID module are not in their acceptable operating ranges (if continuous variables) or configurations (if discrete variables) (as indicated by them being in a RED alarm state), then Steps 1–4 are skipped and the program goes directly to Step 5. It continues in this mode until all variables in the State ID module are considered acceptable again (as indicated by them all being in either GREEN or YELLOW alarm states).

The possible State ID Alarm States need further elaboration depending upon the type of variable being monitored; that is, either discrete or continuous.

Discrete variables are directly defined as State ID variables with a defined Normal Value as the normal configuration status for that variable. This is its GREEN Alarm condition. Whenever, it is not this Normal Value, the variable is in RED Alarm and preempts further FALCONEER™ IV analysis from occurring. There is thus no YELLOW alarm state associated with these State ID variables.

Continuous variables that can be defined in the State ID module analysis are either controlled variables or uncontrolled variables. The following six parameters need to be specified for either type defined in the State ID module:

(1) Minimum measurement limit.
(2) Low Interlock Limit.
(3) Minimum Standard Operating Condition Limit.
(4) Maximum Standard Operating Condition Limit.
(5) High Interlock Limit.
(6) Maximum Measurement Limit.

These parameters are used in the following manner to set the various State ID alarm states for these continuous variables:

GREEN: If the current value of the given variable is less than or equal to Maximum Standard Operating Condition Limit and greater than or equal to the Minimum Standard Operating Condition Limit, the variable is in the GREEN State ID Alarm State and this variable does not preempt further FALCONEER™ IV analysis.

YELLOW High: If the current value of the given variable is greater than the Maximum Standard Operating Condition Limit but less than the High Interlock Limit, the variable is in the YELLOW High State ID Alarm State and this variable does not preempt further FALCONEER™ IV analysis.

YELLOW Low: If the current value of the given variable is less than the Minimum Standard Operating Condition Limit but greater than the

Low Interlock Limit, the variable is in the YELLOW Low State ID Alarm State and this variable does not preempt further FALCONEER™ IV analysis.

RED High: If the current value of the given variable is greater than or equal to the High Interlock Limit but less than or equal to the Maximum Measurement Limit, the variable is in the RED High State ID Alarm State and this variable does preempt further FALCONEER™ IV analysis.

RED Low: If the current value of the given variable is less than or equal to the Low Interlock Limit but greater than or equal to the Minimum Measurement Limit, the variable is in the RED Low State ID Alarm State and this variable does preempt further FALCONEER™ IV analysis.

RED High High: If the current value of the given variable is greater than the Maximum Measurement Limit, the variable is in the RED High High State ID Alarm State and this variable does preempt further FALCONEER™ IV analysis. These alarms are displayed to the user just as RED High alarms are.

RED Low Low: If the current value of the given variable is less than the Minimum Measurement Limit, the variable is in the RED Low Low State ID Alarm State and this variable does preempt further FALCONEER™ IV analysis. These alarms are displayed to the user just as RED Low alarms are.

If any of these parameters are not defined for a particular variable, their values should be set to be equal to the next logical limit that is defined. This will eliminate some of these possible alarm levels. For example, if there is no Maximum Standard Operating Condition Limit then this value should be set equal to the High Interlock Limit. This will eliminate the YELLOW High alarm state as a possible situation for that variable.

5.B.4   STATE IDENTIFICATION ANALYSIS SUMMARY

Again, the motivation for the State ID module is to preempt the further analysis by FALCONEER™ IV when the process is not operating (e.g., shutdown) or is going through startup and has not lined out yet to normal operation (e.g., diverter valves in BYPASS configuration). It thus attempts to identify operating regimes where the corresponding SV&PFA and Virtual SPC analysis is meaningful. Consequently, when none of the State ID variables or other variables enabled in the State ID module are RED (either Low, Low Low, High, or High High), FALCONEER™ IV further performs the appropriate SV&PFA and Virtual SPC Analysis and reports it results with the same Alarm Detail Screens. One interacts with these subsequent alarms in the same manner as with the State ID alarms.

APPENDIX 5.B   REFERENCES

Charniak, E. and McDermott, D. (1985). *Introduction to Artificial Intelligence*, 369–371. Reading, MA: Addison-Wesley Publishing Co., Inc.

Rich, E. (1983). *Artificial Intelligence*, 176–180. New York: McGraw-Hill, Inc.

Stefik, M., Aikins, J., Balzer, R. et al. (1983). The architecture of expert systems. In: *Building Expert Systems* (ed. F. Hayes-Roth, D.A. Waterman, and D.B. Lenat), 97–98. Reading, MA: Addison-Wesley Publishing Co., Inc.

6

PROCESS FAULT DETECTION BASED ON TIME-EXPLICIT KIVIAT DIAGRAM

Shu Xu[1], Mark Nixon[1], Ray C. Wang[2], and Michael Baldea[2]

[1]Process Systems and Solutions, Emerson Automation Solutions, Round Rock, TX, USA
[2]McKetta Department of Chemical Engineering, The University of Texas at Austin, Austin, TX, USA

OVERVIEW

To address the pressing big data challenges encountered by the chemical industry, it is critical to represent data in a lower-dimensional space. In this chapter, a novel time-explicit Kiviat diagram is described to facilitate plant monitoring and fault detection. Three types of processes are simulated in case studies, and the new approach is compared with conventional principal component analysis (PCA)-based approaches, which demonstrates the new approach's superior performance.

Artificial Intelligence in Process Fault Diagnosis: Methods for Plant Surveillance,
First Edition. Edited by Richard J. Fickelscherer.
© 2024 John Wiley & Sons, Inc. Published 2024 by John Wiley & Sons, Inc.

CHAPTER HIGHLIGHTS

- A new time-explicit Kiviat diagram projecting high-dimensional process data into a lower-dimension space to facilitate data visualization.
- Innovative time-explicit Kiviat diagram-based fault-detection procedures tailored to the requirements of different process types – batch, continuous, and periodic.
- Excellent performance obtained by the new approach in all case studies in swiftly identifying process faults while keeping low type I/II error rates.

6.1 INTRODUCTION

The recent rapid development of the Internet of Things (IOT) and smart manufacturing technology has propelled the fourth industrial revolution, where an enormous amount of data is collected from numerous sensors deployed in a plant. This development poses a challenge to process engineers distilling knowledge from a massive amount of data (Qin 2014; Chiang et al. 2017). To overcome this challenge, visualization serves as a preliminary yet critical step for providing powerful insights and process understanding. Front-line control room operators rely heavily on two-dimensional (2D) or sometimes three-dimensional (3D) visual data representation for process monitoring and fault detection. However, for chemical process datasets with hundreds to thousands of variables, a direct visualization of all variables is likely to overwhelm operators and conceal key information that could lead to performance improvement and more effective troubleshooting. Fortunately, a variety of data analytics techniques, mapping the original variables to a lower-dimensional (latent) space, have been developed and deployed to assist process engineers in achieving the goal of visualizing high-dimensional data intuitively (Qin 2012; Becht et al. 2019). For example, principal component analysis (PCA) has become an industrial standard in process monitoring, detecting, and isolating faults associated with individual process variables and units (MacGregor and Kourti 1995; Russell et al. 2001; Ge and Song 2013). Other dimension reduction approaches developed by the machine learning community, commonly used for image and natural language processing, have also been leveraged in process monitoring and fault detection. These include, for example, self-organizing maps (SOMs)

(Alhoniemi et al. 1999), t-distributed stochastic neighbor embedding (t-SNE) (van der Maaten and Hinton 2008; Zhu et al. 2018), and uniform manifold approximation and projection (UMAP) (Joswiak et al. 2019). In this chapter, a new visualization approach – the time-explicit Kiviat diagram – is presented. Section 2 introduces the concept, which provides an intuitive representation of the process data in a lower-dimensional space. In Section 3, specific fault-detection procedures are presented through Kiviat diagrams for three classes of chemical processes: continuous, batch, and periodic. Section 4 provides three industrial case studies corresponding to the same three process types to validate the new approach's efficacy, where performance metrics including fault-detection time, false detection (type I error) rate, and missed detection (type II error) rate are calculated. The new method is compared with the conventional PCA-based approaches in fault-detection performance. Section 5 provides a conclusion.

6.2 TIME-EXPLICIT KIVIAT DIAGRAM

The concept of Kiviat diagrams originates from the parallel coordinates plot used, for example, in computer science for the visualization of software performance (Inselberg 2009). Depicting trends of high-dimensional data with high resolution and low cluttering is beyond the capabilities of the conventional Cartesian system. To break this "curse of dimensionality" and display multivariate data effectively in process industries, parallel coordinates have been recommended (Dunia et al. 2013) as shown in Figure 6.1. It is observed that no abscissa is used in such a plot and an open line corresponding to each multivariate data sample connects the values of each variable. In addition, each of the parallel axes is associated with the ordinate of the Cartesian plot of the corresponding variable.

Parallel coordinates increase visual clutter as more linear segments are added. To mitigate this problem, instead of using parallel axes and representing data samples as open (set of) linear segments, a Kiviat diagram (also referred to as a radar chart or a spider chart) places axes radially around a center point and data samples are rendered as closed polygons. As shown in Figure 6.2, to address time series data frequently encountered in process industries, an additional coordinate is added perpendicularly to the plotting plane that explicitly captures the time dimension which results in a novel time-explicit Kiviat diagram (Wang et al. 2015; Wang et al. 2017).

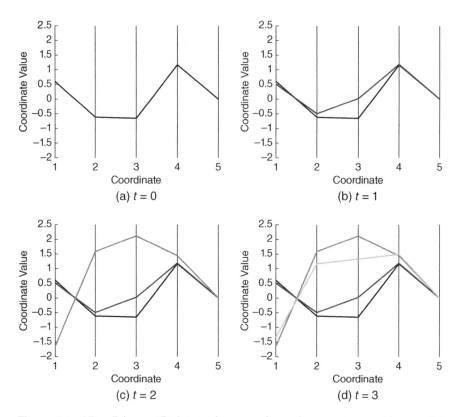

Figure 6.1   Visualizing a 5D dataset in a transformed space spanned by parallel coordinates. Linear segments corresponding to data samples are accumulated as time $t = 0$–3, exacerbating visual cluttering (Wang et al. 2017).

6.3   FAULT DETECTION BASED ON THE TIME-EXPLICIT KIVIAT DIAGRAM

As illustrated in Figure 6.2, a multivariate time series dataset is rendered as a "cylinder." Each slice of the cylinder is a closed polygon and corresponds to the Kiviat diagram at a specific sample time. Its centroid is defined as follows:

$$X_{\text{centroid},j} = \frac{\sum_{i=1}^{n} X_{i,j}}{n} \tag{6.1}$$

$$Y_{\text{centroid},j} = \frac{\sum_{i=1}^{n} Y_{i,j}}{n} \tag{6.2}$$

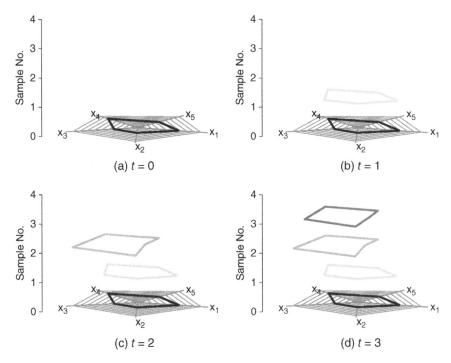

Figure 6.2 Depicting the same 5D dataset in a time-explicit Kiviat diagram, where closed polygons corresponding to data samples are stacked in parallel along the time axis perpendicular to the Kiviat plot plane (Wang et al. 2017).

where $i \in \{1, \ldots, n\}, j \in \{1, \ldots, m\}$, and m, n represent the number of samples and variables of datasets, respectively. The equations above indicate that the centroids are the geometric centers of the polygons.

At a given time, the process state represented by such a multivariate data slice is condensed into a single point, i.e., the centroid, which facilitates detecting changes in the processes (see Figure 6.3).

If datasets are pre-processed so that each variable has zero mean and unit standard deviation, the centroids of a steady-state region will be centered around the intersection point of the axes of the Kiviat plot, with slight deviations from point to point due to noise. "Normal operating regions" in time-explicit Kiviat diagrams are defined and visualized to distinguish between normal and abnormal operation (i.e., faults) and mitigate the noise impact. Fault-detection procedures based on the time-explicit Kiviat diagram for three types of processes are elaborated in the sections below.

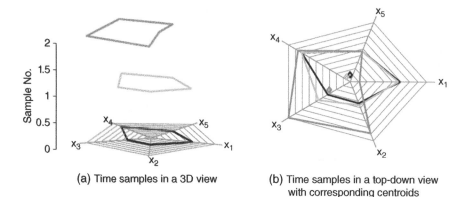

(a) Time samples in a 3D view

(b) Time samples in a top-down view
with corresponding centroids

Figure 6.3 Process state shown by a multivariate data slice is condensed into a single point – the centroid. (b) A bird's-eye view of (a), adding centroids depicted by diamonds (Wang et al. 2017).

6.4 CONTINUOUS PROCESSES

Continuous processes are characterized by continuous material inflow and outflow during the whole operation at or near steady states. Often, the process switches between several steady-state operating points. In this section, for simplicity, it is assumed that the process only features a single steady state. The following steps are carried out to establish a confidence ellipse of centroids representing the "normal" steady state of the process (Wang et al. 2015).

(1) Assuming that the process data matrix $X \in R^{m \times n}$ (which contains m measurements of n variables) reflects a normal steady-state operation with desirable performance, the covariance matrix $\Sigma = XX^T$ and its eigenvalues λ_i and eigenvectors v_i, $i \in \{1, \ldots, n\}$ are computed by

$$\lambda v = \Sigma v \tag{6.3}$$

(2) λ, Σ, and v values computed above are used to define an n-dimensional ellipsoid around the steady-state operating region, which can be described by the following equation:

$$\left(x - \overline{x}\right)^T \Sigma^{-1} \left(x - \overline{x}\right) = 1 \tag{6.4}$$

(3) The length of the axis l_i is determined by the eigenvalues of the covariance matrix and scaled using the critical value κ of the χ^2

distribution that corresponds to the desired confidence level of the ellipsoid shown below:

$$l_i = 2\sqrt{\kappa\lambda_i} \ \forall i \in \{1...n\} \tag{6.5}$$

(4) Extremes of the n-dimensional ellipsoid are projected on the Kiviat diagram (see Figure 6.4a).

   (1) Within the annular region between the extremes of the n-dimensional ellipsoid projected on the Kiviat diagram, random polygons are generated based on sampled values uniformly

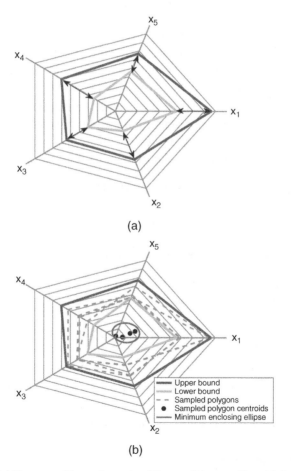

(a)

(b)

Figure 6.4   (a) Upper and lower bounds of the confidence ellipsoid. Arrows indicate individual limits of each variable; (b) minimum enclosing ellipse of centroids (dots) is generated by sampling data points (dashed lines) within the annual region (Wang et al. 2017).

distributed within the bounds of each variable (see Figure 6.4b).
To prevent random polygons close to the edges from falling
outside of the ellipsoid, a verification procedure is conducted
including two simple steps:

(2) Transforming the coordinates Υ of the randomly generated
polygon using the inversion matrix W^{-1} to obtain transformed
coordinates Z:

$$Z = \Upsilon W^{-1} \tag{6.6}$$

where

$$W = v\sqrt{\lambda}.$$

(3) Comparing the norm $D = \|\mathbf{Z}\|$ with the radius of the unit sphere:
if $D \leq 1$, the randomly generated polygon will be kept because
of its close tie to a point within the confidence ellipsoid; otherwise, we discard the polygon and generate a new one.

(4) Repeat step 4 shown above until the number of random polygons reaches a prescribed threshold (typically, 5,000), and calculate the minimum-area enclosing the confidence ellipse of
center c by solving the following optimization problem:

$$\min_{A,c} \log\left(\det\left(A\right)\right)$$

$$s.t. \left(P_i - c\right)^T A \left(P_i - c\right) \leq 1, i = 1,2 \tag{6.7}$$

where P is the matrix of centroid locations.

After the confidence ellipse of centroids is obtained following Steps (1)–(5),
fault detection is performed in the following manner:

(1) Calculate the corresponding polygon and centroid in the Kiviat
diagram for every new data sample.

(2) Check if the centroid lies outside of the confidence region.

(3) Mark the sample as faulty if it lies outside of the confidence
region. A separate criterion (e.g., two consecutive samples are
identified as faulty) can be implemented to raise a process fault.

6.5 BATCH PROCESSES

Unlike continuous processes, batch processes, commonly implemented in
pharmaceutical industries, do not allow for material exchanges with the sur-
roundings during the operation except for the start and the end (fed-batch

process is not considered here). In addition, batch processes, being dynamic, rarely reach a steady state. Such distinguishing characteristics of batch processes call for an alternative approach for computing a single confidence ellipse in the time-explicit Kiviat diagrams. Replacing a single confidence ellipse with multiple confidence regions can achieve the goal of capturing the entire trajectory of the normal batch operation, and the corresponding steps are shown below (Wang et al., 2018).

(1) Unfold batch data (with dimensions I batches $\times J$ time samples $\times K$ variables) into a $J \times IK$ two-dimensional array using time-wise unfolding, as shown in Figure 6.5.

(2) Leverage collective data sets corresponding to normal batches for training and represent them on the same Kiviat diagram.

(3) Calculate centroids together with their confidence regions for each sample at time t in the batch based on multiple batches, following the same procedure shown in Section 3.1.

(4) Stack the confidence ellipses in a way similar to the polygons in Kiviat diagrams in Figure 6.2 to allow for better visualization of the trajectory of the batch (see Figure 6.6).

By comparing the centroids of new batch samples against the confidence regions at each sample time, it is possible to identify when a fault occurs in a batch run and diagnose the root cause of such a fault. It is worth pointing out that such a fault-detection method can not only be used in real time, but can also be leveraged to evaluate the batch operation after its completion.

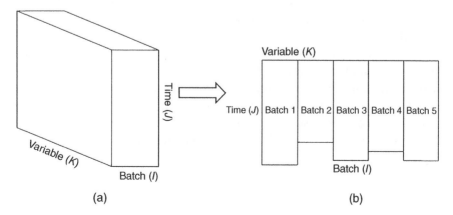

Figure 6.5 Unfolds a 3D batch data shown in (a) with regard to time (b).

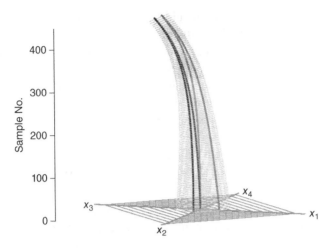

Figure 6.6   A 3D tube composed of confidence regions at each sample drawn for an illustrative batch process data set (Wang et al. 2017).

6.6   PERIODIC PROCESSES

In this section, a distinctive class of chemical operation – the periodic process – is studied (Wang et al. 2017). Unlike its continuous or batch counterparts, a periodic process is characterized by cyclical steady states during normal operation where the cycles typically begin and end in the same state. To take advantage of such characteristics, the inter-cycle operation is treated as a continuous process and the intra-cycle behavior is treated as a batch process. As a result, the time-explicit Kiviat diagram-based fault-detection approach is tailored to meet the requirements of a periodic process, and it consists of two steps: an inter-cycle step that leverages the oscillatory steady state to detect abnormal cycles and an intra-cycle step that strives to pinpoint the locations of deviations in those irregular cycles. Specifically, in the inter-cycle step, a confidence ellipse is defined based on cyclic centroids associated with normal operation cycles and is used to find faulty cycles with cyclic centroids outside such an ellipse. In the intra-cycle step, confidence ellipses for individual samples across cycles of normal operation are calculated, and by comparing the new samples against the confidence ellipse at each sample time, we can identify when a fault occurs within a cycle (see Figure 6.7).

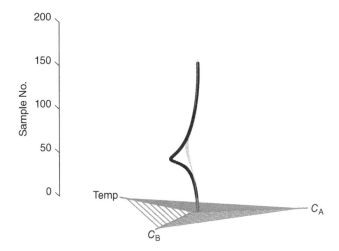

Figure 6.7 Fault hidden in an abnormal cycle is detected. An intra-cycle confidence region is used to differentiate normal data (lying inside) from faulty ones (falling outside) (Wang et al. 2017).

6.7 CASE STUDIES

In this section, case studies are conducted on simulated continuous, batch, and periodic processes to validate the effectiveness of the proposed time-explicit Kiviat diagram in fault detection.

6.8 CONTINUOUS PROCESSES

The benchmark Tennessee Eastman Process (TEP) (Downs and Vogel, 1993) (see Figure 6.8) is used in the continuous process case study, and the new time-explicit Kiviat diagram-based fault-detection method is compared with PCA-based fault-detection methods (Russell et al. 2001; Tamura and Tsujita, 2007). The MATLAB version of the TEP simulator (Ricker 2015) has provisions for applying 20 faults and 11 out of them are listed in Table 6.1. These faults include changes in flow rate/temperature/pressure/composition, drifts of reaction kinetics, and sticking valves.

For each fault listed in Table 6.1, training data indicative of steady-state operation of the process were obtained by simulating the process for 12 hours (720 min) and injecting the fault at $t = 300$ min. Random noise was

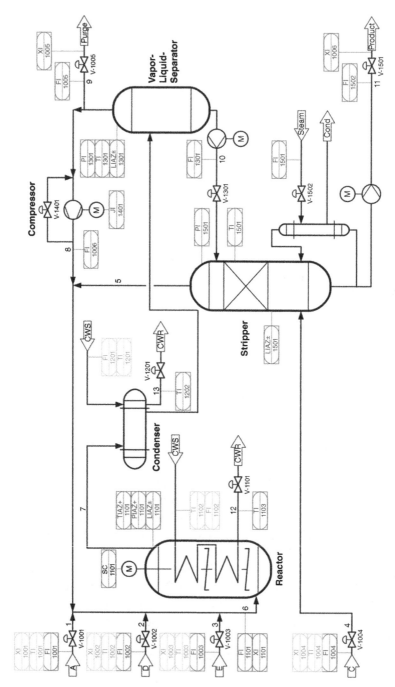

Figure 6.8 Schematic diagram of the Tennessee Eastman Process (Yao et al. 2021).

Table 6.1 Faults simulated in the Tennessee Eastman Process simulator.

Fault no.	Description	Type
1	A/C feed ratio, B composition constant (Stream 4)	Step
2	B composition, A/C ratio constant (Stream 4)	Step
5	Condenser cooling water inlet temperature	Step
6	A feed loss (Stream 1)	Step
7	C header pressure loss – reduced availability (Stream 4)	Step
8	A, B, C feed composition (Stream 4)	Random variation
10	C feed temperature (Stream 4)	Random variation
11	Reactor cooling water inlet temperature	Random variation
12	Condenser cooling water inlet temperature	Random variation
13	Reaction kinetics	Slow drift
14	Reactor cooling water valve	Sticking

added to every run. Principal component analysis (PCA) was leveraged for dimension reduction where nine principal components were used to capture 70.1% of the total variance. The confidence level used to compute the confidence ellipse is set to 95%.

Three metrics are used to evaluate the performance: fault-detection time (i.e., time required to detect fault after its injection in the process), missed detection (type II error) rate, and false detection (type I error) rate. Such metrics are compared against T^2 and Q metrics of PCA and dynamic PCA (Russell et al. 2000). The results are summarized in Tables 6.2–6.4. It is observed that in most cases, the new approach can detect the faults faster while reducing the type I/II error rates compared to conventional PCA and DPCA methods.

Table 6.2 Fault-detection time for the Tennessee Eastman Process.

	Fault-detection time (minutes) (lower is better)				
Fault no.	Proposed method	PCA T^2	PCA Q	DPCA T^2	DPCA Q
1	3	21	9	18	15
2	8	51	36	48	39
5	2	48	3	6	6
6	2	30	3	633	3
7	2	3	3	3	3
8	46	69	60	69	63
10	52	288	147	303	150
11	39	912	33	585	21
12	14	66	24	9	24
13	124	147	111	135	120
14	8	12	3	18	3

Table 6.3 Missed detection rates for the Tennessee Eastman Process.

	Missed detection rates (lower is better)				
Fault no.	Proposed method	PCA T^2	PCA Q	DPCA T^2	DPCA Q
1	0.018	0.008	0.003	0.006	0.005
2	0.018	0.020	0.014	0.019	0.015
5	0.002	0.775	0.746	0.758	0.748
6	0.005	0.011	0	0.013	0
7	0.007	0.085	0	0.159	0
8	0.102	0.034	0.024	0.028	0.025
10	0.138	0.666	0.659	0.580	0.665
11	0.171	0.794	0.356	0.801	0.193
12	0.069	0.029	0.025	0.010	0.024
13	0.283	0.060	0.045	0.049	0.049
14	0.040	0.158	0	0.061	0

Table 6.4 False detection rates for Tennessee Eastman Process.

False detection rates (lower is better)				
Proposed method	PCA T^2	PCA Q	DPCA T^2	DPCA Q
0.032	0.014	0.16	0.006	0.281

6.9 BATCH PROCESSES

The PenSim bioreactor simulator (Ündey et al. 2002) is used in this comparison. Figure 6.9 shows the process schematic.

Sixteen process variables measured in the PenSim process are listed in Table 6.5, where process input variables include the aeration rate, agitator power, and glucose feed rate. Model predicted variables include the concentrations of biomass, glucose, penicillin, dissolved oxygen, and carbon dioxide. Table 6.6 summarizes nine faults simulated in this case study, including step/ramp changes in the inputs. Training data are collected from 20 batches indicative of a normal operation and are used to establish the sample-wise confidence ellipses following Steps (1)–(4) in Section 3.2. Each testing batch data is obtained by running the simulator with a fault listed in Table 6.6 lasting for 30 hours from $t = 100$ h to $t = 130$ h.

Fault-detection time and false detection (type I error) rate are used in evaluating the performance of fault detection. The new approach is compared against the online multiway PCA (MPCA) T^2 and Q statistics

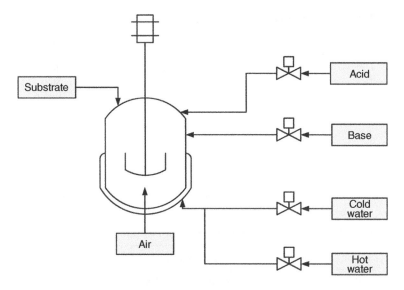

Figure 6.9 Schematic diagram of the PenSim process (Wang et al. 2017).

Table 6.5   List of variables measured in the PenSim process.

Variable number	Variable description
x_1	Aeration rate (L/h)
x_2	Agitator power (W)
x_3	Substrate feed rate (L/h)
x_4	Substrate temperature (K)
x_5	Substrate concentration (g/L)
x_6	Dissolved oxygen concentration (g/L)
x_7	Biomass concentration (g/L)
x_8	Penicillin concentration (g/L)
x_9	Culture volume (L)
x_{10}	Carbon dioxide concentration (g/L)
x_{11}	pH
x_{12}	Temperature (K)
x_{13}	Generated heat (cal)
x_{14}	Acid flow rate (mL/h)
x_{15}	Base flow rate (mL/h)
x_{16}	Cooling/heating water flow rate (L/h)

(Lee et al. 2004) commonly used in batch analytics. Table 6.7 and Table 6.8 summarize the results. It is observed that the proposed approach obtains a fault-detection time comparable to MPCA and also excels in reducing false alarms.

Table 6.6   Faults added in the PenSim process.

Fault no.	Description	Type
1	10% increase in aeration rate	Step
2	20% increase in aeration rate	Step
3	1.5 L/h increase in aeration rate	Ramp
4	20% increase in agitation power	Step
5	40% increase in agitation power	Step
6	0.015 W increase in agitator power	Ramp
7	20% increase in substrate feed	Step
8	40% increase in substrate feed	Step
9	0.12 L/h increase in substrate feed	Ramp

Table 6.7   Fault-detection time for the PenSim process.

	Fault-detection time (hours) (lower is better)		
Fault no.	Proposed method	MPCA T^2	MPCA Q
1	0.5	4	3.5
2	0.5	9.5	9.5
3	13	13	13
4	1.5	2.5	3
5	9	7	7.5
6	15.5	11.5	12.5
7	20	1.5	2
8	14.5	6	
9	12.5	10.5	

Blank cells indicate that no fault was detected.

Table 6.8   False detection rates for the PenSim process.

	False detection rates (lower is better)		
Fault no.	Proposed method	MPCA T^2	MPCA Q
1	0.11	0.075	0.1
2	0.025	0.07	0.085
3	0.01	0.095	0.07
4	0.03	0.105	0.07
5	0.16	0.11	0.07
6	0	0.105	0.035
7	0.07	0.105	0.02
8	0.085	0.105	
9	0.07	0.105	

Blank cells indicate that no fault was detected.

6.10   PERIODIC PROCESSES

In this section, an air separation system for separating oxygen from air via pressure swing adsorption (PSA) is studied. The PSA system was simulated using the gPROMS gML Separations-Adsorption model library (Process Systems Enterprise 2021). The model is established for a two-bed, four-step isothermal process (see Figure 6.10), whose periodic operation follows the switching strategy described in Table 6.9. The simulation parameters are listed in Table 6.10. A total of 26 variables including flow rates of the feed, pressure, and concentrations in and across both beds were recorded. The simulated data were contaminated with white noise with a signal-to-noise ratio of 30. The process was run for 10,000 s in total, and the observed

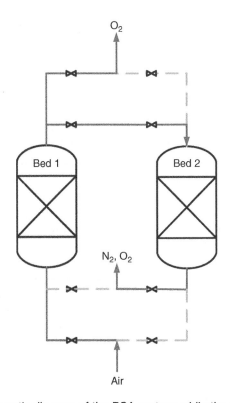

Figure 6.10   Schematic diagram of the PSA system: while the gas flow pathway is represented by the blue solid lines, the inactive piping in the cycle is denoted by red dashed lines. While Bed 1 stays active, Bed 2 is undergoing regeneration (Wang et al. 2017).

Table 6.9   Switching strategy for the pressure swing adsorption (PSA) process.

Duration (s)	Bed 1 state	Bed 2 state
2	Pressurization	Blowdown
60	Adsorption	Desorption
2	Pressure equalization	Pressure equalization
2	Blowdown	Pressurization
60	Desorption	Adsorption
2	Pressure equalization	Pressure equalization

Table 6.10   Parameters for the PSA model simulation.

Parameter	Parameter value
Feed flow rate	0.00364 mol/s
Temperature of feed	298.15 K
Length of bed	0.35 m
Radius of bed	0.0175 m
Particle radius	0.003175 m
ϵ(void fraction)	0.4
P_{feed}	300 KPa

period of a single cycle was 150 s. Faults were simulated by disturbing the temperature and pressure of the feed flow into the beds at $t = 5,000$ s. In this case study, the new hybrid approach is compared with the same dynamic PCA and multiway PCA as shown before in fault-detection performance, and the fault-detection time is used to evaluate fault-detection performance. As seen from the summary of results in Table 6.11, the hybrid approach outperforms both DPCA and MPCA by detecting faults faster.

Table 6.11   Fault-detection time for the PSA system data.

		Fault-detection time (seconds) (lower is better)				
Case	Fault description	Proposed method	DPCA T^2	DPCA Q	MPCA T^2	MPCA Q
1	Increased temperature feed by 5 K in Beds 1 and 2	89	120	115	118	74
2	Increased temperature feed by 5 K in Beds 1 and 2	99	51	54	116	54
3	Pressure drop in Bed 1 by 10%	59	52	103	118	116
4	Pressure rise in Bed 2 by 10%	61	122	116	116	173

6.11 CONCLUSIONS

In this chapter, a brief review on data analytics techniques is provided to address the challenge of visualizing multivariate datasets encountered in process industries. In addition, a new visualization approach – the time-explicit Kiviat diagram – is described, which facilitates the representation of process data in a lower-dimensional space and serves as a structure to define specific visualization and fault-detection techniques for three types of chemical processes: continuous, batch, and periodic. Furthermore, the simulated case studies demonstrate the superior performance of the proposed approach in quickly detecting faults while maintaining low type I/II error rates.

ACKNOWLEDGMENT

This work was supported by the Center for Operator Performance (COP).

REFERENCES

Ündey, C., Çinar, A., and Birol, G. (2002). A modular simulation package for fed-batch fermentation: penicillin production. *Computers & Chemical Engineering* 26 (11): 1553–1565.

Alhoniemi, E., Hollmén, J., Olli, S., and Vesanto, J. (1999). Process monitoring and modeling using the self-organizing map. *Integrated Computer Aided Engineering* 6 (1): 3–14.

Becht, E., McInnes, L., Healy, J. et al. (2019). Dimensionality reduction for visualizing single-cell data using UMAP. *Nature Biotechnology* 37 (1): 38–44.

Chiang, L.H., Lu, B., and Castillo, I. (2017). Big Data anlaytics in chemical engineering. *Annual Review of Chemical and Biomolecular Engineering* 8 (1): 63–85.

Downs, J. and Vogel, E. (1993). A plant-wide industrial process control problem. *Computers & Chemical Engineering* 17 (3): 245–255.

Dunia, R., Edgar, T.F., and Nixon, M. (2013). Process monitoring using principal components in parallel coordinates. *AIChE Journal* 59 (2): 445–456.

Ge, Z. and Song, Z. (2013, 2013). Distributed PCA model for plant-wide process monitoring. *Industrial & Engineering Chemistry Research* 52 (5): 1947–1957.

Inselberg, A. (2009). *Parallel Coordinates*. New York, New York, USA: Springer.

Joswiak, M., Peng, Y., Castillo, I., and Chiang, L.H. (2019). Dimensionality reduction for visualizing industrial chemical process data. *Control Engineering Practice* 93: 104–189.

Lee, J., Yoo, C., and Lee, I. (2004). Statistical monitoring of dynamic processes based on dynamic independent component. *Chemical Engineering Science* 59 (14): 2995–3006.

MacGregor, J.F. and Kourti, T. (1995). Statistical process control of multivariate processes. *Control Engineering Practice* 3 (3): 403–414.

Process Systems Enterprise. (2021). gPROMS gML Separations—Adsoprtion Model Library. London, United Kingdom. Retrieved November 12, 2021, from https://www.psenterprise.com/products/gproms/process/libraries

Qin, S.J. (2012). Survey on data-driven industrial process monitoring and diagnosis. *Annual Reviews in Control* 36 (2): 220–234.

Qin, S.J. (2014). Process data analytics in the era of Big Data. *AIChE Journal* 60 (9): 3092–3100.

Ricker, N. L. (2015). Tennessee Eastman Challenge Archive. Seattle, Washington, USA. Retrieved November 12, 2021, from http://depts.washington.edu/

Russell, E.L., Chiang, L.H., and Braatz, R.D. (2000). Fault detection in industrial processes using canonical variate analysis and dynamic principal component analysis. *Chemometrics and Intelligent Laboratory Systems* 51 (1): 81–93.

Russell, E. L., Chiang, L. H., and Braatz, R. D., (2001) Fault Detection and Diagnosis in Industrial Systems (Vol. 1). London, United Kingdom: Springer-Verlag.

Tamura, M. and Tsujita, S. (2007). A study on the number of principal components and sensitivity of fault detection using PCA. *Computers & Chemical Engineering* 31 (9): 1035–1046.

van der Maaten, L. and Hinton, G. (2008). Visualizing data using t-SNE. *Journal of Machine Learning Research* 9 (86): 2579–2605.

Wang, R. C., Baldea, M., & Edgar, T. F. (2017). Data visualization and visualization-based fault. Processes, 5. doi:https://doi.org/10.3390/pr5030045

Wang, R.C., Edgar, T.F., and Baldea, M. (2017). A geometric framework for monitoring and fault detection for periodic processes. *AIChE Journal* 63 (7): 2719–2730.

Wang, R.C., Edgar, T.F., Baldea, M. et al. (2015). Process fault detection using time-explicit Kiviat diagrams. *AIChE Journal* 61 (12): 4277–4293.

Wang, R.C., Edgar, T.F., Baldea, M. et al. (2018). A geometric method for batch data visualization, process monitoring and fault detection. *Journal of Process Control* 67: 197–205.

Yao, Y., Zhang, J., Luo, W., and Dai, Y. (2021). A hybrid intelligent fault diagnosis strategy for chemical processes based on penalty iterative optimization. *Processes* 9 (8): 1266.

Zhu, W., Webb, Z., Mao, K. et al. (2018). Generic process visualization using parametric t-SNE. *IFAC-PapersOnLine* 51 (18): 803–808.

6.A

VIRTUAL STATISTICAL PROCESS CONTROL ANALYSIS

Richard J. Fickelscherer, PE[1] and Daniel L. Chester[2]

[1]Department of Chemical and Biological Engineering, State University of New York at Buffalo, Buffalo, NY, USA
[2]Department of Computer and Information Sciences (Retired), University of Delaware, Newark, DE, USA

6.A.1 OVERVIEW

The following is an overview of the calculations and logic behind FALCONEER™ IV's Virtual Statistical Process Control (Virtual SPC) analysis. This technology should allow anyone with the capability of continuously monitoring particular process sensor readings or calculating performance equations (a.k.a., key performance indicators [KPIs] or soft sensors) the ability to determine if they are under control or not. Its analysis is based on exponentially weighted moving averages (EWMAs) of those readings or performance equation calculations over time. This analysis directly allows for out of control sensors or calculated variables to be

Artificial Intelligence in Process Fault Diagnosis: Methods for Plant Surveillance,
First Edition. Edited by Richard J. Fickelscherer.
© 2024 John Wiley & Sons, Inc. Published 2024 by John Wiley & Sons, Inc.

flagged more quickly and at lower deviation levels. It should consequently allow the process operators to intercede with the proper control actions necessary to mitigate the underlying process problems without unduly disrupting process operations. This analysis is considered virtual because it is being done automatically in real-time without the need for the operators to collect and chart any process sensor readings.

The normal standard deviation and mean of each measured variable for which Virtual SPC will be performed upon needs to be evaluated to configure those variables properly. This evaluation is automatically performed in FALCONEER IV on individual data tags with 3- to 6-month worth of normal process operations data. The following describes how FALCONEER IV uses these individual tag standard deviations and means for both controlled and uncontrolled measured variables and performance equation calculated variables.

6.A.2 INTRODUCTION

Statistical Process Control (SPC) is a tool used to assess whether a process is currently under or out of control. Various techniques exist for doing this analysis depending on the nature of the process being monitored. In continuous processes (as opposed to the manufacture of discrete, individual units), process data collected at a particular moment in time is not completely independent of its previous data. This phenomenon is referred to as auto-correlation between the data. SPC techniques used to deal with auto-correlation in the data revolve around calculations of exponentially weighted moving averages (EWMAs). These calculations cancel the effects of auto-correlation and allow small but statistically significant shifts in an observed or calculated variable's value to be readily detected. Calculating EWMAs is the method used in our Virtual SPC software module to determine if a particular sensor reading or performance equation (a.k.a., KPIs or soft sensor) calculation is in or out of control.

In FALCONEER IV, there are two types of measured variables and one type of calculated variable to contend with when configuring the particular Virtual SPC analysis to be performed. Measured variables must be defined as either controlled variables or uncontrolled variables. All calculated variables must be defined as performance equation variables. A different EWMA interpretation is performed for each of these types of variables depending on how the particular variable was configured for Virtual SPC analysis in its corresponding tag editor. Each of these possible interpretations is described below.

6.A.3 EWMA CALCULATIONS AND SPECIFIC VIRTUAL SPC ANALYSIS CONFIGURATIONS

The formula for computing the exponentially weighted moving average (EWMA) is (Montgomery 1991a):

$$\text{EWMA}_t = \lambda * Y_t + (1 - \lambda) * \text{EWMA}_{t-1} \qquad (6.A.1)$$

where Y_t is the current value of the monitored variable, EWMA_t is the currently calculated EWMA, EWMA_{t-1} is the previous calculated EWMA, and λ is the weighting factor. λ needs to be determined for each particular variable, but good results are typically obtained for values of 0.1–0.2. The sampling time also needs to be determined, but good results occur if it is greater than the process time constant of the process.[1] The results of the EWMA calculation are interpreted differently depending on the nature of the variable being monitored (controlled, uncontrolled, or performance equation variables) and the way it was configured for Virtual SPC analysis. All possible configurations are discussed below.

6.A.3.1 Controlled Variables

For controlled variables that have been configured without a manual mean or manual control limits (its standard configuration), the upper and lower control limits (UCL and LCL) on their EWMA Charts are given by

$$\text{UCL} = \mu_Y + 3.0 * \sigma_Y * \text{SQRT}\left(\lambda / [2.0 - \lambda]\right) \qquad (6.A.2)$$

$$\text{LCL} = \mu_Y - 3.0 * \sigma_Y * \text{SQRT}\left(\lambda / [2.0 - \lambda]\right) \qquad (6.A.3)$$

where μ_Y is the controlled variable's (i.e., Y_t's) set-point value and σ_Y is its standard deviation of Y_t about that value. This configuration, thus, allows μ_Y to change if the controlled variable's associated set-point changes during the analysis. For this configuration, the current value of the EWMA (i.e., EWMA_t) is compared to these UCL and LCL to determine its Virtual SPC alarm status. From their definitions, both the UCL and LCL are equidistant from the current set-point value. RED Virtual SPC alarms occur whenever the current EWMA goes beyond either the UCL or LCL and YELLOW

---

[1]The process time constant for a system is the summation of the transport delay times and three times the residence times of any tanks in that system.

Virtual SPC alarms occur whenever it goes beyond 2/3 of the interval between the set-point and these two limits.

A variation on this standard configuration for controlled variables is to define a manual mean. This makes μ_Y constant in the above calculations of the UCL and LCL. These limits thus do not change if the actual set-point for the specific Control variable changes. Controlled variables configured in this manner still have their current calculated EWMA values (i.e., $EWMA_t$) compared to these stationary UCL and LCL to determine their Virtual SPC alarm status. Again, from their definition, both the UCL and LCL are equidistant from the manual mean and the RED and YELLOW alarms occur exactly as they do when the set-point is used as the mean.

A final variation on the standard configuration for controlled variables is to define both a manual mean and manual control limits. This makes μ_Y constant and makes both the UCL and LCL constant but not necessarily equidistant from the manual mean. Controlled variables configured in this manner still have their current calculated EWMA values (i.e., $EWMA_t$) compared to these UCL and LCL to determine their Virtual SPC alarm status as above. However, if the manual mean is chosen to be different from the actual mean of the UCL and LCL chosen, such configurations allow the Virtual SPC YELLOW Alarm intervals to be skewed either high or low, correspondingly.

6.A.3.2 Uncontrolled Variables and Performance Equation Variables

For uncontrolled variables and performance equation variables, the standard configuration is also to define them without setting a manual mean or manual control limits. For these two types of variables, since the EWMA statistic can be viewed as a one-step-ahead forecast of the process mean for the next period, the one-step-ahead predication error, $\varepsilon_1(t)$, can be used to directly establish the upper and lower control limits on their EWMA Charts. The absolute value of the predication error, $\Delta(t)$, is the method used in this case (Montgomery 1991b):

$$\varepsilon_1(t) = Y_t - EWMA_{t-1} \tag{6.A.4}$$

$$\Delta(t) = \alpha * \left| \varepsilon_1(t) \right| + (1-\alpha) * \Delta(t-1) \tag{6.A.5}$$

where α is the first-order filter constant and $\Delta(t-1)$ is the absolute value of the prediction error for the last time period $t-1$.

Since $\sigma \approx 1.25 * \Delta(t)$ for a normal distribution, the UCL and LCL used for uncontrolled variables and performance equation variables are predictive by definition:

$$UCL_{t+1} = EWMA_t + 3.75 * \Delta(t) \tag{6.A.6}$$

$$LCL_{t+1} = EWMA_t - 3.75 * \Delta(t) \tag{6.A.7}$$

The choice of α controls how much of the historical process data is used in estimating the standard deviation of the prediction error. Larger values of α put more weight on recent data, while smaller values of α put more weight on older data. Good results are typically obtained when the value equals 0.01 to 0.1.

For both uncontrolled and performance equation variables defined without a manual mean or manual control limits (neither constant nor fixed distance), their current values (i.e., Y_t) are compared to both the UCL and the LCL based on the past value of the EWMA (i.e., the value of $EWMA_{t-1}$) to determine their Virtual SPC alarm status. From their definitions, both the UCL and LCL are equidistant from the past EWMA value. RED Virtual SPC alarms occur whenever the current value (i.e., Y_t) goes beyond either the UCL or LCL and YELLOW Virtual SPC alarms occur whenever it goes beyond 2/3 of the interval between the past EWMA value and these two limits.

A common situation which can occur when using this Virtual SPC analysis configuration of uncontrolled and performance equation variables is that, if the current values (i.e., Y_t) do not change over time (i.e., remain constant long enough), the values of the UCL and LCL both eventually converge to that value and the Virtual SPC alarms sound. While this is a direct method to detect stuck sensors, it also, however, causes many false alarms when the UCL and LCL shrink down over time and then the current value (i.e., Y_t) does change rather insignificantly. For this reason, other possible configurations can be employed to help ensure only significant changes in the monitored variable are brought to the user's attention. These are described below.

FALCONEER IV also allows for uncontrolled and performance equation variables the ability to set the UCL and LCL values to fixed, constant limits. This is referred to as *PRECONTROL*. For both types of variables, the current EWMA calculated value is compared directly against these limits to determine the Virtual SPC alarm status as above. When configuring these limits, it is also necessary to specify a manual mean for that variable. If the manual mean is chosen to be different than the actual mean of the constant

UCL and LCL chosen, such configurations skews the Virtual SPC YELLOW Alarm intervals either high or low, correspondingly. In practice, PRECONTROL limits are used mostly to set upper and lower alert limits on the interpretation of calculation results of performance equation variables (a.k.a., KPIs). It can, however, also be used for uncontrolled variables which are determined to not have control issues at the automatically calculated limits, making it possible to specify a normal range of operation for those variables based on safety or other concerns, etc. This flexibility in defining alert limits directly allows users to better tune the performance of the Virtual SPC analysis on actual process data for the various variables being monitored.

Other Virtual SPC analysis configurations of uncontrolled and performance equation variables are also possible. One such configuration is referred to as fixed distance. In this configuration, if a manual mean is not set for this variable, the variable's UCL and LCL are defined to be a specified number of standard deviations from the current EWMA calculated value (i.e., $EWMA_t$). The variable's current value (i.e., Y_t) is then compared against those limits to determine its Virtual SPC alarm status. If a manual mean is set for this variable, the variable's UCL and LCL are defined to be a specified number of standard deviations from this manual mean. In this configuration, the variable's current EWMA calculated value (i.e., $EWMA_t$) is then compared against those limits. (This last configuration is equivalent to the PRECONTROL configuration with the exception that the limits are always equidistant from the chosen manual mean and that those constant limits get calculated from the variable's standard deviation.)

A final variation on the Virtual SPC analysis configuration of these two types of variables is to set just a manual mean on the variable. In this case, their current values (i.e., Y_t) are compared to both the UCL and the LCL based on the past value of the EWMA (i.e., $EWMA_{t-1}$) to determine their Virtual SPC alarm status. From their definitions, both the UCL and LCL are equidistant from the past EWMA value. RED Virtual SPC alarms occur whenever the current value (i.e., Y_t) goes beyond either the UCL or LCL. However, now the YELLOW Virtual SPC alarms occur whenever it goes beyond 2/3 of the interval between the manual mean value and these two limits. Depending on the value of this manual mean and the past value of the EWMA, this may skew the YELLOW Alarm region compared to that which would occur if this manual mean were not set. Although allowed as a potential Virtual SPC analysis configuration, in practice it does not usually lead to improvements in determining out of control situations over the case where the manual mean is not set. This configuration is, therefore, not recommended in actual applications.

As described above, depending on how they are configured, uncontrolled and performance equation variables have their Virtual SPC alarm status based on either their current values (Y_t's) or their current EWMA values (i.e., $EWMA_t$). In contrast, control variables always have this status based on their current EWMA values (i.e., $EWMA_t$). In either case, exceeding either the UCL or LCL generates a RED alert (out of control); exceeding 2/3 of the interval defined between either the UCL or LCL and the chosen mean generates a YELLOW alert (going out of control); else it is a GREEN alert (in control). Again, this analysis will always be periodically performed at a frequency that can be defined to be different for each variable being monitored. Properly choosing this frequency for each variable will depend on consideration of the various time constants inherent in the actual process system and the process data sampling rate. This determination is necessary for each variable being monitored since continuous process data are auto-correlated and thus require sufficient time between samples before unique information is forthcoming by the Virtual SPC analysis. When correctly configured, this analysis provides a powerful tool for proactively determining the current status of process operations so that timely corrective actions can be taken to help optimize those operations.

6.A.4 VIRTUAL SPC ALARM TRIGGER SUMMARY

As described in Section 6.3, the particular Virtual SPC alarms that can occur for a particular type of variable depend on how the Virtual SPC analysis is configured for that variable. Depending on these configurations, either the variable's CV (i.e., current value) or current EWMA (exponentially weighted moving average) calculated value (updated at the interval specified in that variable's Virtual SPC analysis' configuration) is compared against the High and Low YELLOW and RED Virtual SPC alarm limits. These limits, again depending on the type of variable and its associated configuration, can either be fixed or have changing values. The following list summarizes which of these possible values (either CV or EWMA) and corresponding limits (fixed or changing) is used for the different possible configurations. This corresponds to the value (either CV or EWMA) reported as the trigger for the Virtual SPC alarms recorded in those reports.

Controlled variables: These variables can be configured for Virtual SPC analysis either as:

(1)   Standard EWMA (analysis occurs as described in Section 6.A.3.1 comparing EWMA with the UCL and LCL centered on the controlled variable's associated set-point value).

(2) Fixed mean (analysis occurs as described in Section 6.A.3.1 comparing EWMA with the UCL and LCL centered on the manual mean). *This configuration is not recommended if the actual setpoint of the controlled variable changes frequently.*

(3) Fixed mean and fixed limits (analysis occurs as described in Section 6.A.3.1 comparing EWMA with the manual UCL and LCL possibly skewed on the manual mean).

Uncontrolled and performance equation variables: These variables can be configured for Virtual SPC analysis either as:

(1) Standard EWMA (analysis occurs as described in Section 6.A.3.2 comparing CV with the UCL and LCL centered on the prior EWMA's value). *This configuration is not recommended if the CV does not change significantly over time – it will, however, eventually flag any stuck sensors or constant calculations.*

(2) Fixed mean (analysis occurs as described in Section 6.A.3.2 comparing CV with the UCL and LCL centered on the prior EWMA's value but YELLOW Alarm limits centered on the fixed mean so they may be skewed). *This configuration is not recommended; use standard EWMA instead.*

(3) Fixed mean and fixed distance (analysis occurs as described in Section 6.A.3.2 comparing EWMA with the UCL and LCL centered on the manual mean by the number of specified standard deviations).

(4) Fixed distance (analysis occurs as described in Section 6.A.3.2 comparing CV with the UCL and LCL equidistant from current EWMA by number of specified standard deviations).

(5) Fixed mean and fixed limits (*PRECONTROL*) (analysis occurs as described in Section 6.A.3.2 comparing EWMA with the manual UCL and LCL with the YELLOW Alarm limits possibly skewed on the chosen manual mean).

This variety of potential configurations for performing the Virtual SPC analysis makes FALCONEER IV extremely flexible for auditing the current performance of process operations and determining whether or not all target monitored variables are currently in or out of control.

6.A.5   VIRTUAL SPC ANALYSIS CONCLUSIONS

Using Virtual SPC to monitor the status of measured variables and performance equation calculations is an additional sophisticated tool for optimizing process operations. It brings to bear a type of analysis shown to be useful with auto-correlated data, hopefully giving the process operators both more timely and more meaningful alerts than those already given by their distributed control system (DCS). This should then allow them to respond with the appropriate control actions sooner.

ACKNOWLEDGMENTS

As a co-founder of FALCONEER Technologies LLC, Doug Lenz strongly advocated including Virtual SPC capabilities in FALCONEER IV; we would like to acknowledge the programing efforts of Lee Daniels required to accomplish both this and the streamlined state identification capability also currently included in FALCONEER IV.

APPENDIX 6.A REFERENCES

Montgomery, D.C. (1991a). *Introduction to Statistical Quality Control*, 299–309. New York: John Wiley and Sons.

Montgomery, D.C. (1991b). *Introduction to Statistical Quality Control*, 341–351. New York: John Wiley and Sons.

7

SMART MANUFACTURING AND REAL-TIME CHEMICAL PROCESS HEALTH MONITORING AND DIAGNOSTIC LOCALIZATION

Rajan Rathinasabapathy[1], M. J. Elsass[2], and J. F. Davis[3]

[1]Department of Chemical and Biomolecular Engineering, UCLA, Currently with Technical Services – Process Optimization, Phillips 66, Los Angeles, CA, USA
[2]Department of Chemical Engineering, University of Dayton, Dayton, OH, USA
[3]Office of Advanced Research Computing and Department of Chemical and Biomolecular Engineering, UCLA, Los Angeles, CA, USA

CHAPTER HIGHLIGHTS

- Development of diagnostic localization (DL) as a form of root-cause diagnosis in which abnormalities are analyzed as they manifest for quicker action.
- Demonstration of DL on an industrial ethylene splitter use case.
- Key methods and their integration that make DL possible: functional representation (FR) as a qualitative modeling construct; causal link assessment (CLA) algorithm to combine topographical and symptomatic assessment; and discretization and single time-step analysis for

Artificial Intelligence in Process Fault Diagnosis: Methods for Plant Surveillance,
First Edition. Edited by Richard J. Fickelscherer.
© 2024 John Wiley & Sons, Inc. Published 2024 by John Wiley & Sons, Inc.

process feature extraction, integration of multiple AI/ML techniques, explainable AI, and human-in-the-loop involvement.

- Human-in-the-loop enterprise diagnosis versus root-cause diagnosis.
- Analysis of various operational scenarios: event manifestation, sensor reliability and malfunction, multiple simultaneous malfunctions, and graceful degradation.
- Discussion of process health and early detection (PHED), smart manufacturing, AI integration, and DL.
- Reusable FR device models and the usefulness of an FR model library.

Smart manufacturing (SM) is the strategic investment in people, technology, and practice that enables manufacturers to extract significantly increased value from their existing assets and resources, at network scale, using digitalization to empower more effective workforce, factories, and supply chains. Real-time data and modeling are monetized in operations by taking advantage of significant untapped productivity (maximum use of resources end-to-end throughout an enterprise, including energy and water, to drive down production costs), precision (ensuring operating assets make the product right the first time), and performance (achieving maximum capability of people, assets, energy, and materials for better, faster, cheaper, and safer manufacturing).

At the level of machine and unit operations within the factory walls, benefits accrue with local impacts like preventive and predictive maintenance, equipment health, quality assurance, and energy and material management. When rolled up to line and factory operations, broader KPIs can be managed and optimized. These include process uptime, process interactions, production management, operational health, and material and energy management. When extended to factory-to-factory supply chains and ecosystems operational interoperability, material, product, and capacity availability can be used to drive supply chain productivity, manage disruptions, increase resilience, and open new market opportunities. Importantly, smart manufacturing addresses multiple KPIs simultaneously at the scale at which benefits accrue together. For example, increased productivity decreases waste, which reduces energy consumption and greenhouse gas emissions. The largest impacts occur at enterprise and supply chain levels, but depend on advanced sensors, equipment devices, data, models, systems, controls, and equipment-based processes on the factory floor that are the cyber and physical elements that make smart manufacturing actionable.

Assessing operational health and taking action early is a proactive functional capability in smart manufacturing that has particular relevance and value when applied to process lines in which causes of problem behaviors and their manifestations can be masked or separated in time and location because of integrated effects. We call this process health and early detection (PHED). Significant attention has been paid to the health and maintenance of the individual devices that make up a process operation. PHED is particularly effective when applied to the overall operational health of the entire process line as an integrated enterprise system. PHED goes beyond the preventive maintenance of unit and machine devices to also encompass the sensors, associated control and management devices, and the overall operational health for each unit operation with respect to the process line in which they are used. In the context of smart manufacturing, PHED encompasses: (i) measurement and analysis for individual device and overall operational assessment; (ii) the ability to predict when an abnormal situation could manifest into a larger impact; and (iii) the ability to localize the cause(s) in order to take timely actions.

In this chapter, we develop and demonstrate how smart manufacturing, AI, data, and modeling are integrated for explainable, normal, and abnormal unit and line operational assessments. A key development is an intermediate flowsheet-based diagnostic called *diagnostic localization* that allows earlier actions to be taken before root causes are fully discernable or manifest. Both capabilities are associated with qualitative modeling, symptom and topographical matching, and search with time and change discretized. Requirements for optimizing windows in time for the required fidelity in time, ways to deal with malfunctioning or missing sensors, how to address integrated operational behavior complexities and search, and how to interpret patterns of analysis over time all become key levers for tuning the performance of the system. Additionally, we address the ways to discern when the model itself is incomplete. We also address reusability of the models and approaches not only for sustained management of the given health monitoring system but also the ability to reuse and extend the approach to other enterprise chemical process line operations. Why various modeling approaches are selected and how they are integrated to meet these requirements for human-in-the-loop diagnostic localization are illustrated with a test bed application implemented on an Ethylene plant. In a previous paper, we described our integrated approach focusing on how an algorithm called causal link assessment (CLA) executes (Rathinasabapathy et al. 2016). This chapter addresses the modeling foundations and how multiple approaches are integrated to produce an explainable, trusted, robust, and reusable smart manufacturing approach for human-in-the-loop diagnostic localization. For the purposes of this chapter, we again use the Ethylene plant use case. We summarize the CLA algorithm but also include several new examples to illustrate these modeling foundations. Since the algorithm is not the primary focus of the chapter, we encourage this reference for additional detail.

7.1 INTRODUCTION TO PROCESS OPERATIONAL HEALTH MODELING

Over nearly 30 years, many different approaches to fault diagnosis have been developed and analyzed. Approaches have largely focused on root-cause diagnosis. Categorically, these can be classified as qualitative, quantitative, or history-based models (Venkatasubramanian et al. 2003a,b,c; Davis et al. 1999). Historically, quantitative and history-based models have been emphasized from the perspective of predictive analytics and model predictive control (MPC) (Chiang et al. 2001, 2015; Christofides et al. 2007; Mhaskar et al. 2006, 2007; Raimondo 2013). These methods encompass smart real-time data analytics, fault tolerance, and integrated optimization. This sense of predictive analytics, diagnostics, and control has tended to emphasize tightly integrated data, simulation, statistical, and/or control and optimization models that, in execution, are largely indistinguishable in task. These models depend heavily on data and/or first principles for modeling abnormal behaviors and conducting diagnoses. Operating data are synchronized for all models together and are defined by the smallest time interval suitable for the intended control or analytical objectives relative to a given set of process dynamics. The ability to do diagnostic analyses by mapping from input numeric sensor data to numeric output for actuation in abnormal situations is essential for closed-loop fault tolerant control. For fault tolerant control, the need to distinguish faults or failures is embedded in the control methodology (Mhaskar et al. 2006, 2007).

We define process health and early detection (PHED) as recognizing and distinguishing potential failure situations when an abnormality occurs, and localizing them with sufficient understanding to take action. This may or may not entail identifying the root cause before taking action.

With this definition, PHED is a key functional capability under the umbrella of smart manufacturing that involves measuring and analyzing process data to pinpoint the cause or causes of abnormal events. The embedded objective is to do this as fast and as accurately as possible so corrective actions can be taken in a timely manner and before abnormalities manifest into larger operational failures. The diagnostic problem of interest is the ability to identify and take appropriate action on abnormal situations in a complex process that occur above and beyond just identifiable component level situations. These are abnormal situations that are the result of interactions and complexities in which causes and manifestations become separated in time and location. In these situations, root causes can be masked, there can be multiple simultaneous malfunctions, the sensors themselves may not be working properly, and more difficult root causes like build-up, degradation, and wear could be in play. For these kinds of abnormal situations, first-principles models may not be possible because the phenomena are not understood and data-centered models are not possible because there are no enough data. It is impossible to account for all possible failures, and there may be failure modes that have never been previously experienced.

From a practical diagnostics standpoint, there is generally a need for an abnormal situation to manifest itself in some way. One or more variables being measured have to experience an excursion from normal. The diagnostic objective is to take action before the abnormal event progresses into an event that can no longer be corrected or addressed, which can result in larger losses or even process shutdown. The longer a situation manifests itself the more information there is to pinpoint the cause; however, the act of letting the situation manifest itself increases the potential for larger losses. There is, therefore, an important decision involving when and what to do while balancing impact versus available information. This decision making, which always has uncertainty, can be balanced with the concept of diagnostic localization (DL), which accounts for the fact that the precise root cause does not necessarily need to be known before taking action.

DL entails recognizing potentially abnormal excursions as early as possible, doing cause-and-effect analysis because root causes can manifest as excursions in upstream and downstream operations, sorting through multiple root causes, which may not yet be distinguishable because of cause-and-effect time lags potentially needing to wait for the abnormalities to develop so root locations can be identified, and making decisions before the root cause(s) are fully known. Additionally, the sensors themselves can be a root cause, and there is a need to recognize when there is a new, unimagined root cause.

Decision making and actions in PHED are human-centric; however, this requires the right tools to orient and evaluate the abnormal situation for a human-in-the-loop to appropriately intervene. Qualitative diagnostic models are particularly useful for larger scope, enterprise-wide, human-in-the loop decision making. They also offer robustness in capturing diagnostic behaviors when faults and failures have not previously occurred, and there are few actual data on fault conditions that have been experienced. As discussed extensively by the Abnormal Situation Management (ASM) Consortium,[1] the sources of abnormal situations are roughly 40% equipment, 20% process, and 40% people. The involvement of people requires the right tools for orientation and evaluation of a situation and for appropriate intervention. A number of ASM studies can be found on the ASM

[1]https://process.honeywell.com/us/en/site/asm-consortium.

Consortium website for background information on human factors, diagnosis, etc. Appropriate human-in-the-loop diagnosis is supported with methodologies that do data analysis, identification, and prioritization of likely faults and failures. We add diagnostic localization as a key tool.

7.2 DIAGNOSTIC LOCALIZATION – KEY CONCEPTS

Diagnostic localization (DL) is a novel diagnostic methodology focused on presenting a health assessment together with localized and limited feasible hypotheses when there are abnormal situations to enable humans-in-the-loop to take early action when there are multiple faults and failures with effects displaced in time and location. While DL can identify a single root cause for a component operation, it is an enterprise diagnostic function that is geared toward operations with multiple process units, equipment devices, and sensors. DL accounts for the probability of multiple excursions from normal when considering multiple component operations, process degradation, and some number of sensor errors given many sensors across an enterprise operation.

From a workforce standpoint, a key learning by the Abnormal Situation Management (ASM®) consortium is that the operator is more likely to remain trained and engaged in diagnostic decision making if operator involvement is integrated tightly with the decision-support system. This has steered our technical development away from strictly root-cause diagnosis and toward feasible hypothesis generation, event manifestation, and structural or functional localization for operator consideration. Diagnostic localization is, therefore, a tool that assists the operator in thinking more broadly about possibilities and in evaluating process behavior to narrow the focus to those sub-systems and devices or functions that may be in *fault* or *failed* modes. In a nutshell, diagnostic localization is a less granular form of diagnosis focusing on a broadened set of feasible and prioritized malfunction scenarios, information on the nature of an abnormal event, and a diagnostic narrowing to a malfunctioning system. Identification of a failed component is a possible outcome of diagnostic localization but is not the primary focus (Elsass et al. 2004). We emphasize DL as an assessment for recognizing, predicting, and acting on events before they become larger events by keeping human-in-the-loop, to assist with PHED, rather than root-cause diagnosis in a traditional Process Fault Detection connotation.

Diagnostic localization is most useful for:

- Providing information on abnormal events that are not readily modeled numerically.

- Identifying multiple probable hypotheses for presentation to the operator.
- Providing hypotheses as an event manifests itself to broaden operator perspective especially in early stages.
- Providing diagnostic information flexibly at varying levels of plant granularity as appropriate to the situation and the manifestation of an event.
- Providing interpreted explanations (not just numerics) to the operator.

Keeping in mind that the focus of PHED is on abnormal situations beyond the individual components, we consider the applicability of existing methods.

7.2.1 Qualitative Modeling and Symptomatic and Topographic Search

As noted earlier, Process Fault Detection and Diagnostic methods can be classified as qualitative, quantitative, or history-based models (Venkatasubramanian et al. 2003a,b,c; Davis et al. 1999). Historically, quantitative, and history-based models have been emphasized from the perspective of predictive analytics and model predictive control (Chiang et al. 2001, 2015; Christofides et al. 2007; Mhaskar et al. 2006, 2007; Raimondo 2013). Quantitative models require significant data or first-principle descriptions about each fault. While the data and models, including digital twins, may be available for a process component, i.e., a pump, it is far less likely that data and first-principles models or digital twins exist for enterprise or full unit abnormal situation scenarios. For plant operations that are typically robust, adequate historical data on abnormal excursions, even at the component level, are generally limited and very limited for enterprise behaviors that are the result of known integrated effects. For these reasons, PHED with DL foundationally has to depend on Qualitative Modeling. However, we can take advantage of this in that operational health assessment and DL do not need quantitative detail in order for an operator to take action. What is needed is a localized understanding and probability of likely situations that are occurring. If the Qualitative Modeling is done in human interpretable terminology and constructs, it is interpretable and explainable by design for human-in-the-loop actions.

Qualitative modeling is different from quantitative modeling in that it involves cause-and-effect analysis, pattern analysis, and search. Because of the explicit nature of cause-and-effect descriptions, qualitative modeling can also act as an umbrella and distributor of more detailed quantitative analyses. However, as noted throughout the literature, key drawbacks of qualitative modeling, and in particular Signed Directed Graph (SDG) methods, are inconsistencies when performing causal propagation on feedback loops, (Maurya et al. 2003a,b, 2006; Bjorkman 2013; Yim et al. 2006; Hu et al. 2015) combinatorial explosion of possible outcomes due to multiple simultaneous faults, (Bjorkman 2013; Zhang et al. 2005; Chen and Chang 2007) and generation of spurious solutions (Maurya et al. 2007). Search strategies are broadly classified as topographic and symptomatic. In topographic search, a malfunctioning system is searched with reference to a template representing normal operation. The abnormality is identified when there is not a match with an expectation. Because of topographic search, the location within the operation can be identified. All topographic strategies depend on search with reference to a model of normal function. Therefore, topographic search strategies are well suited for identification of abnormalities that cannot be modeled directly. In symptomatic search, a set of observations representing the abnormal state of the system can be used as a search template to find a matching set in a library of known symptoms. The abnormal state of operation must be modeled in order to identify the malfunction. Multiple and novel malfunctions may not be identified.

The DL methodology takes advantage of these methods by applying an inference mechanism to a qualitative plant model (Bjorkman 2013; Maurya et al. 2007; Gabbar 2007; Chen and Chang 2007; Vachhani et al. 2007; Wang et al. 2008; Thambirajah et al. 2009; Hamdan et al. 2012). A pattern matching inference mechanism overcomes the shortcomings of symptomatic and topographic search by using both (Mylaraswamy and Emigholz 2000). A symptomatic search is used to generate a causal process pattern for the entire enterprise operation. Malfunctions are then localized to the process topography using a hierarchical process decomposition.

7.2.2 Functional Representation as a Qualitative Modeling Construct

Functional representation (FR) is a qualitative, structurally organized knowledge formalism that has been extensively developed by Chandrasekaran and Josephson, and studied for malfunction diagnosis in process applications (Elsass 2001). Several related types of qualitative modeling approaches have been used for fault diagnosis, including fault trees (Chang et al. 2002), SDGs (Maurya et al. 2003a,b, 2004; Chen et al. 2007, 2009; Hamdan et al. 2012; Gao et al. 2010; Chang and Chen 2011), bond-graphs (Ould-Bouamama et al. 2012), and Dynamic Flowgraphs (Zhang et al. 2005). All these approaches foundationally use an SDG type causal model that is based on the topology of the process equipment items in a flowsheet. The FR formalism captures causal paths similarly to represent behavior. FR differs and proves a useful choice for DL in how function and structure are captured and linked to these causal paths.

FR distinguishes three main knowledge types. *Structural knowledge* describes what a device is and the topological aspects of the device that include input–output material, energy and information ports, port direction, and the type of ports. *Functional knowledge* captures the intent of the device and/or what a device does with respect to different inputs, i.e., its function. *Behavioral knowledge* describes how a function is achieved. *Behavioral knowledge* is represented as a Causal Process Description (CPD), (Chandrasekaran 1994) which represents the instantaneous state of a device. This state is linked to a function and can be identified as a normal or malfunctioning behavior. The inference mechanism is used to capture a set of CPDs within a window in time. Sets of CPDs over time capture the process dynamics.

An FR Model is best described by an example. Starting with the control valve shown in Figure 7.1, the process can be examined on a device-by-device basis. Figures 7.2 and 7.3 show the FR model for the control valve (the model can handle both fail-closed and fail-open configurations). Structural knowledge of the valve is captured topologically as a set of linked ports that are *signal, material inlet, internal,* and *material outlet.* Models contain process variable state descriptions at each port, and the port states are linked into configurations modeling feasible behaviors. Furthermore, only process variables that are affected by the device are modeled. In the example, the *material inlet* port models flow through the valve. Even though other variables such as temperature and composition are relevant, the control valve model only contains *flow* descriptions because flow is the only variable that is impacted in the valve. In addition to the material inlet and

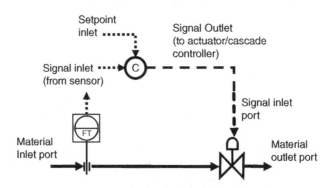

Figure 7.1 Control loop – controller, sensor, and control valve and its ports.

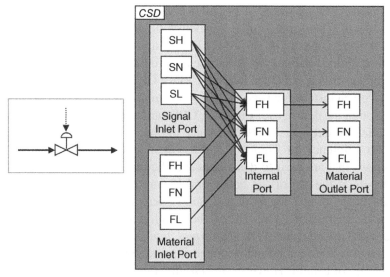

For this form of FR, syntactic primitives are:
Variables: Flow (F), Signal (S)
Qualitative Variable Descriptions: High (H), Normal (N), Low (L)

Figure 7.2   Functional representation model of a "control valve" device.

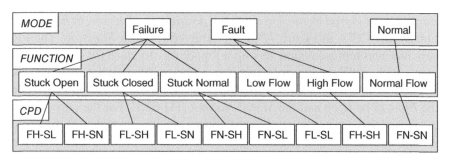

Figure 7.3   Alternative functional representation model of a "control valve" device.

outlet ports, there is a signal inlet port that models the signal (pneumatic, electric, etc.) sent to the valve to change the valve position (more open or more closed). These ports are linked to an *internal* port where the material and signal descriptions are brought together. In devices with mass or energy accumulation, the *internal* port contains a description of the accumulation variable(s). For the control valve, the port states are modeled with a set of qualitative features, *signal* and *flow,* and their possible values, e.g., *signal high* (SH), *signal normal* (SN), and *signal low* (SL), and *flow high* (FH), *flow normal* (FN), and *flow low* (FL). In the case of a valve, there is no accumulation of material, so FH inlet links to FH internal and FH outlet, and so forth. These inlet–outlet behavioral linkages within a device form a *Causal State Digraph* (CSD) of all known and relevant device behaviors. Path behaviors, for example, FH at the inlet to FH at the outlet (FH-FH), are called *Causal Process Descriptions* (CPDs). CPDs describe a device behavior at a given instance in time as opposed to an inlet state causally propagating through the device over time, i.e., CPDs are a snapshot in time depicting the variables and states for that instance in time.

With reference to Figure 7.3, the possible CPDs are grouped into *Functions* that are used to describe a specific behavior or set of behaviors. *Mode* sits at the top of the behavioral hierarchy and is used to delineate "what" behavior is being exhibited, and *Function* is used to describe "how" the Mode is achieved. While FR can be used for other applications, the three Modes that are relevant to DL are *normal*, *fault*, and *failure*. Please note that additional Modes can be incorporated if more granular diagnostic classifications are of benefit. However, for DL, the three Modes have proved to be adequate.

Normal behaviors are intuitive. *Fault* behaviors are those where there is an abnormal state in the device due to abnormal input(s) and *Failure* behaviors model a malfunction in the device. Each *Mode* can encompass multiple behaviors. With respect to the example, *Failure* can occur with the valve *stuck open* or *closed* or *stuck in a normal position*, *Fault* as *low* or *high flow*. Finally, the CPDs in Figure 7.3 are interpreted for each *Function*. The FR structure and representation sets up a top–down structure for causal statements such as "If the control valve is in a *Failure Mode* where the *valve* is *stuck open*, then expect the *Flow* to be *High* and the *Signal Low* (FH-SL), or the *Flow* to be *High* and the *Signal Normal* (FH-SN)." For the purposes of diagnosis, this representation is used bottom up. For example, "If the *Flow* is *High* and the *Signal* is *Low*, then the *valve* is *stuck open* and is in a *Failure Mode*." For the interest of the reader, FR-based modeling has other applications. One relevant example is Goodaker et al. (1995, 1996) applied it for HAZOP.

With the same principles, the FR model for the remaining devices in the control loop in Figure 7.1 – a "Flow Sensor" and a "Controller," are expanded here and shown in Figures 7.4–7.8. The Flow Sensor device model is similar to the Control Valve model. Note that the Flow sensor has the ability to model sensor malfunctions, which occur when the signal state (which represents the sensor measurement) is different from the measured process variable state in the device. The controller has three signal ports: a *signal inlet port*, an *input set-point*, and a *signal outlet port*. Keeping in mind that CPDs model a device state at an instant in time, the controller

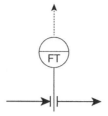

Figure 7.4   Flow sensor and its ports.

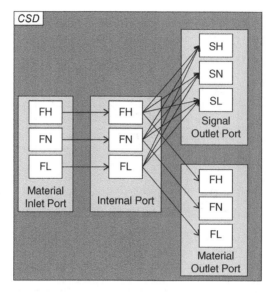

Figure 7.5   Functional representation model of the "flow sensor" device.

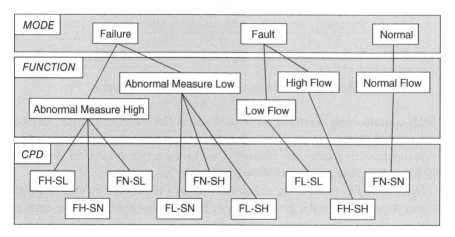

Figure 7.6   Alternative functional representation model of a "flow sensor" device.

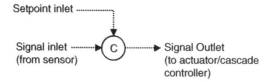

Figure 7.7   Controller and its ports.

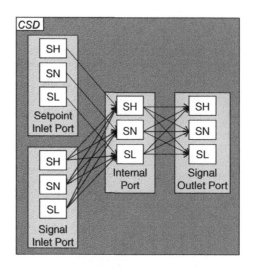

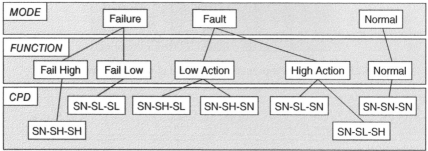

Figure 7.8   FR model of the "controller" and its ports.

CPDs contain every permutation of inlet and outlet states. Controller failure and normal states are straightforward. Fault states cover every possible behavior resulting from any change. Due to the large number of potential device states, only a subset is shown in Figure 7.8.

Individual FR models are connected to an FR process model or plant model depending on the project scope. For our example, the flow sensor device, control valve device, and controller device are assembled by connecting their ports to form a control loop model to form a subsystem in the

FR plant model. An inference mechanism propagates port states through the process model to develop a set of process level behavioral models termed Causal Process Patterns (CPP). The CPPs highlight the devices in *fault* and *failure* states to provide diagnostic information.

7.2.3   Causal Link Assessment for Combined Topographical and Symptomatic Assessment

Diagnostic localization requires an a priori knowledge base (functional representation plant model) and an inference mechanism to perform diagnosis. In our work, we have named the inference mechanism causal link assessment (CLA). The CLA algorithm utilizes the FR plant model together with process data, generally numeric, that have been interpreted and cast into qualitative features to generate plant-wide behaviors (the CPPs) formed from the connected device CPDs where associated data features are examined.

Figure 7.9 shows the high-level framework that lays out how CLA integrates with the different diagnostic localization components. As shown, CLA uses Syntactic Pattern Recognition as the analytic tool in a two-step process that involves causal process pattern (CPP) generation followed by interpretation. Generally, the task of pattern recognition is a data analytic method to automatically recognize patterns in data (numbers, text, images, etc.) using machine learning algorithms. As shown in

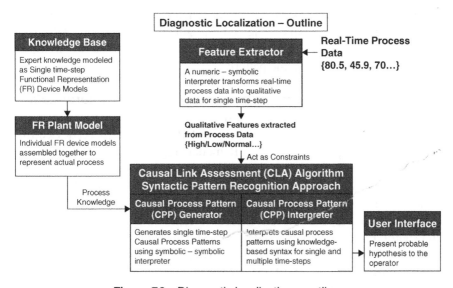

Figure 7.9   Diagnostic localization – outline.

the earlier section, the process knowledge in FR plant models is syntactic by design, and the syntactic pattern recognition deals with structural or textual information in patterns rather than numerical information (Pal and Pal 2001; Fu 1977; Gonzalez and Thomason 1978). These patterns are typically comprised of sub-elements or *primitives* (not arrays of numbers) that are interrelated. In the FR models, these primitives involve the process variables and process descriptions. Rules called *syntactic* or *production* rules govern the relationships among the primitives. The collection of primitives and syntactic rules together form the *pattern grammar* and the classification process in which patterns are analyzed and recognized is called *syntax analysis* or *parsing*. Syntactic pattern recognition involves the following computational components (Gonzalez and Thomason 1978):

- Selection and extraction of a set of primitives.
- Identification of the relationships among the extracted primitives.
- Recognition of the allowable primitive pattern structures.

The CPP Generator is a symbolic–symbolic operator that combines the symbolic process knowledge in FR models with real-time process data abstracted as symbolic process measurement features to form syntactic behavioral patterns. This portion of the algorithm primarily embodies a symptomatic search algorithm that generates and tests hypotheses based on sensor observations. However, simultaneously with generating and testing hypotheses for each device, feasible causal descriptions for each device are linked and accumulated as the process is structurally examined. In this way, CLA combines both symptomatic and topographic assessments. After pattern generation, the CPP Interpreter uses a set of knowledge-based operators to relate and prioritize the patterns.

7.2.3.1 *Causal Process Pattern Generation and Interpretation Illustrated* In this section, we illustrate how CPPs are generated by accumulating feasible device behaviors in an orderly and exhaustive, device-by-device analysis operating on the FR model. Figure 7.10 shows a simple process consisting of a process boundary, temperature sensor, and a simplified control valve. We will focus on CPP generation and show only inlet and outlet features and values without the full FR models.

In this example, the relevant measured features are *temperature, flow,* and *signal*. For each variable, the possible feature values are *High, Normal,* and *Low*. The abbreviations for this example are *Flow* (F), *Signal* (S), *Temperature* (T), *High* (H), *Normal* (N), and *Low* (L). CPP generation can begin with any device. The example starts with the *Boundary* and assumes

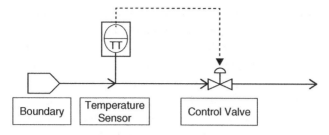

Figure 7.10   Device sequence to illustrate CPP generation.

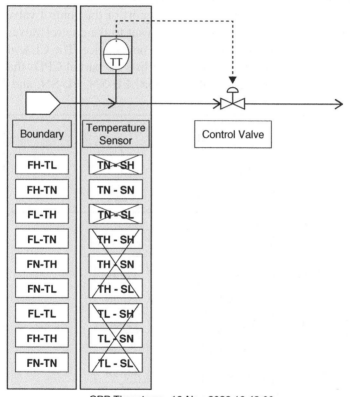

CPP Timestamp: 16-Nov-2022 16:43:00

Figure 7.11   Causal process pattern generation at the temperature sensor.

that the boundary states of *Flow* and *Temperature* are unknown for a defined time step. We, therefore, list all nine possible CPDs that exist at the *Boundary* in Figure 7.11. The *Boundary* is structurally connected to the *temperature sensor* by the material inlet port. With *Flow* unknown at the *temperature sensor*, the flow states are passed on to the next device.

Assume that the temperature sensor is reliable and the temperature sensor is measuring a normal temperature, which constraints the state to be TN-SN (only possible CPD for this time step), as shown in Figure 7.11. If we had not assumed that the temperature sensor to be reliable, all *temperature* feature states are valid and there would have been nine CPDs instead of one. The sensor can be reliable and giving a high or low value (corresponding to high or low temperature), and in this case, the normal temperature is coming from the actual temperature measurement.

The CLA algorithm next examines the Control Valve, the next device in the sequence as shown in Figure 7.12. The flow outlet port of the temperature sensor is connected to the flow inlet port of the control valve model. The temperature descriptions are not relevant to the control valve, and the *temperature* state is passed through to the next device. The CLA algorithm searches the FR control valve model and lists all partial CPDs that have a normal signal (SN) at the signal inlet port: FH-SN, FL-SN, and FN-SN.

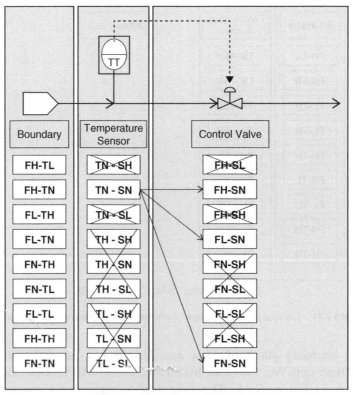

CPP Timestamp: 16-Nov-2022 16:43:00

Figure 7.12 Causal process pattern generation at the control valve.

Since there is no flow sensor to verify the status of flow, all three CPPs propagate until a reliable sensor is available to prune the CPPs.

If there were no temperature measurement upstream of the control valve, all 9 CPPs from the boundary would propagate forward resulting in 81 CPPs (9 temperature feature states × 9 flow feature states) at the control valve. We can see from this example that CPPs can rapidly proliferate but can be pared down with feature states. The approach does allow process assumptions to be applied with discretion. For example, the simplified demonstration ignores possibilities of a leak at the valve.

The CPP interpretation (hypothesis discrimination) is the final step in diagnostic localization to analyze, prioritize, and explain the findings in understandable terms to the end user with useful localized process diagnostic explanations. The CPP Interpreter, as a hypothesis discriminator, uses process-oriented, knowledge-based rules to analyze feasible plant CPPs for usefully localized process explanations of abnormal behaviors. The CPPs are prioritized with respect to *fault* and *failure* situations.

The relation of a CPD, its "function"/"malfunction" categorization, and the *normal, failure,* or *fault* mode are directly available from the FR model of the device, as shown in Figure 7.13. The example is a screen capture from an FR model implemented in Gensym G2. Shown are several CPDs associated with different modes. By searching within and across the CPPs for mode and function matches, the CPP Interpreter has a list of possible "mode" and "function" details for all the process devices. A CPP with at least one device mode in a *failure* status is considered a feasible hypothesis. Referring back

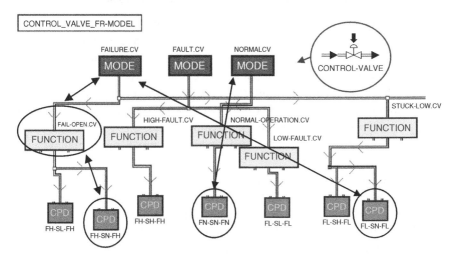

Figure 7.13   FR model and CPP interpretation.

to the CPP generation example, the Control Valve FR Model in Figure 7.13 has two *failure* states (FH-SN and FL-SN) and one *normal* state (FN-SN), which are the malfunction hypotheses for this CPP. From the "function," additional details about the control valve malfunctions can be extracted, which are "FAIL-OPEN.CV," "STUCK-LOW.CV," or "NORMAL.CV." A reliable sensor reading is required to narrow the hypothesis in this example. Overall, a CPP has underlying mode/function/CPD details that are meaningful and valuable to the end user when flagged during DL. The CPP Interpretator also has multi-time-step rules that can prioritize CPPs based on failure progression across time steps.

7.3   TIME

DL is designed to present diagnostic cues to the user. As an abnormal event manifests itself over time, it is critical how time is molded into the system since the cues will be presented across multiple time steps. While most causal approaches involve propagating a malfunction through a process model over time, the FR-based CLA DL system takes a different approach in that a process behavioral snapshot is modeled for a window in time highlighting the fault and failure information. This process is repeated over time to create a set of process snapshots for successive time windows. The composite of these behaviors is used to generate diagnostic information.

7.3.1   Discretization and Single Time-Step Analysis

Discretization in time is a key construct in DL in that it enables CPPs to be generated and analyzed as snapshots in time. With respect to our interest in user interaction, discretization also makes it possible to interpret numeric data as understandable, qualitative features for a window in time. The results of the syntactic pattern analysis and interpretation are similarly understandable in a symbolic form. With the numeric data significantly compressed into symbolic features, the CLA algorithm can computationally focus on a reduced space of a much more manageable syntactic pattern analysis.

In this work, the diagnostic search space is reduced by discretizing continuous process line data into 1-minute intervals, a duration based on behaviors of interests that occur over hours and days for this particular process. The 1-minute behavioral snapshots are computationally feasible well within 1-minute time span, which also enables the analysis and recommendations across time steps to be done in real operational time.

Discretization is a simple global activity where all process variables are sampled simultaneously on a predetermined frequency and predetermined duration with no other rules. Since discretization reduces data, it is certainly possible that valuable information in between collections is lost based on the scan frequency. The choice of discretization frequency is, therefore, important and depends on the time constants of the behaviors of interest. In our process studies, these time constants had already been considered since the data historians were storing measured data at 1-minute frequencies. This work involves diagnosing events that take significantly longer than a minute to propagate manifest though a process line operation. For faster dynamic considerations, the discretization frequency can be modified but the data collection would need to also be increased in frequency.

The example in the previous section (Figures 7.10–7.12) captures how single time-step data are used in CPP generation. As time progresses, each snapshot captures the changing process dynamics and an updated list of potential abnormal events is provided to the operator. If there is no change in variable states between time steps, the patterns are the same. For DL, we focus on when the variable state changes and the patterns are different across time steps. These pattern differences across multiple snapshots, together with when they occur, also allow the system to prioritize and rank malfunctions. Depending on the process dynamics, the time window frequency can be tuned to the process requirements. Single time-step analyses are truly advantageous from the perspective of human interfacing and interaction. Data, information, and analyses are processed in chunks of time making it possible to then follow the trail and reason about the sequence of events. When operators/engineers diagnose process events on the factory floor, it is common practice to analyze an event step-by-step in terms of when it started and how it propagated over time to distinguish cause and effect. The time between snapshot analyses enables the operator to think through the clues presented and take external factors into consideration.

In summary, discretization in fixed windows in time is foundational for many reasons in CLA execution:

- Helps to reduce the diagnostic search to a defined and finite space without any loss of relevant information.
- Aligns with the human thought process, i.e., how humans troubleshoot and diagnose process issues step-by-step, paving the way for human-in-the-loop approach.
- Decouples analyses for a given time step versus across time steps allowing for targeted technical approaches, i.e., pattern analysis and knowledge-based analysis.

- Reduces overall computational complexity making it possible to do causal modeling at a single time step in real-time enabling analyses and human interaction across time steps, also in real time (if the time constants of the behaviors of interest allow).

7.3.2 Dynamics in an Individual Functional Representation

Single time-step discretization also simplifies device dynamics making it possible to consider many device behaviors as input–output effects when their intrinsic dynamics occur quickly relative to behaviors of interest. There are also devices where behavior transition from initial state to final state over a given time period such that the values of inlet and outlet features will be different for different time steps for the same *fault* or *failure* mode. For example, when comparing changes in a control valve in a control loop with a liquid level in a large tank, the control valve changes occur much faster than an appreciable level change in a tank. However, devices like a tank do need to account for their intrinsic dynamics, and it becomes important to ensure that the granularity needed for diagnosis is not lost when performing the single time-step analysis.

In general, diagnostic localization does not require dynamic models. As alluded to, many devices do not need dynamic descriptions at all. When dynamic behaviors of a device are available or needed, these behaviors can be further discretized within the single time-step behavioral descriptions and captured as separate but nested CPDs in the FR model. The sole purpose of accounting for time varying conditions in our CLA model is such that a pattern of inlet and outlet features can link appropriately to the correct *Mode* from time step to time step. There is no need to model dynamic behavior in the sense of a quantitative prediction.

Consider a simple tank level example shown in Figure 7.14. If the inlet flow rate increases, the tank level and outlet flow rate increase over time but

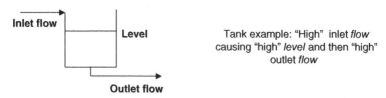

Inlet flow / Level / Outlet flow

Tank example: "High" inlet *flow* causing "high" *level* and then "high" outlet *flow*

CPD	Inlet Flow	Internal (Level)	Outlet Flow
1	Normal	Normal	Normal
2	High	Normal	Normal
3	High	High	Normal
4	High	High	High

Figure 7.14 Tank example – discretized process dynamics.

not instantaneously. The time required to observe *high level* and *high outlet flows* depends on conditions specific to the tank system. In this example, we capture the time-dependent pattern behavior as four discrete CPDs where each is applicable depending on the tank level and flow state at the time step of interest. In this simple tank example, CPD 1 is the *normal* state value pattern with normal at all state descriptions. CPDs 2 and 3 allow for the *inlet Flow* to be *High* while the *Level* is still *Normal* and the *Level* to be *High* while the *outlet Flow* is still *Normal*. Lastly, CPD 4 shows how the *High inlet Flow* has propagated through the system with high *Flow* at the inlet and outlet and a high *Level* in the tank. Not accounting for these pattern possibilities leads to unexplained state patterns during some time steps.

While higher granularity dynamic modeling is not precluded, the CLA approach recognizes explicitly that feature/value pattern fidelity is needed, not the fidelity of value prediction in the sense of a quantitative model. As shown for the tank example, only one additional level of pattern granularity is needed for level and flow out. Note also that *time* is not explicitly incorporated in any of the CPDs. Actual time and the effects of the tank configuration are accounted for in the data abstraction procedure that maps the numeric sensor data to the qualitative feature values. For example, the actual time that the *level* becomes *high* is reflected in the time step when this change is first observed. Only the combinations of *inlet Flow, Level,* and *outlet Flow* need to be accounted for. It has been our experience for complex and simple devices that accurate qualitative diagnostic analysis does not require much dynamic granularity (Elsass 2001).

7.3.3   Time Window and Feature Extraction

Feature extraction is a tool used in various scientific methods including qualitative content analysis in which constituent data, for example, numeric sensor data, image, sound, time series, snapshots, and human observation, are interpreted or abstracted into a higher level context that is indicative of a classification or pattern. For this work, feature extraction from time series data (it could be any form of data) is defined as determining the state of a given measurement relative to its context in the process operation for a series of measurements within a specified time window. The features of interest are determined by the input and output conditions of every FR component model to be used in the CLA process. Within the context of CLA, the feature extraction process interprets streaming data for a window in time so that it can be analyzed by CLA algorithm within the same time step. Figure 7.15 provides a graphical picture of the aspect of the diagnostic localization workflow.

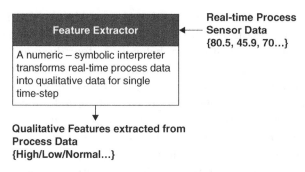

Figure 7.15   Feature extraction.

Specifically, for the ethylene process used as a test bed for this work, every measurement used across all of the component FR models used *High*, *Low*, and *Normal* feature values when interpreting the numeric data. Feature extraction, therefore, involved segmenting the time series data as it streamed in during the window in time and executing a feature interpretation analysis that determined the status of every measurement within that window in time. The feature extraction process is repeated for every time window. In this work in which data were collected on 1-minute time intervals, it was computationally possible to complete the feature extraction for every measurement and run the CLA algorithm within the 1-minute duration for the time window.

The feature extraction methodology is a machine learned algorithm that interprets data based on a range of what is expected when the operation is running normally. It is by design closely tied to human thinking and interpretation, i.e., troubleshooting that is typically based on qualitative features "high temperature," "low pressure," etc. These same feature values are explainable in DL and can be processed in DL for symptomatic and topographical assessment.

In general, Davis et al. (1999) classify process data extraction into two primary tasks: *data analysis* and *data interpretation*. *Data analysis* transforms numeric data into some other numeric data rejecting the irrelevant information, while *data interpretation* transforms numerical data into symbolic data that can be assigned meaningful labels. These labels can include state descriptions, trends, landmarks, and fault descriptions. In this document, we use the term *feature extraction* to include both the *data analysis* and *data interpretation* techniques that label numerical process data as one of three symbolic values – *normal, low,* and *high*. These labels are subjective and highly dependent on a continuous designation of a reference band of normal operation for each sensor in the given process. The *feature extractor* used in our work is a *3-sigma control chart* (Aft 1986), with mean values extracted from a set of training data. The *control chart* is cross checked with annotated data, event descriptions, and direct process expertise to make sure abstractions match the state of the process. For our case study, the 3-sigma control chart proved to deliver results that were consistent with labeling done by process experts. Feature extraction for all the variables could be done at once.

Processes can certainly have different operating regimes and the reference band can change continuously as process conditions change. This is critical to the performance of the CLA since an incorrect feature state will result in an incorrect diagnosis. There is an interface for operators or engineers to be able to flexibly update the reference band. The control chart for the Ethylene process did require review and updating about every 2 weeks to account for operational changes that changed the numeric ranges. Thresholds set by the operators worked well. As future work, to improve feature extraction with changing process conditions, we recommend exploring the use of Principal Component Analysis (PCA) or AI/ML tools and/or leverage Process Fault Detection tools. A few references of interest are Qualitative Trend Analysis (QTA) of sensor trends used to augment causal reasoning models (Maurya et al. 2007; Hamdan et al. 2012; Gao et al. 2010), Chang and Chen's fuzzy logic approach in sensor data evaluation (Chang and Chen 2011), Vedam and Venkatsubramanian's use of PCA in the fault detection task of an SDG fault diagnosis system (Vedam and Venkatasubramanian 1999), and Whiteley and Davis's extensive work on non-linear pattern recognition systems (Whiteley and Davis 1992).

Figure 7.16 illustrates the abstraction of a process data set for a pressure measurement. If the data fall within the upper and lower limits within a time

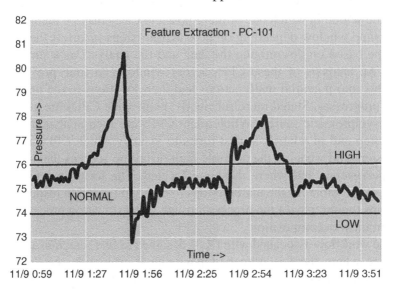

Figure 7.16 Control chart – feature extraction.

step, the pressure is labeled "normal," and outside the limits, the pressure is determined to be either "high" or "low" labeled accordingly.

While limit checking in this way is simple, the approach has proven to be quite robust for accurately characterizing the qualitative interpretation of the sensor measurements relative to describing the actual state of the plant. As seen in the control chart, there could be times when the process data can bounce back and forth across the threshold as an event is manifesting itself. Setting the threshold for repeated features across time steps is important to account for the inevitable noise in the system. Reference bands together with repeated CPP threshold proved effective.

7.4   THE WORKFLOW OF DIAGNOSTIC LOCALIZATION

Recapping the earlier discussion, diagnostic localization involves a priori knowledge representation of the plant function in *normal* and *failure* modes. In this work, a qualitative plant model composed of component functional representations is presented. The first step in developing the DL system is building the offline FR plant model using a P&ID or a Process Flow Diagram, assembling the individual FR device models, and building the topography of the unit into the diagnosis. This FR-based plant model is aligned with an inference mechanism called causal link assessment that searches for the possible syntactic patterns called Causal Process Patterns (CPPs). CLA uses a hybrid symptomatic and topographic search in which one or more possible CPPs may be identified for each time window depending on how malfunctions manifest themselves in time based on discretizing the data and the analysis as a single time step. An analysis of these CPPs across windows in time provides the insights into potential malfunctions and their locations in the process as time progresses. Stated more succinctly, plant-wide CPDs are generated by a symptomatic search. Whole plant CPPs are then analyzed for that time step and in relation to past times to identify which CPPs are more persistent and what are the possible fault or failure sources that are causing persistence and where these are located. The location information makes it possible for the malfunctions to be organized with respect to the systems topography of the process through a classification process called hierarchical decomposition.

The workflow associated with DL is designed to first search in a window of time where there is primary topographical evaluation that combines symptomatic and topographic assessments. The FR modeling process is component based and captures the structural, modular, functional, and behavioral knowledge of each individual device in the process. When the FR device models are assembled to build the FR plant model, the topography of the process is captured. As discussed in previous sections, for a given window in time, the feature abstractor examines the data (smoothed or harmonized with past data) for every measurement and determines the measured state of the plant as a vector of all interpreted values for the measurements. The causal pattern generator establishes the possible malfunctions by identifying the possible causal interactions and discarding those that are not possible. Hierarchical classification is applied to the CPPs identified within a time window to examine all the levels of process systems in which the malfunctions are occurring to provide point-in-time systems view of what malfunctions and their locations may be manifesting. The CPPs and this malfunction location assessment are examined across time steps for persistence to determine likely malfunctions and their location over time. As malfunctions manifest over time, the persistence of any individual malfunctions for root-cause diagnosis is monitored together with the persistence of location, which stabilizes at higher levels of system abstraction and then narrows as the possible malfunctions become clearer with additional features that manifest over time. Diagnostic localization occurs because a persistent location at higher systems level abstraction can occur before a malfunction becomes fully persistent. Location information can be useful for taking action before root-cause diagnosis is fully resolved.

DL does not explicitly model malfunctions nor is it dependent on malfunction models. Even if a specific malfunction is not identified and labeled, there can still be a causal path for a yet unidentified situation. This statement is premised on the concept that the behaviors of individual devices can be accounted for when carrying many combinations of CPD's forward until they can be reliably pruned. In general, these three independent DL capabilities – device FR models, CLA for plant-wide causal linkages, and feature extraction based on actual measured excursions from expected normal behavior – combine to negate the dependency on only malfunction templates and full completeness of malfunction modeling. This provides DL the ability to still recognize the possibility of unaccounted-for malfunctions and provide localization information. DL is designed to keep the human-in-the-loop and to provide cues to the operator. If full resolution is not possible, the cues and information provided still help take action, as well as point to situations, that require a fuller understanding. As demonstrated, the level of detail in the FR model is sufficient for the information content of the data, the causal analysis, and the DL and causal information provided to the operator.

Whole plant process patterns (behaviors), CPPs, are each unique and represent a feasible behavioral state of the process constrained by the real-time process data. The CPP Generator discussed is a symbolic – symbolic operator that combines the symbolic process knowledge in FR models with interpreted symbolic real-time measurement features to form syntactic behavioral process patterns using a set of predefined rules. This is also referred to as hypothesis generation in machine learning. However, the CPP generator in CLA works in reverse by using a syntactic pattern recognition technique to *compose* the CPPs from a defined set of primitives rather than decomposing patterns. Detailed governing rules for CPP generation can be found in Rathinasabapathy's thesis (Rathinasabapathy 2005).

Each FR device model contributes only one CPD to any one CPP. The number of CPDs in a CPP always matches the number of devices in the process. If a device model has more than one feasible CPD, CPP branching occurs at that point, creating separate unique CPPs. As the symptomatic search progresses through all of the devices, the CPP generation can increase exponentially. Measurement features from open-loop sensors are used as constraints to prune the CPPs. Also, closing loops when moving through the process models prune the branching. In general, the more the ports on a device with known behaviors, the fewer the resulting CPPs. For instance, when moving to a control valve (three ports), if the states at one port (say material inlet) are known, then the system has to create branches for every signal state; however, if both the signal and inlet port states are known, then there is not any branching. The potential can be higher for devices with many ports. This is shown in Figure 7.24. This requires confidence in the sensor measurement. If sensor reliability is in question for any of the sensors, those CPPs can be left unconstrained. Cross-device causal relationships and expert process knowledge can also be used as constraints as discussed in detail by Rathinasabapathy (2005).

Figure 7.17 illustrates a CPP for a FR plant model ("overhead section" of the ethylene splitter) for a time window. The FR plant model is on the left

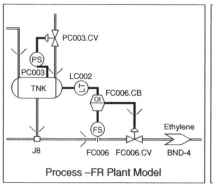

Process –FR Plant Model

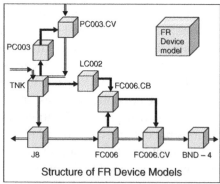

Structure of FR Device Models

Figure 7.17   Example – FR plant model topography.

and on the right is a depiction of the individual FR device models from which CPPs can be generated. Given the process topology and the FR plant model match, the hierarchical nature of the process topography is maintained throughout the process of generating the CPPs. *Material lines* (double lines) are connected to *material lines,* and *signal lines* (solid lines) are connected to *signal lines.* This structure ensures the process variables and their states are logically matched at the connections of the CPDs when the CLA algorithm is executed. The arrows show the direction of material/signal in the process, not necessarily the path of the CLA or CPP generation. The CPP path can be different and optimized by device sequencing, which is discussed in a later section of this chapter.

This is illustrated in Figure 7.18, which shows a closer look at a portion of the overhead section FR model with fewer devices and higher resolution. This can be thought of as a CPP space. The italicized text in the blue shows a set of interpreted measurement values for a point in time. These are common to the linked CPDs from the adjacent devices (F – *Flow*, T – *Temperature*, S – *Signal*, L – *Level,* and N – *Normal*). The connection between two CPDs from adjacent connected devices can happen only when the CPDs have the same variable states for common variables. For example, a Temperature Normal cannot be connected to a Temperature High or a Signal Normal cannot be connected to a Flow Normal. The text in red shows the possible CPDs.

There are two reasons for multiple CPDs for a device. First, DL recognizes three modes: *normal, fault,* and *failure* (shown in Figure 7.19). For each mode to exist for a device, there must be at least one Function which in turn requires a CPD. There is only one CPD for each function, but there can be multiple Functions for each mode and therefore multiple CPDs.

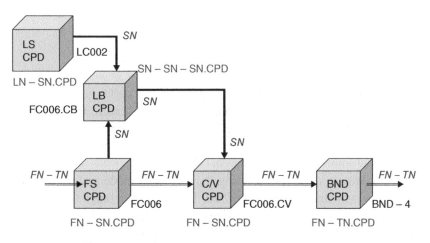

Figure 7.18   Causal process pattern – a closer look.

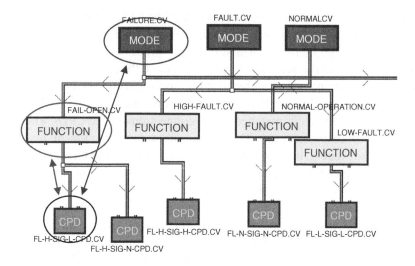

Figure 7.19 CPD – mode – function relation available from FR model.

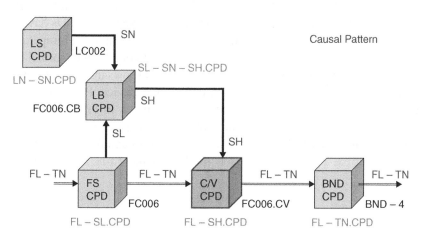

Figure 7.20 Causal process pattern.

When a device is analyzed, CLA findings are prioritized based on *fault* and *failure* modes. The relation of a CPD, its "function"/"malfunction" categorization, and the *normal, failure,* or *fault* mode are directly available from the FR model of the device as shown in Figure 7.19 (a ScreenCapture from an FR model implemented in Gensym G2).

Another way to interpret DL using the CPP with mode and function patterns is shown by the CPP in Figure 7.20. This CPP has a failed Control Valve Function – FC006.cv, which can be viewed within mode and function patterns in Figures 7.21 and 7.22. In Figure 7.21, the "C/V Mode" is in *failure* status,

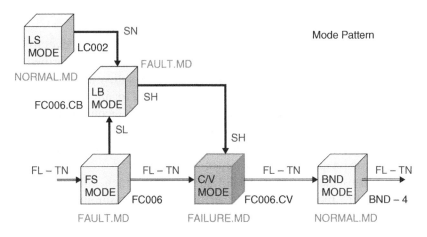

Figure 7.21 Mode pattern.

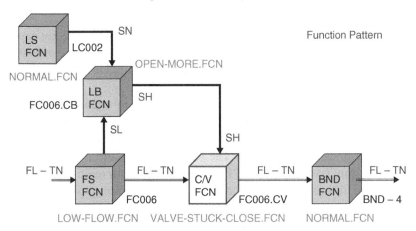

Figure 7.22 Function pattern.

which is the malfunction hypothesis for this CPP. From Figure 7.22, the "function pattern" details about the malfunction can be extracted, which is "VALVE-STUCK-CLOSE.FCN." The underlying mode/function details in these syntactic patterns are valuable and more meaningful to the end user.

In an abnormal situation, there are typically one or more CPPs containing a *failure* mode hypothesis. This subset of CPPs is then interpreted as a group to prioritize the most probable failure hypotheses using a set of rules, typically a mix of process knowledge, possibility, and probability. Rathinasabapathy (2005) discusses these rules, hypothesis generation, and prioritization in detail. The prioritized hypotheses are presented to the operator at a level of granularity defined by the user. The level of detail is customizable for each end user based on role. Devices, malfunctions, modes, or a combination can be presented topically.

There are several key reasons why DL works:

- DL is defined as diagnosis to a sufficient system or area level to take appropriate action even if a fault or failure has been not clarified at a point in time. The advantage comes from DL's ability to identify faults and failures as they are manifesting in time and location.
- Adopting a single time-step methodology avoids inconsistencies when trying to propagate cause-and-effect behaviors over time. Persistence of behaviors and/or changes in diagnostic patterns across multiple time steps can be used to prioritize recommendations.
- CLA uses the FR plant model structure in which *fault* and *failure* modes are modeled using all of the sensor measurement states for every device in the process plant flow sheet for a fixed time horizon. Complexity in enterprise behaviors results from connected equipment items and the possible interactions of *fault* and *failure* modes not from a complex simulation.
- DL does not require high fidelity models. FR models indirectly capture dynamics as inlet–outlet causal states, so essential dynamic effects are not lost. Overall, process dynamics are not required for DL eliminating loop back issues.
- DL has an advantage with a distributed structure organized by work-flow, i.e., Feature extractor, FR plant model, CPP generator, CPP interpreter, and a user interface to bring different computational environments and methodologies together. The independent nature and flexibility of the Feature Extractor and CPP Interpreter allow for readily changing to include different methods and knowledge for different processes.
- The FR model captures the topography of the device in the process providing the ability to provide topographical assessment knowledge to the end user.
- The Feature abstractor works independently assessing the state of the process and providing the data necessary for symptomatic assessment. The frequency of assessment can be adjusted to every process differently as required. The FR knowledge construct and the independent components working together are key to why and how DL works.
- DL has the flexibility to select the right process variables to manage CPPs. By localizing specified variables to sub-sections of the process, the number of CPPs can be better managed (Rathinasabapathy 2005). DL has the flexibility for variable reduction by localizing variables.

In this work, once specified, the device sequence for the purposes of CPP generation remains fixed, but this can be a future area of improvement. The optimum device sequence maximizes the application of state constraints and minimizes the degrees-of-freedom for variables to take on values. Governing rules for device sequencing and process variable localization is discussed in detail by Rathinasabapathy (2005).

7.5 DL-CLA USE CASE IMPLEMENTATION: NOVA CHEMICAL ETHYLENE SPLITTER

The CLA methodology was demonstrated in an actual diagnostic situation on a Nova Chemical ethane-ethylene splitter (C2 splitter). The relevant C2 splitter process plant description is available in the literature (Nelson 1954; Schutt and Zdonik 1956).

As shown in the plant flow sheet in Figure 7.23, an ethane–ethylene feed mixture is fractionated in a high-pressure C2 splitter tower to get high purity ethylene product in the overhead. Ethane is the bottoms product and is typically recycled. A heat pump is used as a cooling source for the overhead vapor and as a heat source for the reboiler. The overhead ethylene vapor leaving the top of the column is compressed, cooled, and condensed in the reboiler and partially refluxed to the top of the column after pressure reduction through a valve. The bottoms product is vaporized and its refrigeration capacity is utilized in different sections of the process.

The C2 splitter was modeled in the FR formalism with 49 devices that include 14 sensors and 2 split devices. The fractionator and the heat pump, the two more complex devices, were each modeled in two parts to take advantage of devices already available in an FR device database from previous work (Elsass 2001). As shown, the fractionator was modeled as a stripper *STR* and an enriching section *ENR*. The heat pump was modeled as two heaters, *H1* and *H2*. Other devices were reused from an FR device library built in previous work. FR device modeling for the C2 splitter is discussed in detail by Rathinasabapathy (2005).

Feature extraction was run continuously by configuring a data file of all filtered/averaged sensor values for a given time step and looking for any excursions from normal, e.g., high or low values. When an excursion was detected, data abstraction was used to generate a file of the state qualitative feature values. CPP generation was then initiated starting at the ethane product outlet with boundary values, BND-3. The following device sequence was then executed: BND-3, TC002, TC002.CV, J3, CLR, J2, COM, PC001, H2, H1, CC001, ENR, STR, PMP, CC002, EEJB, FC003, FC003.CB, FC003.CV, VAP, PC002, J4, FC005, REB, FC005.CV, J7, TNK, J8, FC006, FC006.CB, FC006.CV, LC002, LC001, PC002.CV, TC001, TC001.CB, FC005.CB, FC004, FC004. CV, FC004.CB, FT007, FT007.CV, FC002.CV, FC002, FC001.CV, FC001, PC003, PC003.CV, and J9. The device sequence is user specified. In our E2

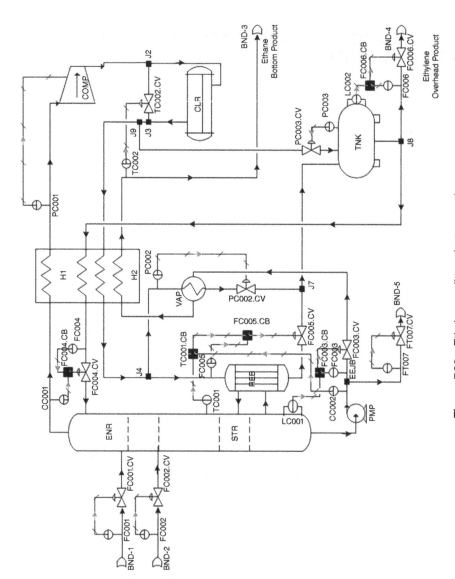

Figure 7.23 Ethylene splitter plant operation.

splitter use case, the same sequence was used for all subsequent time steps. While the overall results do not depend on the device sequence, the CPP computational load does. The effects of different starting points and selections are discussed in the analysis section. CPP interpretation was conducted by analyzing multiple CPPs for each time step for persistence across time steps.

7.5.1 CPP Generation

Figure 7.24 illustrates the full scale and scope of generating a single full plant CPP. In the figure, each row corresponds to a device and each box corresponds to a CPP accumulated to the point of the particular device in

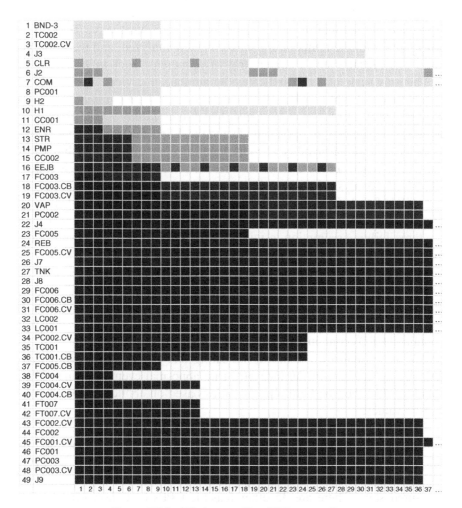

Figure 7.24   Ethylene splitter CPP generation.

the row. With reference to the previous illustrations, CPPs branch and accumulate when there are multiple CPDs and reduce when applying state features and reasonable knowledge-based assumptions as the procedure progresses as well as when the progression comes to a device with multiple constrained port descriptions. The total number of CPDs associated with a box increases as the devices are analyzed in sequence, moving down the chart. As a result, the row number is the sequence but also the number of accumulated devices analyzed. The number of columns corresponds to the number of current CPPs at that point in the process; however, column placement does not relate to any specific branching. With reference to Figure 7.23, the procedure begins with BND-3 with nine state feature values shown in row 1. The temperature sensor TC002 is then analyzed and the number of partial CPPs is 3. The control valve TC002.CV associated with the temperature sensor is then analyzed, and the accumulated CPPs for the sensor and control valve increase back to 9.

The gray scale of a box indicates the number of devices in *normal* and *failure* modes. Light gray boxes indicate devices that are *normal*. A medium gray box indicates when at least one device is in *failure* mode. Dark gray indicates two devices in *failure* mode and black indicates three devices. We assumed that the probability of more than three simultaneous failures was not likely. When setting the failure threshold (three in this case), it is possible during analysis that more than three malfunctions could exist in a partial CPP before being rejected later. Application of the assumption is a way of avoiding significant branching if in reality it is not necessary. In this case, the three-failure assumption terminated CPP branching if there were three *failure* modes in any CPP.

In the initial four rows, all boxes are light gray indicating that none of the first four devices are in *failure* mode in any CPP. Even though malfunctions appear as dark gray boxes sporadically in rows 5 through 7, the eighth row shows no *failures*, e.g., all light gray boxes. This indicates that the CPPs that contained CPDs containing a *failure mode* were rejected along the way. All boxes in the final row are black, indicating that all CPPs have three malfunctioning devices. In the face of a true abnormal event, the final CPPs should reflect the true failure possibilities based on application of all state feature values.

For the particular time step analyzed, the CPP Generator produced 36 feasible CPPs as shown in row 49. Table 7.1 shows the full CPP for just 1 of the 36, e.g., one of the boxes in row 49. As illustrated in the table, a CPP is a collection of a feasible causal pathway for 49 CPDs that explains the measured state. There are 35 other feasible CPPs (not shown here). The highlighted CPDs are explained in the following.

Table 7.1 One full CPP from the C2 splitter case study.

No.	CPD	Device
1	F-H-T-L-CPD	BND-3
2	TL-SL-CPD	TC002
3	FN-SL-CPD	TC002.CV
4	FN-TN-CPD	J3
5	ALL-N-CPD	CLR
6	FN-PH-CPD	J2
7	FNPH-CI-SH-FNPH-CO-CPD	COM
8	P-H-SIG-H-CPD	PC001
9	FH-BPI-CPD	H2
10	FH-RI-FTH-RO-CPD	H1
11	CH-SH-CPD	CC001
12	FH-VAP-FEED	ENR
13	FHTH-LI-FL-RO-FH-CO	STR
14	FH-IN-OUT-CPD	PMP
15	ALL-N-CPD	CC002
16	FH-BI-FH-PO-FN-EEO	EEJB
17	FH-SH-CPD	FC003
18	HI-HST-NO-CPD	FC003.CB
19	FH-SN-CPD	FC003.CV
20	FHTN-SI-CPD	VAP
21	P-H-SIG-H-CPD	PC002
22	FH-TN	J4
23	FL-SL-CPD	FC005
24	FLTN-SI-FLTL-SO-CPD	REB
25	FL-SL-CPD	FC005.CV
26	FL-TN	J7
27	ALL-N-CPD	TNK
28	FN-TN-J8	J8
29	FN-SN-CPD	FC006
30	NI-NST-NO-CPD	FC006.CB
31	FN-SN-CPD	FC006.CV
32	IN-N-OUT-N-CPD	LC002
33	IN-H-OUT-H-CPD	LC001
34	FH-SH-CPD	PC002.CV
35	IN-N-OUT-N-CPD	TC001
36	NI-NST-NO-CPD	TC001.CB
37	LI-NST-LO-CPD	FC005.CB
38	FH-SH-CPD	FC004
39	FH-SN-CPD	FC004.CV
40	HI-HST-NO-CPD	FC004.CB
41	FN-SN-CPD	FT007
42	FN-SN-CPD	FT007.CV
43	FL-SL-CPD	FC002.CV
44	FH-SH-CPD	FC002
45	FN-SN-CPD	FC001.CV
46	FN-SN-CPD	FC001
47	IN-N-OUT-N-CPD	PC003
48	IN-N-OUT-N-CPD	PC003.CV
49	FN-SN-CPD	J9

7.5.2 CPP Interpretation

Referring back to the ethylene splitter use case, a single CPP, as shown in Table 7.1, contains a CPD for each device in the process for a particular time step. Each CPP generated is unique. To further interpret, each CPP is analyzed for the devices in which the *failure* mode occurs. As shown in Table 7.1, there are three devices, H2 (heater 2), ENR (enriching section), and EEJB (emergency ethane) highlighted (refer to the plant flow sheet) for this CPP. All other CPDs are either in *normal* or *fault* modes. The other 35 CPPs are similarly analyzed, and the repeated occurrence of *failures* and the devices in which they occur are determined. The CPDs across multiple time steps were similarly analyzed for *failures* and device location. In the interpretation scheme used for the C2 splitter, we first considered *failure* modes that persisted over time. Those that persisted across multiple time steps were ranked higher than those appearing sporadically. Next, the predominance of *failures* in multiple CPPs, those appearing in a greater number of CPPs, was then ranked higher. The user can readily adjust. In developing this interpretation method, we note the similarity to approaches that involve generating an exhaustive set of causal maps and matching them to a current process state (Chang and Chen 2011).

In reviewing the CLA generation results for this test case, CPP generation produced 36 unique feasible CPPs of which one is illustrated in Table 7.1. Each of the 36 contained the same three malfunctioning devices as illustrated with the highlighted rows 9, 12, and 16. Device 9 in the CPP corresponds to the heater device and the function description FH-BPI-CPD.·H2 is associated with the *Heater Plugged* FR Function. Device 12 is the enriching device and its CPD is linked to the *High Feed Vapor Enriching FR* Function. Device 16 is a pipeline splitter and its CPD is linked to the *emergency ethane manual valve kept open* Function.

At the time of the demonstration test, it was not known that there were three malfunctions occurring and the recommendations that emerged were not known until the conclusion of the CPP generation and interpretation tasks. In the CPP interpretation task, the 36 CPPs were searched for common malfunctions that revealed the common pattern for three simultaneous failures.

These CLA results were subsequently confirmed with operators as correctly representing the abnormal state of the process. A detailed post event analysis revealed multiple simultaneous events occurring with overlapping effects. When fully understood after applying CLA, one abnormal event was a heat-pump heater plug and propanol had been injected to free it. The bad actors establishing the event were high pressure upstream of the heater and high temperature downstream of the vaporizer. With increases in both the pressure and the temperature, a blockage downstream of the pressure sensor was deduced. The propanol injection unplugged the heater, and when the pressure dropped back into the normal range, the heater plug was confirmed.

In addition to a heat-pump heater plug, there was an increased vapor flow to the column due to an upstream unit upset and an abnormally high bottoms ethane product flow due to the emergency ethane line left open for operational reasons. The vapor feed change resulted in a change in feed composition and temperature. To maintain the column temperature, the reboiler controls reduced heating (ethylene HP discharge), which caused more ethylene flow through the vaporizer causing a reboiler control action to increase the system pressure. To reduce system pressure, PC002.CV was opened more resulting in high flow through the vaporizer with no heat exchange. The increase in overhead composition was, therefore, linked to the increase in vapor feed flow.

7.5.3 Diagnostic Localization

The section above focused on the detailed failure analysis. Failures, however, can also be grouped by system or subsystem. In doing so, we enable the ability to account for the possibility that failure modes may have not yet resolved for the fault situation producing them but the location may have. For example, if all of the failures are concentrated in a particular system, this is valuable diagnostic information. System localization is what is meant by diagnostic localization. For user presentation and visualization, there can be a large number of feasible failure modes especially at the onset of an event. Presenting localized diagnoses at the system or subsystem level helps to narrow the diagnostic focus and reduce the number of feasible hypotheses presented to the user. Regardless of the granularity, the user can always drill down to specifics in the CPP if necessary.

The figure shows the G2 user interface developed for the prototype based on the FR plant model, a diagnostic-cues window at the top right, a control panel at the top with options to start/stop DL, and a CPP window at the bottom right. The CPP window allows the user to follow the progression of the investigation of devices and analyze CPPs. The malfunctioning devices are highlighted in various colors in the FR plant model with malfunction possibilities listed in the diagnostic-cues window. The user interface is designed for diagnostic localization by displaying results topographically and in terms of systems and subsystems. *Functional,* *behavioral,* and *mode* details are in another window.

7.6 ANALYZING POTENTIAL MALFUNCTIONS OVER TIME

Most AI-based Process Fault Detection technologies provide Process Operators and Engineers with valuable process excursions that become the basis for doing a diagnosis to detect abnormal events. Operators then monitor the behaviors of individual sensors and the manifestation of an abnormal event over time. Diagnostic localization is premised on the same philosophy of providing rolling information to the operations by providing measurement behaviors and device failure localization hypotheses as events manifest themselves.

Early in an event that requires time for effects to propagate and develop throughout a process, diagnostic hypotheses will typically include multiple and unrelated malfunction hypotheses. Diagnostic localization presents the operator with initial rank-ordered hypotheses at the onset of an event and then updates these hypotheses as the event manifests itself. The intent is to provide the operator with a broader perspective of possibilities early in the event and throughout the event. As a malfunction event manifests itself, DL can begin to localize a malfunction within sub-systems such as the "overhead section" of the "separations unit." As time progresses, the event manifests itself further, more information becomes available, sensor information stabilizes in malfunction states, and hypotheses narrow to the malfunctioning devices. It is worth noting that diagnostic localization compares diagnostic results between multiple time steps for consistency and not for the relation between the previous and current results.

At each window in time, the diagnostic hypotheses consist of CPPs that contain malfunctions, along with their device locations within the topography of the plant model. As time windows accumulate, the malfunctions in successive CPPs start to stabilize. Diagnostic localization findings are therefore based on recency and persistence of malfunctions in CPPs across multiple time steps. A simple counting process makes it possible to offer an explainable sense of confidence. Hypotheses that repeat over multiple time steps are ranked higher than those that were generated less frequently or that died out. For instance, if the system has run for 10 time steps, a hypothesis that is generated in 8 of the time steps would be ranked higher than a hypothesis generated in 3 time steps. Furthermore, those hypotheses that are more recent take precedence. If there are 10 time steps with a hypothesis generated for just the first 5 time steps and a hypothesis generated for just the last 5 time steps, the hypothesis generated in the last 5 time steps would be ranked higher. If a hypothesis is not generated every time step, it has a smaller probability. Fewer counts from recent hypotheses take priority over higher counts from past hypotheses. As an abnormal event propagates through the process, the changing feature states act as constraints, tapering the possible CPPs, leading to a smaller number of hypotheses but with greater probability.

Consider a simple multi-time step example. A simple process example is shown in Figure 7.25 that has a process boundary, a flow sensor, a cooler, and a temperature sensor. Assume the states at the boundary are known (Flow Normal and Temperature Normal) and that the sensors are all reliable. During 329 minutes of operations (time = 0 to 329), the flow and temperature states at the boundary, the flow sensor (FT), and the temperature sensor (TT) are all showing normal. This results in "Normal Cooling" in the Cooler. The only valid CPD at the Temperature sensor is TN-SN (temperature normal–signal normal). However, at time = 330 min, the temperature sensor TT reads high as shown in the control charts in Figure 7.26.

The high temperature is monitored for an additional 10 minutes during which every time step analysis produces the same CPP with the CPD TH-SH at the temperature sensor and "Poor Cooling" malfunction at the Cooler device. For this process, 10 minutes of repeated time step patterns is considered sufficient for confirming the probability that poor cooling

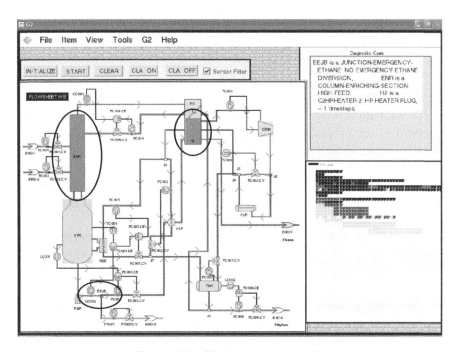

Figure 7.25 Final case study results.

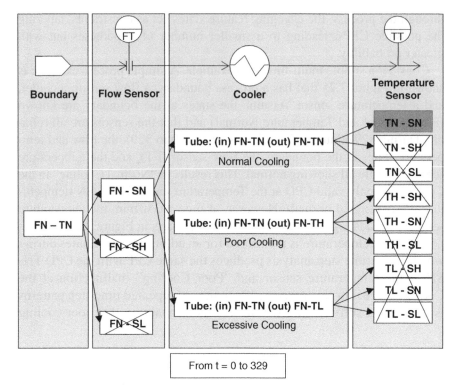

Figure 7.26   CPP example from time = 0 to 329 intervals.

in the exchanger may be a problem. The DL algorithm does allow the operator to set a threshold on how many time steps with a repeated CPP need to occur before a malfunction is considered flagged. Please note that when using the system in operation, observed CPPs with malfunctions will flag off and on across several time steps before the CPP stabilized. Setting the threshold of time steps is an important knowledge-based operational parameter.

This example offers a view into how DL works when an abnormal event first manifests itself. The approach allows for more than one abnormal event to be in play. If the abnormal events are interrelated, then the CPP for the entire process will stabilize. However, it is possible for localized events to stabilize during one time period and for another independent event to appear and stabilize during a different time period. This example can also be thought through using the tabular method in Table 7.1 for analyzing multi-time-step malfunctions to rank and prioritize.

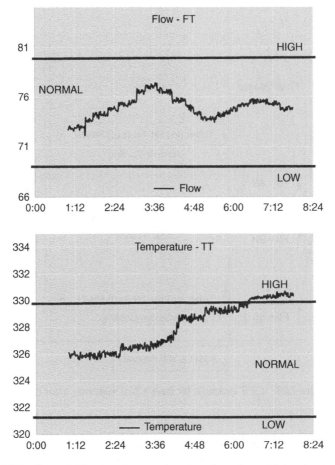

Figure 7.27 Control Charts – temperature trending above high reference band.

7.7 ANALYSIS OF VARIOUS OPERATIONAL SCENARIOS

7.7.1 Event Manifestation, Sensor Reliability/Sensor Malfunctions

Chemical processes can be considerably different in number of devices and topographical and integration complexity. There are additional complexities with measurement instrumentation location, availability, timeliness, and reliability. There can be operational complications with multiple simultaneous malfunctions versus a single malfunction, and there are certain process behaviors that are not associated with a physical device and cannot be defined in a regular FR device model. Examples include human operations (opening

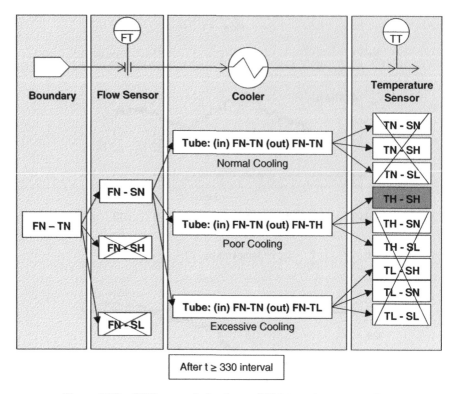

Figure 7.28 CPP example for time ≥ 330 intervals, poor cooling.

a field valve or adjusting a burner in the furnace), human DCS entry errors, logic blocks, pipeline mixing, splitting, plugging, etc. These types of issues that are not associated with physical devices still need to be captured in the form of CPDs. As a result, the concept of Hypothetical devices or Function-Only Devices was introduced.

Diagnostic localization is also subject to the possibility of an unmodeled device behavior, an unavailable feature state, or an incorrect FR model. However, DL is inherently a reliable system in that the Causal Link Algorithm (CLA) is not searching for normal and malfunctioning devices. Rather it is only searching behaviors and is always allowing for multiple behaviors. The CLA algorithm is inherently stable and simply degrades with more possible CPPs. Even in the case of an unmodeled behavior or a missing device model, DL does not fail abruptly. The causal patterns degrade gracefully by presenting faulty device behaviors with a wider range of options but still useful in providing meaningful insights for a person to continue the diagnosis manually. This form of graceful degradation is a strength in that it makes it possible to identify multiple malfunctions and account for malfunctions but looking at the CPPs. This algorithmic tactic also allows us to consider some number of sensor malfunctions with limits.

7.7.2 Hypothetical and Function-Only Devices

Hypothetical devices are not physical devices in the process and do not have a *function*. For example, a "set-point" device makes it possible to capture an operator set-point entry as a CPD and organize it under *normal* or *failure* modes. A "set-point" device does not have an "inlet" port but does have an "outlet signal" port whose *signal* is an input to a "control valve" or a "computation block."

Function-Only devices are physical elements in the process that do not have functions in the sense of a process operation that is a direct step in the manufacturing of the product. For example, pipes are conduits and not modeled as FR devices as long as there are no process changes in them. In another example, pipeline junctions split or merge process streams in some cases with process state changes. For CLA, a pass-through device, i.e., Function-Only device, is required to pass the variables and feature states through the junction. In a "splitter," a single process stream splits into two or more streams with the same process conditions except for *flow*. In a "mixer," all the process variable states can change.

7.7.2.1 *Feed Change Effects*   In the ethylene splitter case study, during the heater plug event, the vapor feed to the distillation column changed due to disturbances from the upstream plant. In order to distinguish between the overlapping effects of the "feed change" and the in process effects of a heater plug and a closed manual valve, it was necessary to model "feed change" behaviors. In this case, a feed change is a known operation or process disturbance and not a process behavior that can be associated with any device in the process. CPDs reflecting the "feed change" were created and added to the "distillation column" FR model. If the feed changes to the unit can be inferred upstream of the process, such information can be incorporated into the model.

7.7.2.2 *Manual Control Valve*   The third malfunction in the ethylene splitter heater plug event was the untimely closure of "FT007.CV," a manually operated "control valve." The "column bottoms" has an emergency ethane diversion stream that is operated manually. The emergency ethane diversion stream had been open before the plugging event causing ethane

flow through this stream. The "valve" was closed during the event for unknown reasons, but it caused a "high" ethane *flow* through the "heater."

The manual "control valve" has a "*material inlet*" port and a "*material outlet*" port but no "*signal input*" from a control system in a traditional sense; however, there is a "signal input" as an Operator action. The emergency ethane junction block (EEJB) FR model was modified to infer the manual valve operation by analyzing the *flow* changes from the different streams. As a device, EEJB was not only constructed as a "splitter" but also as a hypothetical device to infer the "manual control valve" operation. This is an example of operational customization details not available in P&IDs and can be acquired only from process experts operating the unit, i.e., automatic valves used as manual valves.

Another example stems from a primary assumption that *flow–pressure* interactions are often negligible. However, in the ethylene splitter process, a *flow–pressure* interaction was observed in the "reboiler loop" that was confirmed by process experts. The "reboiler" changes the ethylene *flow* through the "tube side," depending on the "column" *heat* requirements. The *flow* change on the "tube side" immediately changes the *pressure* in the upstream, i.e., the "compressor" downstream *pressure* and eventually the whole system *pressure*. The *flow–pressure* interaction in the "reboiler loop" cannot be explained by CLA without accounting for the variable interaction. In the ethylene splitter prototype, the "splitter" device "J4" is modified as a hypothetical device that accounts for the variable interaction.

Hypotheses generated throughout the manifestation of an event ultimately depend on the plant models, the data, and the users, similar to any other decision-support system. Given the high degree of modularity, the FR formalism is conducive to systematic construction with device models. Modeling error is minimized because models are constructed in small, well-defined chunks. Annotated case studies are vital to testing and validating the model. The most common adjustments to the FR model during the construction and test phases were incorrect mode categorizations and missing behaviors.

7.7.3 Unaccounted Malfunctions, Graceful Degradation, Multiple Malfunctions, Etc.

An incorrect categorization under *failure* mode will result in a progressive type of analysis problem. If a malfunction is identified with device failure instead of device fault, the device will be interpreted incorrectly as malfunctioning. All CPDs with this flawed characterization will be carried forward increasing the number of malfunctions in a CPP. A malfunction can never be diagnosed if its CPD is wrongly categorized under *fault* or *normal* mode. Also, a missing *behavior* (or an incorrect selection of variables) results in no diagnosis. Consider the example of a "control valve" FR model in Figure 7.3. If the malfunction "Stuck Normal" is not accounted for, the CPDs "FN-SH-FN" and "FN-SL-FN" cannot be constructed. Without these CPDs, all CPPs that have a combination of "FN" and "SL or SH" will terminate at the "control valve."

7.7.4   Sensor Availability and Reliability

Reliable sensors are important in providing the required feature states. There is also the importance of independent, open-loop sensors, which are critical to diagnosability. Process units typically do not have adequate independent sensors and measurements, making it impossible to distinguish all malfunctions. There can also be discrete sensors such as analyzers and off-line labs in which data are not available when needed. Since sensor availability and reliability are keys to pruning the search results, any opportunity to use alternate sensors, etc. dynamically should be taken. From a DL perspective, the availability of open-loop sensors significantly helps manage the total number of CPPs during execution and in the end results.

As an example, a process stream with process variables *flow* and *temperature*, with no validation of the actual process state through the abstracted sensor data, CLA has to progress with 9 CPPs instead of 2. The end user is also presented with additional possibilities.

Not surprisingly and consistent with diagnosability studies, the availability of key open-loop sensors can be used to trim the CPPs with the assumption that they are reliable. However, an unreliable open-loop sensor will cause the CLA to misdiagnose. CLA has a built-in ability for the end user to validate all or some of open-loop sensors as reliable. If a sensor is marked unreliable, then an erroneous reading is carried forward as a possible CPD and all CPPs involving that sensor are carried forward. This produces many more CPPs for the end user to consider (i.e., the end user does not trust the sensor). The flexibility to switch on/off an open-loop sensor provides the end user the ability to perform a *what-if analysis*, which can be a powerful training tool or help with HAZOP sessions.

If the diagnosis involves malfunctioning sensors and these are known a priori, feature states cannot be used as constraints to prune CPPs and sensor malfunctions within limits can be identified. While it is possible many sensors can fail simultaneously, due to a failed DCS motherboard or due to a cold snap, it is unusual all or most sensors malfunction simultaneously for other reasons. When there is a suspicion of a failed or a malfunctioning sensor, DL provides the flexibility to drop those sensors from CPP pruning. This would increase the search space but allows DL to diagnose a sensor malfunction. The value of open-loop sensors is demonstrated on a simulated case study involving CSTRs. The inclusion of open-loop sensor information resulted in considerably fewer CPPs, better diagnostic resolution, and significantly improved computational performance (Rathinasabapathy 2005). The downside of having many sensors tagged as unreliable will generate too many CPPs and will quickly fill up the three simultaneous malfunction limit.

7.7.5 Process Complexity

Two processes have been modeled in FR for the purpose of diagnostic localization, the ethylene splitter introduced previously and an earlier demethanizer process. Both prototypes were used to correctly diagnose the process malfunction. The Demethanizer unit was modeled after an existing industrial unit and contained less devices and less complex devices compared to the ethylene splitter. Sensor malfunctions were of particular concern in the Demethanizer prototype built by Elsass (2001). The complex devices we refer are those with several ports that manipulate multiple process variables. The ethylene splitter is a complex process with heat integration between the "column overheads" and the "column bottoms." Compared to the Demethanizer, the larger number of devices increased the computational load in terms of sheer number of CPPs that needed to be calculated, analyzed, and tracked. The process complexity coupled with devices with multiple ports leads to an escalation in the number of CPPs. The application of the CLA algorithm to the ethylene splitter had to be developed to better manage active and inactive ports when going through the devices in a non-sequential manner. Better "device sequences" and "selective process variable localization" to particular sub-sections of the process are the strategies employed.

In the Demethanizer DL prototype, one long sequence of devices could be identified. Given a single long sequence of devices, a particular process stream and its process variables could be readily tracked. However, due to the complexity of the Ethylene Splitter, a device-by-device analysis of active and inactive ports was required.

A single long sequence of devices for the ethylene splitter was not possible:

- Process integration, i.e., the "bottom product stream" and the "overhead product streams" are heavily intertwined.

- The overhead and other streams split and re-combine at different locations in the process.
- The feed streams and the emergency ethane streams were included in the model.

The combination of these necessitated the identification of small device-sequence groups and then the ability to group and combine information. One important benefit of small sequence groups is the ability to localize variables. By localizing *level, signal,* and *composition* to small group analyses and not carry these variables through all devices, the number of CPPs is significantly reduced. Decomposing into device-sequence groups made it necessary to switch between process streams at a common device to connect groups. These computational methodologies were not needed for the Demethanizer due to the linear nature of the model.

Figure 7.29 shows the major device sequences of the ethylene splitter prototype. The device sequence 1 starts from "BND-3" and ends at "TC002. CV." *Flow* and *temperature* are the only variables considered in this sequence. Device sequence 2 starts at "TC002.CV," the common device between sequences 1 and 2. Sequence 1 ends at the "signal port" and

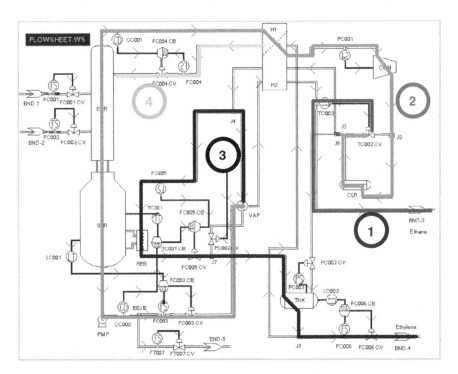

Figure 7.29 Ethylene splitter major device sequences.

sequence 2 starts at the "material port" of "TC002.CV." Sequence 2 is the longest in the ethylene splitter prototype with the "cooler," "compressor," "HP heater," "distillation column," and "column bottom" devices. *Pressure* and *composition* are additional variables considered in this sequence. Sequence 2 ends on the "shell side" of "VAP." Sequence 3 starts on the "tube side" of "VAP." Sequence 4 covers the devices between "heater" and "column" on the "reflux loop." It should be noted that sequence 4 is not connected to sequence 3. All the remaining devices are formed into small sequences and integrated in the same way as described here.

7.7.5.1 Complex Events

The ethylene splitter event with multiple simultaneous device malfunctions is spread out in process, diverse and complex in character. This led to a number of new considerations. In this section, we discuss two major considerations to manage CLA outcomes with process knowledge – the "*maximum simultaneous malfunction*" threshold and the "*sensor – control valve combinations*" that need to be resolved in order to diagnose multiple malfunctions.

7.7.5.2 "Maximum Simultaneous Malfunctions" Threshold

A user-defined threshold for the maximum number of simultaneous malfunctions has proved to be very useful in helping to manage the number of CPPs during the CLA analysis. The threshold is a process-determined number that reflects how many devices are likely to be simultaneously malfunctioning. The setting allows any CPPs with malfunction numbers greater than the threshold to be rejected. In the Demethanizer DL system, the user-defined threshold was set at 3 meaning that it was very unlikely to see three devices failing at the same time.

For the smaller Demethanizer process and the single malfunction case study, this threshold of three worked fine. Because the ethylene splitter heater plug event involved three malfunctioning devices, the event could not be diagnosed completely without relaxing the threshold. The prototype was tested for different threshold settings. A larger threshold is not a computational issue. However, analyzing large number of CPPs by a human is not practical.

7.8 DL-CLA INTEGRATION WITH SMART MANUFACTURING (SM)

SM at its core is about data-driven and model-based manufacturing intelligence at scale for end-to-end management of operations. The focus of this chapter is end-to-end diagnostics of a process plant operation. SM is open not only to data-centered quantitative modeling but also to the perspective of tasks in execution purposely discretized to form an explicit workflow. Each task can act on data synchronized for that step, while the overall workflow is monitored for state, execution, and completeness (Davis et al. 2012, 2015). Such a workflow-based approach offers the advantage of accommodating hybrid modeling approaches. Discretization also readily accommodates human-in-the-loop interpretation and action but supported and in coordination with computation-based data interpretation and analysis. In this way, SM encompasses not only the roles of tightly coupled, model-based control, but also control in a systems-of-systems approach that includes applications and people as collaborative in real-time operation (Madni and Sievers 2013). This workflow-based viewpoint when provisioned as a service offers additional technical advantages in that operating data can be "taken" to applications running on different execution platform infrastructures than that of the control system.

From this SM perspective, the CLA methodology is structured for use and scaling on infrastructure specifically for qualitative modeling and CPP generation that are different from those needed for control. This makes it possible to invest in one data structure for control but use the same data structure to also solve the diagnostic problem without having to invest in a separate infrastructure for data abstraction or CPP generation, interpretation, and presentation. The plant side data structure provides the input to a data abstractor as a service to generate qualitative features as a service. Qualitative modeling tools accessible in SM infrastructure can support FR model building and workflow implementations of the CLA components and the human-in-the-loop interfaces. Infrastructure costs and support that have been typically difficult for qualitative modeling environments are dramatically lowered, and both infrastructure and applications can be readily changed. Importantly for diagnosis, it affords progressive growth of diagnosis for increasing data asset utilization opportunities that can begin at an operational level but could scale to larger site-wide enterprises.

The authors foresee the DL tool as an app that comes with the components required to run a real-time process fault diagnosis. The user can build the FR plant model using the process devices from the FR model library and would have the flexibility to use any data abstractor component (simple standard deviation limit tools or complex machine learning tools) available in an SM marketplace to quantify data and additional components to trigger CLA. Operational data, collected, ingested, and contextualized across the operation and the control systems are readily available through SM edge and cloud infrastructure.

7.9 AN FR MODEL LIBRARY

Building knowledge-based models is a time-consuming and knowledge-intensive task that depends on both the formalism of the knowledge representation and the acquisition of specific process information in the context of that formalism. There is significant incentive to reduce the engineering effort by making these knowledge models reusable (Pos et al. 1996). Qualitative models in which the quantitative features are resolved separately make this possible.

FR device models have the same functions and behaviors in many cases and can be readily reused without any changes. In our work, FR device models constructed previously are stored in a reusable *FR Device Library*[2] as shown in Figure 7.30. Each FR model in the device library has the same structure of *mode, function,* and *behavior* as previously discussed. The figure shows the library with drag and drop features in which the model, function, and behavior are all carried with the device.

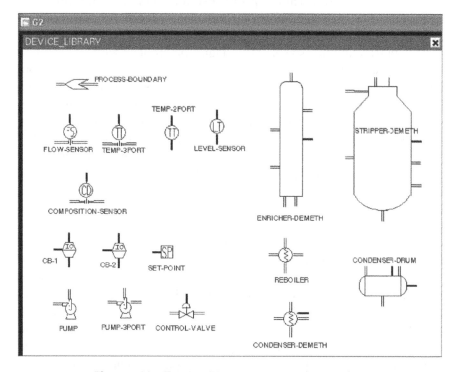

Figure 7.30 Existing FR Device Library (built in G2).

[2]Simply referred as "device library" or "library" in the rest of the document.

7.9.1   Reusing FR Device Models

The Ethylene DL project began with an existing library of device models developed for the Demethanizer prototype (Elsass 2001).

Completely reusable FR models are used as-is and are process-independent. These are models of devices that perform the same function or malfunction and exhibit the same behavior irrespective of the process. They have the same structural knowledge, same number of ports, port direction, and type of ports. Their behaviors can be explained with the same variables and feature states. Most sensors (pressure, flow, temperature, etc.), control valves, and pumps are examples of completely reusable devices. FR models can be used interchangeably too – for example, there are different types of flow sensors with different working principles but if they have the same structure (ports, type), mode, function, and CPDs, they are completely interchangeable in the FR plant model.

Partially reusable FR models cannot be used as-is but are reusable with modifications while still process-dependent. A difference in the structure of the device (ports, type), how it is connected in the process, and functional or behavioral differences can all lead to not use as-is. For example, different types of heat exchangers (coolers, heaters, reboilers, etc.) have the same structure and the same high-level function of heat exchange, but their behaviors are completely different, i.e., one cools versus another one heats. These FR models cannot be reused, but it is often easy to clone an FR heater device in the library and modify it to be a cooler by just changing functions and CPDs to account for the differences rather than building a new model from scratch as shown in Figure 7.32. And every instance of a novel/modified device model is added to the library. Over time, a rich library of devices with more specific functions will be created.

7.9.1.1   *Heat Exchanger – Example of a Partially Reusable Device*   As shown, Rathinasabapathy (2005) modifies an FR reboiler (heat exchanger) model available in the FR model library into a FR Cooler model needed for the new process. The structural knowledge is identical, i.e., both are shell and tube heat exchangers with two inlet and outlet ports.

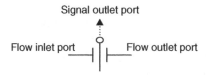

Figure 7.31   Flow sensor – completely reusable device.

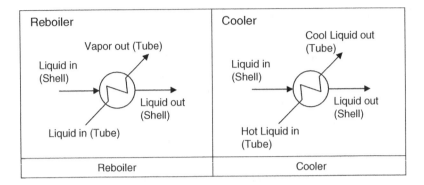

Figure 7.32   Reboiler and cooler – partially reusable devices.

The behavior can be described with the same variables *flow* and *temperature*. The mode of operation is identical. The overall function is the same, but the type of heat transfer is different leading to differences in function, malfunctions, and behaviors. The reboiler function is to heat and has a phase change on the tube side versus the cooler where there is no phase change and the function is to cool down. The malfunctions are different, low heat transfer results in poor reboiling versus poor cooling. The behavioral knowledge is different, a low heat transfer results in low vapor traffic at the reboiler tube outlet versus a low heat transfer results in high temperature at the cooler tube outlet. Because the variables and ports are the same and both FR models have the same CPDs, the modification required simple reorganizing of CPDs and some relabeling of the functions from *normal heat transfer, low heat transfer,* and *high heat transfer* to *normal cooling, poor cooling,* and *excessive* cooling. Now, a new FR *cooler* model is constructed and it becomes a new FR model in the library. A tabular checklist method using expert process knowledge was used to capture the similarities and differences before the model modification.

7.9.1.2 Column Stripper – Example of a Partially Reusable Device   Another example of a partially reusable FR model is shown with a distillation column stripping section. This model is already available in the FR device library. The ethylene splitter distillation column stripping section has an additional port, a level-sensor, an additional variable *level* must, therefore, be added to the model; otherwise both strippers have the same function and behaviors. Adding variable *level* with three variable states "high," "low," and "normal" increases the CPDs by threefold. The existing CPDs are replicated for the new variable and variable states and the functions labeled to create a new FR model.

Table 7.2   Reboiler and cooler checklist.

No.	Knowledge	Reboiler	Cooler	Notes
1.	Functional knowledge	Reboil tube side liquid to vapor. Phase change	Cool tube side liquid. No phase change	Different
2.	Potential malfunctions	"Low" heat exchange (poor reboiling), "high" heat exchange (excessive reboiling)	"Low" heat exchange (poor cooling), "high" heat exchange (excessive cooling)	Different
3.	CPDs	4 ports, 2 variables, 3 variable states	Same number of CPDs	CPDs need reorganized[a]
4.	Behavioral knowledge	Behavior is different. Example: "Low" heat transfer results in "low" *temperature* at tube outlet	"Low" heat transfer results in "high" *temperature* at tube outlet	Different

[a] The total number of CPDs is same, but they need to be reorganized under their appropriate malfunctions.

7.9.2   Complex FR Models

Complex devices are unique, process-specific, and they do not have similar enough models in the library to simply modify. Consider an example of a hydrocracker reactor, a Fluidized Catalytic Cracker (FCC) reactor, or a Selective Catalytic Reduction reactor. These are unique devices designed for specific purposes, and a similar device model is likely unavailable in the library until the library has grown sufficiently or been used to model these devices. These devices are complex and have to be built new.

A *complex device* typically has numerous ports, many process variables, and the behavior descriptions can lead to numerous CPDs. Managing, pruning, and organizing the CPDs can become complicated. To manage these complications, we break *complex devices* into smaller sub-devices. The sub-devices are relatively easier to model, and the increased modularity makes it possible to better manage the number of causal links in CPDs between sub-devices without losing any device knowledge. The CPD surges are managed at the CLA layer rather than the model. Also note, not every sensor in the complex device needs to be modeled in the FR model. More detail is better, but unless demanded by the process diagnostics, it is advisable to simplify modeling. Detailed models increase the model size and the search space during run time.

Elsass (2001) discusses the decomposition of complex devices into sub-devices and gives an example from a distillation column that can be

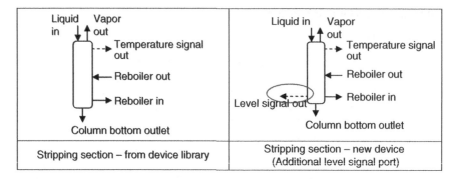

Figure 7.33   Stripping section example.

decomposed into an enricher and a stripper. From a process perspective, the enricher and stripper can be thought of as independent devices because dynamics are slow. None of the overall column behaviors are lost because CPDs capture the interaction between the enricher and the stripper at the connecting ports. After decomposition, the total number of ports dealt with for each device is fewer than the original number of ports for an entire device. This is the key reason that the decomposition process helps with model building. In the distillation column example shown in Figure 7.33, the total number of ports for the column as a whole is 7. When decomposed, the resulting two devices have five or six ports, respectively. The *level* variable need not be modeled as a variable in the enriching section. In another view, the combination of ports and variables is important. In simple devices, there could be just one variable (flow in a valve for instance). More ports with more process variables that need to be accounted exponentially increases the number of CPPs and the complexity of the device. A device with 10 ports that only models Flow would not be too complex.

Building complex devices by decomposition not only enhances model building but also results in reusable sub-devices. For example, the next project could have a distillation column with the same enricher but a different stripper. The FR model needs to be modified only for the stripper. Some columns could have multiple temperature sensors, as noted earlier, unless they add additional value in diagnosing they do not have to be modeled.

The ethylene splitter heat pump (HP) is another example of a complex device as shown in Figure 7.35, with nine ports and two variables (*temperature* and *flow*). Process understanding from industry experts with the ethylene splitter unit revealed that there are two different streams in the HP with negligible interaction. With that process expertise, the HP can be decomposed into two sub-devices and modeled separately. The decomposition

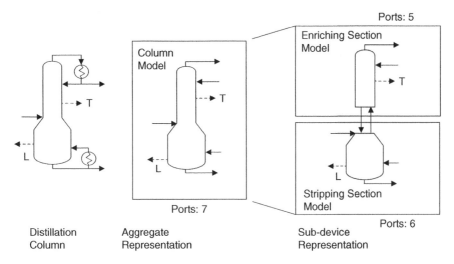

Figure 7.34   FR distillation column decomposition.

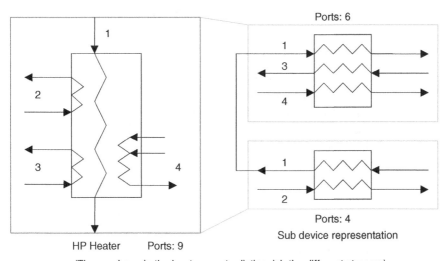

(The numbers, in the heater, are to distinguish the different streams)

Figure 7.35   HP heater decomposition.

provided sub-devices with fewer ports. Additionally, a reusable sub-device is already available in the device library. By reorganizing the CPDs and relabeling the functions, an existing heat exchanger FR model was reused. Device decomposition helps to build very complicated FR models and ease complicated CLA runtime process as well.

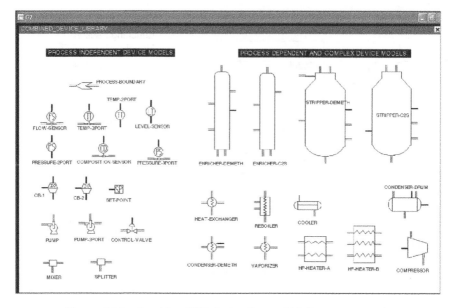

Figure 7.36   Combined FR device library (G2 Screen capture).

7.9.3   Analysis and Results

The newly constructed FR models for the ethylene splitter FR plant model were added to the existing device library increasing the number of devices in the library from 17 to 30 as shown in Figure 7.36. We did a comparative analysis of the time and effort in building new FR device models versus reusing models from the FR device library (Rathinasabapathy 2005). Our calculation showed a total time and effort savings of about 50%. Our observation is that in any process completely reusable, devices constitute at least 50% of the total devices. We continued our reusability analysis on a simulation case study by building a FR plant model for a CSTR example that had 12 devices. The reusable FR models helped us save about 66% of time and efforts (Rathinasabapathy 2005).

7.10   CONCLUSIONS

This chapter develops the integrative AI concepts and methods, for a form of root-cause diagnosis called diagnostic localization (DL). DL narrows the possible malfunctions and uses the system structure of the process to localize them, as abnormalities manifest over time and throughout an enterprise operation. DL allows earlier actions to be taken before root causes are fully discernable or manifest. This concentrates on the conceptual, theoretical, and implementation detail on how AI concepts and methods, data and information models, and multiple forms of quantitative and qualitative modeling are integrated with workflow for real-time, explainable, normal, and abnormal assessments of PHED. We further describe a DL approach that integrates human-in-the-loop action with probable and explainable malfunction propositions that are localized to process systems in greater and greater detail over time, as the effects of malfunctions develop. DL's ability to take earlier actions before root causes are fully discernable or manifest is demonstrated for an Ethylene plant use case.

DL's capabilities are associated with qualitative modeling, symptom and topographical matching, and search with time and change discretized even when there is insufficient data or models in malfunctioning situations. Why various modeling approaches are selected and how they are integrated are illustrated with the Ethylene plant test bed application. The chapters go into more depth with new use case examples on the foundations of FR modeling, a causal link assessment (CLA) algorithm, and causal process pattern (CPP) generation and interpretation than have been published previously. Requirements for optimizing windows in time for the required fidelity in time, ways to deal with malfunctioning or missing sensors, how to address integrated operational behavior complexities and search, and how to interpret patterns of analysis over time all become key levers for tuning the performance of the DL system. Additionally, we have addressed ways to discern when the model itself is incomplete.

Finally, this chapter brings together the modeling foundations and describes in detail how the multiple approaches are integrated and how they work together to produce an explainable, trusted, robust, and reusable smart manufacturing approach for human-in-the-loop DL. An in-depth analysis describes DL's ability to diagnose simultaneous multiple malfunctions and how it deals with sensor reliability, sensor malfunctions, and graceful degradation. The chapter addresses the reusability of the models and approaches not only for sustained management of a given health monitoring system but also for extending the approach to other enterprise chemical process line operations.

REFERENCES

Aft, L.S. (1986). *Fundamentals of Industrial Quality Control*, 146. Addison-Wesley.

Bjorkman, K. (2013). Solving dynamic flowgraph methodology models using binary decision diagrams. *Reliability Engineering & System Safety* 111: 206–216.

Chandrasekaran, B. (1994). Functional representation: a brief historical perspective. *Applied Artificial Intelligence* 8: 173–197.

Chang, C.-T. and Chen, C.Y. (2011). Fault diagnosis with automata generated languages. *Computers & Chemical Engineering* 35 (2): 329–341.

Chang, S.-Y., Lin, C.-R., and Chang, C.-T. (2002). A fuzzy diagnosis approach using dynamic fault trees. *Chemical Engineering Science* 57 (15): 2971–2985.

Chen, J.Y. and Chang, C.-T. (2007). Systematic enumeration of fuzzy diagnosis rules for identifying multiple faults in chemical processes. *Industrial & Engineering Chemistry Research* 46 (11): 3635–3655.

Chen, J.Y. and Chang, C.-T. (2009). Development of fault diagnosis strategies based on qualitative predictions of symptom evolution behaviors. *Journal of Process Control* 19 (5): 842–858.

Chiang, L.H., Braatz, R.D., and Russell, E. (2001). *Fault Detection and Diagnosis in Industrial Systems*. London: Springer.

Chiang, L.H., Jiang, B., Zhu, X. et al. (2015). Diagnosis of multiple and unknown faults using the causal map and multivariate statistics. *Journal of Process Control*. 28: 27–39.

Christofides, P.D., Davis, J.F., El-Farra, N.H. et al. (2007). Smart plant operations: vision, progress and challenges. *AIChE Journal* 53 (11): 2734–2741.

Davis, J.F., Piovoso, M.J., Hoo, K.A., and Bakshi, B.R. (1999). Process data analysis and interpretation. *Advanced Chemical Engineering* 25: 1–103.

Davis, J.F., Edgar, T., Porter, J. et al. (2012). Smart manufacturing, manufacturing intelligence and demand-dynamic performance. *Computers & Chemical Engineering* 47: 145–156.

Davis, J.F., Edgar, T., Graybill, R. et al. (2015). Smart manufacturing. *Annual Review of Chemical and Biomolecular Engineering* 6 (1): 141–160.

Elsass MJ. Multipurpose Sharable Engineering Knowledge Repository. PhD Dissertation The Ohio State University, 2001.

Elsass MJ, Saravanarajan, James F. Davis, Dinkar Mylaraswamy, dal Vernon Reising, John Josephson, "An integrated decision support framework for managing and interpreting information in process diagnosis". Process Systems Engineering 2004, B. Chen and A.W. Westerberg (editors), Elsevier Science B.V.

Fu, K.S. (1977). Introduction to Syntactic Pattern Recognition. In: *Syntactic Pattern Recognition, Applications*, Chapter 1, pp. 5 (ed. K.S. Fu). Springer-Verlag.

Gabbar, H.A. (2007). Improved qualitative fault propagation analysis. *Journal of Loss Prevention in the Process Industries* 20 (3): 260–270.

Gao, D., Wu, C., Zhang, B., and Ma, X. (2010). Signed directed graph and qualitative trend analysis based fault diagnosis in chemical industry. *Chinese Journal of Chemical Engineering* 18 (2): 265–276.

Gonzalez, R.C. and Thomason, M.G. (1978). *Syntactic Pattern Recognition – An Introduction*. Addison-Wesley Publishing Company Inc.

Goodaker, A.W., M.J. Elsass, J.F. Davis, D.C. Miller and J.R. Josephson, "Sharable Engineering Knowledge Databases: Application to HAZOP", Paper – 175b, AIChE Fall meeting, Miami, 1995.

Goodaker, A.W., Davis, J.F., Josephson, J.R. et al. (1996). *Design of Multipurpose Knowledge Databases for Process Plants*. Technical Report: The Ohio State University. Computers and Chemical Engineering.

Hamdan, I.M., Reklaitis, G.V., and Venkatasubramanian, V. (2012). Real-time exceptional events management for a partial continuous dry granulation line. *Journal of Pharmaceutical Innovation* 7 (3–4): 95–118.

Hu, J., Zhang, L., and Wang, Y. (2015). A systematic modeling of fault interdependencies in petroleum process system for early warning. *Journal of Chemical Engineering of Japan* 48 (8): 678–683.

Madni, A.M. and Sievers, M. (2013). System of systems integration: key considerations and challenges. *System Engineering* 17 (3): 330–347.

Maurya, M.R., Rengaswamy, R., and Venkatasubramanian, V. (2003a). A systematic framework for the development and analysis of signed digraphs for chemical processes. 1. Algorithms and analysis. *Induatrial & Engineering Chemistry Research* 42 (20): 4789–4810.

Maurya, M.R., Rengaswamy, R., and Venkatasubramanian, V. (2003b). A systematic framework for the development and analysis of signed digraphs for chemical processes. 2. Control loops and flowsheet analysis. *Industrial & Engineering Chemistry Research* 42 (20): 4811–4827.

Maurya, M.R., Rengaswamy, R., and Venkatasubramanian, V. (2004). Application of signed digraphs-based analysis for fault diagnosis of chemical process flowsheets. *Engineering Applications of Artificial Intelligence* 17 (5): 501–518.

Maurya, M.R., Rengaswamy, R., and Venkatasubramanian, V. (2006). A signed directed graph-based systematic framework for steady-state malfunction diagnosis inside control loops. *Chemical Engineering Science* 61 (6): 1790–1810.

Maurya, M.R., Rengaswamy, R., and Venkatasubramanian, V. (2007). A signed directed graph and qualitative trend analysis-based framework for incipient fault diagnosis. *Chemical Engineering Research and Design* 85 (10): 1407–1422.

Mhaskar, P., Gani, A., El-Farra, N.H. et al. (2006). Integrated fault detection and fault-tolerant control of process. *AIChE Journal* 52: 2129–2148.

Mhaskar, P., Gani, A., McFall, C. et al. (2007). Fault-tolerant control of nonlinear process systems subject to sensor faults. *AIChE Journal* 53: 654–668.

Mylaraswamy D, Emigholz K, Bullemer National Petrochemical and refiners association. Computer Conference. Fielding a Multiple State Estimator Platform. Chicago; 2000.

Nelson, W.L. (1954). *Petroleum Refinery Engineering*, 4the. New York: McGraw-Hill.

Ould-Bouamama, B., El-Harabi, R., Abdelkrim, M.N., and Gayed, M.K. (2012). Bond graphs for the diagnosis of chemical processes. *Computer Chemical Engineering* 36: 301–324.

Pal, S.K. and Pal, A. (2001). *Pattern Recognitions from Classical to Modern Approaches*. Calcutta, World Scientific: Indian Statistical Institute.

Pos, A., Borst, P., Top, J., and Akkermans, H. (1996). Reusability of simulation models. *Knowledge Based Systems* 9: 119–125.

Raimondo, D.M. (2013). Conference on control and fault-tolerant systems (Systol). In: *Proceedings of the 2nd International Conference on Control and Fault-Tolerant Systems. Nice*, 444–449.

Rathinasabapathy, Rajan., Diagnostic Localization in Chemical Process Industries Using Plant-Wide Causal Patterns. PhD Dissertation, University of California Los Angeles, 2005.

Rathinasabapathy, R., Elsass, M.J., Josephson, J., and Davis, J. (2016). A smart manufacturing methodology for real time chemical process diagnosis using causal link assessment. *AIChE Journal*, https://doi.org/10.1002/aic.15403.

Schutt, H.C. and Zdonik, S.B. (1956). How to recover ethylene from hydrocarbon mixtures. *Oil Gas Journal* 54 (65): 171–174.

Thambirajah, J., Benabbas, L., Bauer, M., and Thornhill, N.F. (2009). Cause-and-effect analysis in chemical processes utilizing XML, plant connectivity and quantitative process history. *Computer Chemical Engineering* 33 (2): 503–512.

Vachhani, P., Narasimhan, S., and Rengaswamy, R. (2007). An integrated qualitative–quantitative hypothesis driven approach for comprehensive fault diagnosis. *Chemical Engineering Research Design* 85 (9): 1281–1294.

Vedam, H. and Venkatasubramanian, V. (1999). PCA-SDG based process monitoring and fault diagnosis. *Control Engineering Practical* 7 (7): 903–917.

Venkatasubramanian, V., Rengaswamy, R., Yin, K., and Kavuri, S.N. (2003a). A review of process fault detection and diagnosis. Part I: Quantitative model-based methods. *Computer Chemical Engineering* 27 (3): 293–311.

Venkatasubramanian, V., Rengaswamy, R., and Kavuri, S.N. (2003b). A review of process fault detection and diagnosis. Part II: Qualitative models and search strategies. *Computer Chemical Engineering* 27 (3): 313–326.

Venkatasubramanian, V., Rengaswamy, R., Kavuri, S.N., and Yin, K. (2003c). A review of process fault detection and diagnosis. Part III: Process history based methods. *Computer Chemical Engineering* 27 (3): 327–346.

Wang, H., Chen, B., He, X. et al. (2008). A signed digraphs based method for detecting inherently unsafe factors of chemical process at conceptual design stage. *Chinese Journal of Chemical Engineering* 16 (1): 52–56.

Whiteley, J.R. and Davis, J.F. (1992). Knowledge-based interpretation of sensor patterns. *Computer Chemical Engineering* 16 (4): 329–346.

Yim, S.Y., Ananthakumar, H.G., Benabbas, L. et al. (2006). Using process topology in plant-wide control loop performance assessment. *Computer Chemical Engineering* 31 (2): 86–99.

Zhang, Z.Q., Wu, C.G., Zhang, B.K. et al. (2005). SDG multiple fault diagnosis by real-time inverse inference. *Reliability Engineering System Safety* 87 (2): 173–189.

8

OPTIMAL QUANTITATIVE MODEL-BASED PROCESS FAULT DIAGNOSIS

Richard J. Fickelscherer, PE[1] and Daniel L. Chester[2]

[1]Department of Chemical and Biological Engineering, State University of New York at Buffalo, Buffalo, NY, USA
[2]Department of Computer and Information Sciences (Retired), University of Delaware, Newark, DE, USA

OVERVIEW

As presented in this chapter, it is our contention that evaluating accurate quantitative engineering models of normal process operation with real-time process data is the most promising and powerful means for directly identifying underlying process operating problems. Doing so generates an unimpeachable source from which to logically infer the current state of the process being modeled. Performing this inference continuously on-line enables these programs to perform "intelligent process supervision" of the daily operations of their associated process systems. It thus allows the most fundamental understanding of a given process system's design and operation to be directly utilized in evaluating its current operating conditions.

Artificial Intelligence in Process Fault Diagnosis: Methods for Plant Surveillance, First Edition. Edited by Richard J. Fickelscherer.
© 2024 John Wiley & Sons, Inc. Published 2024 by John Wiley & Sons, Inc.

CHAPTER HIGHLIGHTS

- Innovative patented Fuzzy logic diagnostic algorithm which simplifies and optimizes automated process fault diagnosis based on quantitative models of normal process operation.
- Requires only the same fundamental process knowledge originally used to design a given process.
- Continuously fully leverages such knowledge to intelligently supervise current process operations.
- Thoroughly analyzes live process data to more fully extract all pertinent and timely diagnostic information.
- Helps directly reduce operating costs and substantially improves both process safety and efficiency.
- Therefore, helps optimize, in real-time, process operations throughout all of the processing industries for those target processes of which the required quantitative process models of normal operation can be formulated.

8.1 INTRODUCTION

A general method for performing automated process fault diagnosis in chemical and nuclear processing plants has been actively sought ever since computers were first incorporated into process control systems (Fickelscherer 1990). The motivation for this search has been the enormous potential such automation would have for improving both process plant safety and productivity. However, for a variety of reasons, past attempts at automating process fault analysis have not proven to be very successful. They were much more expensive to develop, in terms of the amount of effort, resources and time, than was originally anticipated. More importantly, the resulting computer code performed below expectations and proved to be very difficult to maintain as the target process systems evolved over time. More robust diagnostic strategies and computer implementations are thus still being actively sought (Ma and Jiang 2011; Das et al. 2012; Venkatasubramanian et al. 2003a,b,c).

A quantitative model-based process diagnostic strategy called the **Method of Minimal Evidence (MOME)** (Fickelscherer 1990, 1994; Fickelscherer and Chester 2013, 2016; Fickelscherer et al. 2003, 2005; Skotte et al. 2001) that overcomes many of the practical limitations encountered in other strategies is described here. It evolved from the experience gained during the development, verification, and implementation of a fully automated process fault analyzer for a commercial adipic acid plant. This strategy has since been converted into two generalized algorithms, one based on Fuzzy logic reasoning (Chester et al. 2008; Fickelscherer and Chester 2013; Fickelscherer et al. 2005; Zadeh 1988) and another based on assumption state differences (Fickelscherer and Chester 2016), both of which have been completely automated[1]: all the users now has to specify are the quantitative engineering models describing normal process operation and their associated statistical parameters computed from sufficient amounts of actual normal operating data. This allows the more difficult problem of automated process fault analysis to be reduced to the much simpler problem of process modeling. This treatment describes both the quantitative model-based diagnostic reasoning employed by MOME and the patented Fuzzy logic algorithm (Chester et al. 2008) incorporating it. Its use directly allows highly effective online, real-time **Sensor Validation and Proactive Fault Analysis (SV&PFA)** to be performed continuously. First, some of the concepts associated with automated process fault analysis as used throughout the following discussion will be defined.

8.2   PROCESS FAULT ANALYSIS CONCEPT TERMINOLOGY

The terms **fault** and **fault situation** will refer to the actual underlying cause that creates either a potential or an actual process operating problem. The term **symptom** refers to the observable manifestations of fault situations. Symptom will also refer to the observable manifestations of non-fault situations. For example, a stuck valve will be considered a fault situation, whether or not it affects process operation in any observable way. Furthermore, if this stuck valve, say, causes a high temperature alarm, that alarm will be considered a symptom of that fault situation. It might also well be a symptom of many other possible fault and non-fault situations.

Diagnostic evidence refers to any symptom that supports the plausibility of one or more fault hypotheses. This evidence may also support the plausibility of any number of non-fault hypotheses. Diagnostic evidence is based upon information obtained directly from the current operating behavior of the target process system. This real-time process information is refined into diagnostic evidence through the application of knowledge about the process system's normal operating behavior. It will be assumed that only the values

---

[1]The patented Fuzzy logic algorithm version of the MOME diagnostic strategy is currently implemented in our commercial software package called FALCONEER™ IV; the Assumption State Differences algorithm version exists as a proof of concept prototype.

of the process variables directly measured by process sensors at a constant sampling frequency will be used to create current diagnostic evidence continuously. This assumption is made for convenience and will not affect the generality of the following discussion whatsoever.

The **target process system** is that portion of the entire processing plant complex being actively analyzed for process fault situations by the given process fault analyzer.

A **process operating event** refers to any occurrence which affects the target process system's operating behavior sufficiently enough to generate symptoms. Such occurrences can be caused by actual process fault situations, such as pump failures, stuck valves, etc. or non-fault events, such as normal process startups and shutdowns, production rate changes, etc. Any combination of these various possible disturbances can likewise create such occurrences.

A process fault analyzer's **intended scope** is that subset of potential process operating events that the fault analyzer is explicitly designed to correctly identify and classify the cause of. A fault analyzer's intended scope by definition includes those caused by its **target fault situations**, which are those potential process fault situations that the process fault analyzer is specifically designed to detect and diagnose. This intended scope is constrained by the process fault analyzer's **intended operational domain**, which specifies the process operating states that that analyzer can meaningfully analyze for its target fault situations, for example, all possible process production levels, steady and unsteady state operation, etc.

Correctly classifying a process operating event means that the process fault analyzer can properly discriminate the cause of that event from all the other possible causes; that is, the process fault analyzer will not misdiagnose that cause as another. According to this definition, the process fault analyzer does not necessarily have to explicitly identify a unique fault hypothesis to operate properly: it just has to not misidentify that fault. Thus, in order to correctly classify the cause of a current process operating event, the process fault analyzer has to either correctly diagnose the fault which is actually occurring as one of the current possible plausible fault hypotheses or misdiagnose it by not making any diagnosis whatsoever, that is, by remaining silent. Performing in this conservative manner is advantageous because it will not distract or confuse the process operators during potentially highly dangerous emergency situations.

A process fault analyzer's **competence** refers to how correctly it classifies the causes of the various possible process operating events contained within its intended scope. A process fault analyzer that can correctly classify the cause of all of these events is considered **completely competent**.

A process fault analyzer's competence can be quite different from its robustness. **Robustness** refers to how correctly a process fault analyzer can classify the cause of the various process operating events that actually occur during its target process system's operation. Consequently, a process fault analyzer's competence is related to how well its performance lives up to its design specifications while its robustness is an indication of how useful those design specifications actually are for the given target process system. A process fault analyzer's true competence approaches its robustness as its intended scope includes more of the process operating events that can actually occur during the target process system's operation.

A process fault analyzer's **diagnostic resolution** is the level of discrimination between a particular fault hypothesis and all other possible contending fault hypotheses. **Perfect resolution** refers to those fault hypotheses that can be uniquely discriminated from all of the other possible potential fault hypotheses (i.e., those required to correctly classify the causes of all possible process operating events contained within the process fault analyzer's intended scope).

A process fault analyzer's **diagnostic sensitivity** for a particular fault situation refers to the minimum values of that fault situation's magnitude or rate of occurrence that can be correctly detected and diagnosed. These sensitivities depend directly upon the current process operating conditions and the degree of diagnostic resolution sought. Fault situations that have magnitudes and/or rates of occurrence at or above the process fault analyzer's minimum diagnostic sensitivity will be considered **significant** with respect to that analyzer. Please notice that this definition of significance has no correspondence to the potential severity of the consequences that that particular process fault situation could have on safe process operations.

A process fault analyzer's **diagnostic response time** is the time interval that elapses between the onset of a significant process operating event and the classification of its cause by the process fault analyzer. It is thus the time interval required to express the complete pattern of diagnostic evidence used by the process fault analyzer to diagnose each of its associated target fault situations.

Finally, a process fault analyzer's **utility** is a measure of the actual usefulness derived from its diagnoses by the process operators. Utility is thus a composite metric of the process fault analyzer's robustness along with its possible diagnostic resolution, sensitivity, and response time for identifying the cause(s) of each of the process operating events that actually occur.

8.3 MOME QUANTITATIVE MODELS OVERVIEW

Quantitative model-based diagnostic reasoning is a highly systematic and powerful means for deriving plausible hypotheses as to the causes of abnormal process behavior. It was used to create the diagnostic knowledge bases of the original FALCON (Fickelscherer 1990, 1994), original FALCONEER (Fickelscherer et al. 2003; Skotte et al. 2001), and various FALCONEER™ IV (Fickelscherer and Chester 2013; Fickelscherer et al. 2005) **Knowledge-Based Systems (KBSs)**. The following describes exactly what constitutes effective Primary and Secondary models of normal process operation.

A **Primary model** is defined as any mathematical equation that is useful for describing normal process behavior.[2] It must contain at least one sensor variable and it must not contain any process variables which are not directly measured by process sensors and which are not adequate parameters. It may contain any number of adequate parameter variables and may also be time dependent (e.g., include differential terms, integral terms, time delays, etc.). We assume that the Primary models are linearly independent of each other, since if one is linearly dependent on some of the others, it does not provide any additional information for the process being modeled beyond what the additional possible Secondary models (defined below) provide.

Modeling assumption variables are defined as any process variables about which assumptions must be made to derive Primary models. Examples of such assumptions include that the current value of a sensor measurement used as the value of the variable within the model is correct or that a value of an unmeasured parameter used within the model is equal to its normal and/or extreme value. These assumption variables are explicitly represented as definite terms within the Primary model. Deviations between these assumption variable values and the actual process state generate the residuals (defined below) used by MOME to do SV&PFA.

Modeling assumption variables thus become invalidated by a significant deviation between the actual process state and that assumed by the Primary model. For most of the assumption variables, these deviations can occur either in the positive or in the negative direction. Those caused by a particular

---

[2]The following terms will be used throughout our discussion: a **sensor variable** is defined as any process variable that is measured directly by a process sensor (e.g., such as a process temperature being measured by a corresponding thermocouple); a **parameter variable** is defined as any process variable which is not directly measured, but which can be adequately characterized by either a standard value and/or an extreme value under normal process operating conditions (e.g., such as the density of a particular process stream); and a **modeling assumption variable** is defined as any sensor or parameter variable required to derive a given quantitative process model capable of describing normal process operating behavior.

assumption variable deviation which is both significant and positive will cause those particular assumption variables to be defined as being invalid and high (identified for assumption variable deviation a_i as a_i^H).[3] Likewise, deviations which are both significant and negative will cause those particular assumption variables to be defined as invalid and low (identified for assumption variable deviation a_i as a_i^L).

The ith Primary model describing normal process operation can be characterized as an equation having the following form:

$$0 = f_i \left(\text{Primary model } i \text{ modeling assumption variables, time} \right)$$
$$- B_i \left(\text{process operating conditions} \right) \tag{8.1}$$

where f_i is a mathematical function and the modeling assumption variables are either specific sensor measurements or standard and/or extreme values of specific parameters and where B_i is the normal offset of model i (referred to as its *Beta*) which could be a function of process operating conditions such as current production rate, etc.[4]

Once the modeling assumption variables are instantiated with real-time process sensor data or standard and/or extreme values of unmeasured parameters and its current estimated *Beta* computed, a residual results (i.e., ε_i below) which is just a function of process sensor noise and any currently occurring modeling assumption variable deviations, that is:

$$\varepsilon_i = f_i \left(\text{current values of Primary model } i \text{ modeling assumption} \right.$$
$$\left. \text{variables and time} \right)$$
$$- B_i \left(\text{current process operating conditions} \right) \tag{8.2}$$

---

[3]For assumption variables representing sensor variable measurements, high deviations mean that the current sensor reading is higher than the actual value of the phenomenon; for a parameter variable just the opposite is true (i.e., the parameter variable's value used in the model is lower than the phenomenon's actual value). Low deviations of the assumption variables mean just the opposite for both sensor and parameter variables.

[4]*Betas* can arise for a variety of reasons. Sometimes one or more of the sensor variables used in a residual calculation needs recalibration because of an ongoing measurement bias. This was true for the particular placement of redundant pH meters in one of the FMC FALCONEER™ IV applications. *Betas* can also arise directly from the particular models used to describe normal operation. In the original FALCON KBS at DuPont, the overall mass balance on the LTC Recycle loop had a −7% *Beta* because the HTC Pre-Heater feed stream contained 7% offgas, fooling the flow sensor on that stream by that amount. Other causes of *Betas* are from legitimate physical phenomena. In the two FALCONEER™ IV KBS's used at FMC, using the Antoine Equation to model the Pressure/Temperature relationship in the crystallizers resulted in a *Beta* equal to the actual boiling point elevation of the mother liquor.

If the **residual** (i.e., ε_i) of an evaluated Primary model is **significantly high or low** (significantly higher or lower than zero), then it can be inferred that at least one or more of the possible modeling assumption variable deviations (i.e., possible process operating problems such as faults which could cause such a residual) is occurring.

If the **residual** (i.e., ε_i) of an evaluated Primary model is **not significant**, then either: (1) there are no modeling assumption variable deviations; (2) one or more such deviations are occurring but at magnitudes or rates of change which are below the sensitivity of that model to discriminate such deviations (i.e., those deviations are not currently considered significant for that particular model); or (3) two or more significant assumption variable deviations are interacting in an opposing fashion. Which of these three possible cases is currently true is determined by logically interpreting the complete pattern of all real-time residual values.

Assumption variables are classified for each model according to how that model's residual changes with deviations of that particular assumption variable. Modeling assumption variable deviations that cause proportional changes in that given model's residual are defined to be **linear assumption variables**. This is an important distinction because the magnitude of the underlying assumption variable deviation can be directly calculated from the model's current residual.[5] If the modeling assumption variable is linear in two or more models, these independent estimates of those magnitudes, if they agree, can be used as additional evidence for the diagnosis that that assumption variable is indeed deviating; else it is direct evidence against that diagnosis. Alternatively, additional models can be created from pairs of Primary models, which directly eliminate those common linear assumption variables. The behavior of these additional linearly dependent (referred to as **Secondary**) **models'** residuals also directly generates such additional diagnostic evidence. The exhaustive use of Secondary models' residuals behavior is directly exploited by the MOME strategy.

Briefly, the **MOME diagnostic methodology** for fault diagnosis compares patterns of the Primary and Secondary residual behavior expected to occur during the various possible assumption variable deviations with the real-time patterns of those residuals currently present in the target process system. It uses the minimum unique patterns required for correctly doing this analysis, allowing for many of the possible multiple assumption

---

[5] That is, there exists analytical redundancy in the Primary model for such linear assumptions.

variable deviations (i.e., **multiple faults**) to be directly identified. This is important because this methodology does not discern only process faults, but all possible causes of abnormal operating behavior, fault and non-fault situations alike. The logic is designed to venture diagnoses only if it is highly certain of the cause(s) of the underlying behavior. This conservative behavior is advantageous because it should not confuse its users with implausible diagnoses at times when the actual target process system's operating state is in flux.

The **patterns of expected residual behavior** that result from applying this method (i.e., the **SV&PFA diagnostic rules**) contain the minimum patterns required to diagnose each of the possible fault situations. This directly maximizes the possible sensitivity of the fault analyzer for these various faults, maximizes the possible resolution (discrimination between various possible faults and/or non-fault situations) of that analysis, and optimizes its overall competence when confronted with multiple assumption variable deviations. The strategy can be further used to directly determine the strategic placement of process sensors for better performing fault diagnosis and the shrewd division of large processing plant complexes for distributing independent process fault analyzers.

As described above, all possible **combinations of two Primary models** that result from **eliminating common linear assumption variables** constitute the set of linearly dependent **Secondary models**. The resulting possible Secondary models will provide useful diagnostic evidence for discriminating between assumption variable deviations k and l if and only if the following inequality holds (Fickelscherer 1990):

$$\left(\partial \varepsilon_i / \partial a_k \right) / \left(\partial \varepsilon_j / \partial a_k \right) \neq / \left(\partial \varepsilon_i / \partial a_l \right) / \left(\partial \varepsilon_j / \partial a_l \right) \tag{8.3}$$

ε_i, ε_j = residuals of Primary models i and j, a_k, a_l = linear modeling assumption variables k and l.

The residuals of the resulting useful Secondary models can be calculated directly from the residuals of the Primary models via the following formula (Petti et al. 1990):

$$\varepsilon_{j+k,i} = \varepsilon_j - \left((\partial \varepsilon_j / \partial a_i) / (\partial \varepsilon_k / \partial a_i) \right) * \varepsilon_k \tag{8.4}$$

ε_j, ε_k = residuals of Primary models j and k,
$\varepsilon_{j+k,i}$ = residual of Secondary model formed by combining Primary models j and k, eliminating assumption variable i,
a_i = linear modeling assumption variable i.

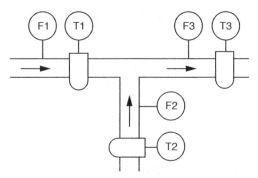

INSTRUMENTED PROCESS MIXING TEE

Figure 8.1   FMC ESP process barometric condenser water feed.

The variances of the resulting useful Secondary models can be calculated directly from the variances of the Primary models via the following formula (Kramer 1990) for the sum of variances:

$$v_{j+k,i}^{2} = v_{j}^{2} - \left(\left(\partial \varepsilon_{j} / \partial a_{i} \right) / \left(\partial \varepsilon_{k} / \partial a_{i} \right) \right)^{2*} v_{k}^{2} \tag{8.5}$$

v_{j}, v_{k} = variances of Primary models j and k.
$v_{j+k,i}$ = variance of Secondary model formed by combining Primary models j and k, eliminating assumption variable i.
ε_{j}, ε_{k} = residuals of Primary models j and k.
a_{i} = linear modeling assumption variable i.

Example 8.1   Example of MOME secondary models.
Consider the instrumented mixing tee depicted in Figure 8.1. The tee blends flows from two heat exchangers (HTX1 and HTX2) and passes them to a condenser (COND). The following mass and energy balances (Primary models P_{1} and P_{2} below, respectively) can be derived for this system:

Primary model 1.   (P_{1}) below is a mass balance on the mixing tee.

P_{1} $\varepsilon_{P1} = \left(p1 * \left(F1 - L1 \right) + p2 * \left(F2 - L2 \right) - p3 * \left(F3 + L3 \right) \right) * 60 \text{ min /hour}$

$p1 = p2 = p3 = \text{density of water} = 8.34 \text{ lbs./gal}$

Its derivation requires the following six modeling assumptions:

Assumption variables a_i for Primary model P_1:

a_1)   Linear—F1 (H_2O flow rate from HTX1) flow sensor is correct: $\partial \varepsilon_{P1} / \partial F1 = +500$

a_2)   Linear—F2 (H_2O flow rate from HTX2) flow sensor is correct:
$\partial\varepsilon_{p1}/\partial F2 = +500$

a_3)   Linear—F3 (H_2O flow rate into COND) flow sensor is correct:
$\partial\varepsilon_{p1}/\partial F3 = -500$

a_4)   Linear—L1 = 0; no process leaks in HTX1 H_2O Return Line:
$\partial\varepsilon_{p1}/\partial L1 = -500$

a_5)   Linear—L2 = 0; no process leaks in HTX2 H_2O Return Line:
$\partial\varepsilon_{p1}/\partial L2 = -500$

a_6)   Linear—L3 = 0; no process leaks in COND H_2O Feed Line:
$\partial\varepsilon_{p1}/\partial L3 = -500$

Primary model 2. $\mathbf{P_2}$ below is an energy balance on the mixing tee.

$$\mathbf{P_2} \quad \varepsilon_{P2} = \left(p1 * cp1 * (F1-L1)*(T1-0) + p2 * cp2 * (F2-L2)*(T2-0)\right.$$
$$\left. - p3 * cp3 * (F3+L3)*(T3-0)\right) * 60\,\text{min/hour}$$

$p1 = p2 = p3 = \text{density of water} = 8.34\,\text{lbs./gal}$

$cp1 = cp2 = cp3 = \text{heat capacity of water} = 1.8\,\text{BTU / lb.°C}$

Its derivation requires the following nine modeling assumptions:

Assumption variables a_i for Primary model P_2:

a_1)   Linear—F1 (H_2O flow rate from HTX1) flow sensor is correct:
$\partial\varepsilon_{p2}/\partial F1 = +900*T1$

a_2)   Linear—F2 (H_2O flow rate from HTX2) flow sensor is correct:
$\partial\varepsilon_{p2}/\partial F2 = +900*T2$

a_3)   Linear—F3 (H_2O flow rate into COND) flow sensor is correct:
$\partial\varepsilon_{p2}/\partial F3 = -900*T3$

a_4)   Linear—L1 = 0; no process leaks in HTX1 H_2O Return Line:
$\partial\varepsilon_{p2}/\partial L1 = -900*T1$

a_5)   Linear—L2 = 0; no process leaks in HTX2 H_2O Return Line:
$\partial\varepsilon_{p2}/\partial L2 = -900*T2$

a_6)   Linear—L3 = 0; no process leaks in COND H_2O Feed Line:
$\partial\varepsilon_{p2}/\partial L3 = -900*T3$

a_7)   Linear—T1 (H_2O temp. out of HTX1) is correct:
$\partial\varepsilon_{p2}/\partial T1 = +900*(F1-L1)$

a_8)   Linear—T2 (H_2O temp. out of HTX2) is correct:
$\partial\varepsilon_{p2}/\partial T2 = +900*(F2-L2)$

a_9)   Linear—T3 (H_2O temp. into COND) is correct:
$\partial\varepsilon_{p2}/\partial T3 = -900*(F3+L3)$

From these two Primary models, the following three unique Secondary models (Secondary models S_1, S_2, and S_3 below) can be derived by eliminating the flow rates F2, F1, and F3, or leaks, L2, L1, and L3, respectively[6]:

S_1 $\varepsilon_{S1} = \left((F1 - L1)^* (T1 - T2) - (F3 + L3)^* (T3 - T2) \right)$

S_2 $\varepsilon_{S2} = \left((F2 - L2)^* (T2 - T1) - (F3 + L3)^* (T3 - T1) \right)$

S_3 $\varepsilon_{S3} = \left((F1 - L1)^* (T1 - T3) - (F2 - L2)^* (T3 - T2) \right)$

The usefulness of the possible Secondary models depends upon how well Inequality (3) holds. The limitations placed upon the usefulness can be demonstrated as follows:

Consider the assumption variables stating that flow meters F1 and F2 are operating properly. Taking the ratios of the partial derivatives as indicated by the Inequality (3) results in the following inequality:

$$\left(\partial \varepsilon_{P2} / \partial F1 \right) / \left(\partial \varepsilon_{P1} / \partial F1 \right) \neq \left(\partial \varepsilon_{P2} / \partial F2 \right) / \left(\partial \varepsilon_{P1} / \partial F2 \right)$$
$$+900 * T1 / +500 \neq +900 * T2 / +500$$
$$T1 \neq T2$$

Thus, as these process operating temperatures T1 and T2 approach one another, the usefulness of Secondary models S_2 and S_1 for providing useful additional evidence rapidly diminishes (i.e., the same diagnostic evidence given by the mass balance's residual (ε_{P1}) would also be given by the energy balance's residual (ε_{P2})). This makes any further discrimination between errors in flow measurements F1 and F2 impossible.

The diagnostic logic behind quantitative model-based process fault diagnosis is straightforward: any significant residual of a given process model can be used directly as evidence that one or more of its associated modeling

[6]Forming identical Secondary models as shown here by eliminating two or more unique linear modeling assumption variables from the same parent Primary models is referred to as coupled assumption elimination. This occurs because the grouping of coupled assumption variables exists in both parent Primary models as multiples of identical terms. Although these multiple copies of the same Secondary model that result from the coupled assumption elimination do not provide additional evidence to discriminate between the coupled assumption variable deviations, they still provide evidence for the other possible assumption variable deviations. Consequently, although redundant, they do not interfere with or impair the subsequent process fault analysis based on them and the other Primary and Secondary model residuals. In FALCONEER™ IV all possible Secondary model residuals are serially created and their results exhaustively inferred upon.

assumption variables' current values which can cause such behavior is deviating from its true value. Consequently, whenever a model becomes violated, all of its associated modeling assumption variables' current values that can cause such a violation are suspected to be deviating. Plausible hypotheses as to which assumption variable or variables are actually currently deviating can be inferred by logically interpreting the entire pattern of the relevant residuals' current statuses.[7]

Consequently, if any of the modeling assumptions truly required for deriving a given Primary model is overlooked, or if the above statistical parameters are calculated incorrectly, the competency of the resulting process fault analyzer will be jeopardized. This has important implications for the usefulness of potential quantitative process models as Primary models: as explained below, their proper derivation must be highly vetted by the developers of the process fault analyzer.

Representing the **declarative knowledge** (also commonly referred to as **domain knowledge**) about the target process system's normal operating behavior as a set of Primary models has a major advantage when it comes to verifying that knowledge's validity and accuracy. This is because each of the Primary models can be verified independently of all of the others. Thus, a team of process engineers familiar with the target process system can scrutinize each model independently to ensure: (1) that all of the modeling assumptions required to derive that model are taken into account; (2) that the standard and/or extreme values of all parameter variables contained within that model are reasonable; and (3) that that model accurately predicts normal process behavior over its entire intended operational domain. If any discrepancies are discovered, that model can be modified: (1) by changing its form through the addition or deletion of terms; (2) by adjusting the standard/extreme values of its associated parameter variables; (3) by constraining its intended operational domain through the modification, addition, or deletion of modeling assumption variables; or (4) by adjusting its normal variance and/or *Beta*. The Primary models resulting from this phase of the verification procedure should always accurately predict the actual process behavior during normal process operation. Such highly vetted models are deemed to be **well-formulated**.

At this stage in the process fault analyzer's development, the possible diagnostic information derived from the declarative knowledge still needs to be refined further into a SV&PFA diagnostic knowledge base. This is accomplished by creating SV&PFA diagnostic rules: patterns of diagnostic evidence capable of discriminating between the various possible process

---

[7]Precisely which residuals are considered relevant depends upon the specific diagnostic strategy employed.

operating events, especially the various possible fault situations. However, as will be demonstrated below, the procedures required to create these diagnostic rules are algorithmic in nature, and therefore it was possible to automate these procedures to ensure that they are always properly applied. Consequently, the competency of the resulting process fault analyzer now depends entirely only upon the correctness of its declarative knowledge base, that is, the process knowledge contained within the various Primary models. This in turn depends entirely upon how well the normal operating behavior of the target process system is understood and represented by those various Primary models. **Thus, it is possible to guarantee that the process fault analyzer will always perform competently if and only if all of the Primary models are guaranteed to be well-formulated.** This indicates that the main investment of effort during the development and verification of a model-based process fault analyzer should be in deriving and verifying those models and then evaluating their correspondingly required statistical parameters from sufficient normal process operating data. We typically recommend using 3–6 months worth of such data to perform the required statistical analysis of these Primary models.

8.4 MOME QUANTITATIVE MODEL DIAGNOSTIC STRATEGY

In this section, a general diagnostic strategy for creating optimal model-based process fault analyzers called MOME is presented. This methodology uses the minimum amount of diagnostic evidence necessary to uniquely discriminate between a significantly deviating modeling assumption variable (e.g., an assumption which assumes the absence of a particular process fault situation) and all other valid modeling assumption variables. Moreover, it ensures that the resulting process fault analyzer will always perform competently and optimizes the diagnostic sensitivity and resolution of its diagnoses. Diagnostic knowledge bases created according to MOME are also conducive for diagnosing many multiple fault situations, for determining the strategic placement of process sensors to better facilitate process fault analysis, and for determining the shrewd distribution of independent process fault analyzers within large processing plants. It has been demonstrated to be competent in both an adipic acid plant (Figure 8.2A and B) formerly owned and operated by DuPont in Victoria, Texas and an electrolytic persulfate plant (Figure 8.3) formerly owned and operated by FMC in Tonawanda, New York. The basic logical inference utilized by MOME is described in detail below.

As described previously, diagnostic evidence is inferred from the real-time residuals of evaluated Primary and Secondary models. The values of the

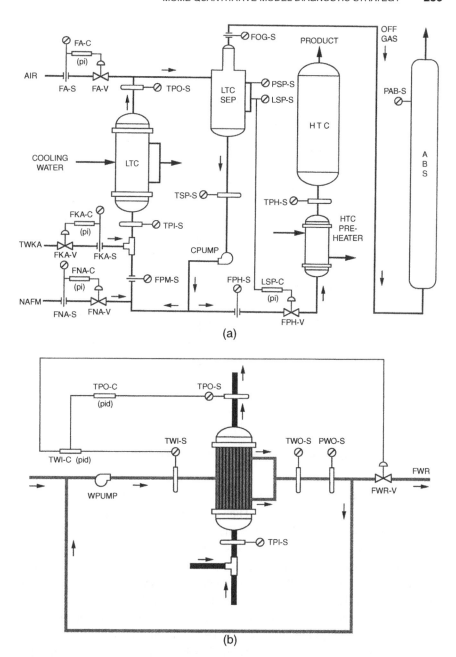

Figure 8.2 (a) DuPont adipic acid process low temperature converter (LTC). (b) DuPont adipic acid LTC cooling system.

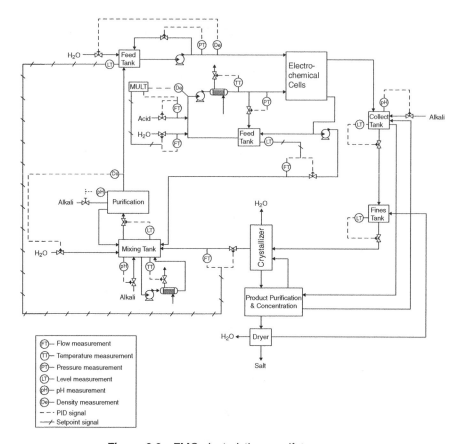

Figure 8.3 FMC electrolytic persulfate process.

residuals directly indicate which models are currently being satisfied and which are currently being violated. The violation of a well-formulated Primary or Secondary model directly implies that at least one of its associated modeling assumption variables that can cause such a violation is deviating significantly. In contrast, the satisfaction of a well-formulated Primary or Secondary model implies either that all of its associated modeling assumption variables are valid (i.e., at least not deviating significantly for that model), or that two or more of those assumption variables are deviating significantly but are interacting in such a way as to cause that model to appear satisfied.

Once the Primary and Secondary models have been evaluated, plausible hypotheses as to the underlying cause or causes of the observed abnormal process behavior can be derived by logically interpreting the resulting patterns of these model satisfactions and violations. Fault diagnosis performed in this manner is said to follow a quantitative model-based diagnostic strategy. Currently, a wide variety of such diagnostic strategies exist (Kramer and Mah 1994; Venkatasubramanian et al. 2003c). However, none of these other strategies simultaneously optimize the diagnostic sensitivity and resolution of diagnostic rules they create while guaranteeing that the resulting process fault analyzer will always perform competently, as MOME does.

The MOME diagnostic strategy for process fault analysis compares patterns of residual behavior expected to occur during the various possible assumption variable deviations with the patterns of those real-time residuals currently present in the target process system. Any matches directly identify the underlying possible assumption variable(s) deviations currently occurring there. MOME uses the minimum unique patterns required for correctly doing this analysis, allowing for many of the possible multiple assumption deviations to also be directly identified. This is important because this methodology does not discern only process faults, but also the causes of all possible process operating events contained in the process fault analyzers intended scope, fault, and non-fault situations alike. The logic is designed to venture unique diagnoses only if it is highly certain as to the possible cause(s) of the underlying problem. Such conservative behavior is advantageous because it should not confuse its users with logically implausible diagnoses at times when the actual target process system's operating state is in flux.

The patterns of expected residual behavior resulting from applying this method (e.g., the **SV&PFA diagnostic rules**) contain the minimum patterns required to diagnose each of the possible fault situations. This directly maximizes the sensitivity of the fault analyzer for these various faults, maximizes the resolution (discrimination between various possible faults and non-fault events for the actual magnitude of fault occurring) of that analysis, and optimizes its overall competence when confronted with multiple assumption deviations. The following example demonstrates these capabilities.

Example 8.2:   Example of MOME SV&PFA diagnostic rules' logic. Consider further the instrumented process mixing tee shown in Figure 8.1 and modeled in detail in Example 8.1. This present example of the MOME diagnostic strategy demonstrates some of the patterns of expected model residuals (i.e., **SV&PFA diagnostic rules**) from which the current operating state of this mixing tee process can be directly inferred.[8]

---

[8]These engineering models were contained in the Sensor Validation and Proactive Fault Analysis (SV&PFA) module of the *FALCONEER™ IV* application at FMC's Tonawanda Electrolytic Sodium Persulfate (ESP) plant. *FALCONEER* is an acronym for **F**ault Ana**L**ysis **CON**sultant via **E**ngineering **E**quation **R**esiduals.

The various possible combinations of residual behavior that can occur whenever any of the nine linear modeling assumption variables used to derive Primary models P_1 and P_2, and Secondary models S_1, S_2, and S_3 above are deviating significantly can be predetermined and written down as SV&PFA diagnostic rules to identify those invalid assumption variables. The pattern of residual behavior determined from real-time sensor data can then be compared against these expected patterns to derive a diagnosis of the current process state. The nomenclature and logic symbols used in the following diagnostic rules are defined as follows[9]:

MOME Diagnostic Rule Nomenclature

P_X^Y Primary model X residual is in State Y

S_X^Y Secondary model X residual is in State Y

a_X^Y modeling assumption variable X is in State Y

where[10]: State **Y** can be either: **H**—means the given quantity is considered significantly **high**, **S**—means the given quantity is considered neither significantly high nor significantly low; e.g., it is considered normal or in other words **satisfied**, **L**—means the given quantity is considered significantly **low.**

LOGIC Symbols

∧ represents conjunction—"logical AND"
∨ represents disjunction—"logical inclusive OR"
⊕ represents disjunction—"logical exclusive OR"
¬ represents negation—"logical NOT"
→ represents implication—"logical implication"
↔ represents equivalence—"logical equivalence"

---

[9]Please note: the following analysis and subsequent discussion is based on Boolean logic. This does not limit its generality in any way. This analytical method will be further formalized and expanded to non-Boolean reasoning below, specifically to rely upon Fuzzy logic reasoning and certainty factor calculations.

[10]The value of State Y for a given assumption variable depends upon whether that variable is for the accuracy of either a sensor measurement or a constant parameter in the relevant Primary and Secondary models. This state for assumption deviations caused by inaccurate sensor measurements that violate models depending upon them is in the same direction as the signs of those models' first derivatives with respect to those assumptions. This state for assumption deviations caused by incorrect values of constant parameters is just the opposite of those signs. Such assignments are demonstrated in this current example.

Consider the situation where all the residuals of the two Primary and three Secondary models above are satisfied.[11] For single fault situations, this would normally result in the inference that all nine of the associated assumption variables are satisfied (i.e., not deviating sufficiently to violate any of the models). The smallest value of the given assumption deviation that will violate at least one of these two Primary models directly defines the minimum sensitivity of the resulting fault analyzer to each of the above nine modeling assumption variables. Such minimum deviations for any given model are defined as **minimum assumption variable deviations** (Fickelscherer 1990). They are typically different for each Primary and Secondary model. The corresponding sensor validation diagnostic rule would thus be as follows:

Sensor Validation Diagnostic Rule:

$$P_1^S \wedge P_2^S \wedge S_1^S \wedge S_2^S \wedge S_3^S \rightarrow a_1^S \wedge a_2^S \wedge a_3^S \wedge a_4^S \wedge a_5^S \wedge a_6^S \wedge a_7^S \wedge a_8^S \wedge a_9^S.$$

Patterns of violated and satisfied residuals (i.e., the other possible SV&PFA diagnostic rules) that would be expected when a given assumption variable deviated can be readily constructed for this instrumented process mixing tee. The following diagnostic rules were systematically created by the following MOME procedure:

First, the states of the violated residuals when particular significant assumption variable deviations occur are determined. This is accomplished by initially evaluating the sign of the first derivative of each model to each assumption variable. The formulas for these first derivatives (i.e., $\partial \varepsilon_i / \partial a_j$) are listed in the following table (Table 8.1).

For argument's sake, in this example assume the flow rate F1 is greater than flow rate F2 (i.e., F1 > F2) and temperature T1 is greater than T2 (i.e., T1 > T2).[12] These assumptions are arbitrary but are required in

---

[11]All 5 models are considered **relevant** in the following analysis for all 15 possible assumption variable deviations. The MOME methodology does its analysis only on relevant Primary and Secondary models. All Primary and Secondary models directly depending on a given assumption variable are considered relevant to that possible assumption variable deviation. So are all Secondary models formed by directly eliminating that given assumption variable. Likewise, so are any Primary models that are required to create Secondary models dependent upon the given assumption variable but that do not require that assumption variable themselves. The set of all other possible Primary and Secondary models used to describe normal operating behavior not included in the relevant set are consequently considered irrelevant in our analysis for deviations of that given modeling assumption variable.

[12]From the physics of this unit operation, these two assumptions directly mandate that F3 > F1 > F2 and T1 > T3 > T2.

Table 8.1 $\partial \varepsilon_i / \partial a_j$.

Model	P_1	P_2	S_1	S_2	S_3
Assumption variable					
a_1 (i.e., F1) sensor	+500	+900*T1	T1−T2	0	T1−T3
a_2 (i.e., F2) sensor	+500	+900*T2	0	T2−T1	T2−T3
a_3 (i.e., F3) sensor	−500	−900*T3	T2−T3	T1−T3	0
a_4 (i.e., L1) parameter	−500	−900*T1	T2−T1	0	T3−T1
a_5 (i.e., L2) parameter	−500	−900*T2	0	T1−T2	T3−T2
a_6 (i.e., L3) parameter	−500	−900*T3	T2−T3	T1−T3	0
a_7 (i.e., T1) sensor	0	+900*(F1−L1)	F1−L1	F3+L3− F2+L2	F1−L1
a_8 (i.e., T2) sensor	0	+900*(F2−L2)	F3+L3− F1+L1	F2−L2	F2−L2
a_9 (i.e., T3) sensor	0	−900*(F3+L3)	−F3−L3	−F3−L3	L2−F2− F1+L1

order to definitively evaluate signs of the above first derivatives. They do not affect the generality of the following discussion in any manner. These assumptions directly lead to the following table of the signs of these derivatives (i.e., **sign** $(\partial \varepsilon_i / \partial a_j)$).[13]

This determination of signs is required for the following reason. If the sign of this derivative is positive, the associated model residual will go high for significantly high assumption deviations of measured variables (sensors above) and go low for significantly high assumption deviations of unmeasured variables (parameters above). This residual will likewise go low and high, respectively, for the case of significantly low assumption deviations. Just the complete opposite is true if the sign of this derivative is negative. Furthermore, if the derivative evaluates to 0, the associated residual is unaffected by the indicated assumption variable deviations. To summarize this **assignment logic**:

[13]FALCONEER™ IV computes these signs with current real-time process data on each analysis cycle and adjusts its inferencing accordingly.

IF
sign($\partial \varepsilon_i / \partial a_j$)   AND a_j
IS   IS A   THEN

+	Sensor	$a_j^H \rightarrow \varepsilon_i^S \oplus \varepsilon_i^H$ & $a_j^L \rightarrow \varepsilon_i^S \oplus \varepsilon_i^L$
+	Parameter	$a_j^H \rightarrow \varepsilon_i^S \oplus \varepsilon_i^L$ & $a_j^L \rightarrow \varepsilon_i^S \oplus \varepsilon_i^H$
−	Sensor	$a_j^H \rightarrow \varepsilon_i^S \oplus \varepsilon_i^L$ & $a_j^L \rightarrow \varepsilon_i^S \oplus \varepsilon_i^H$
−	Parameter	$a_j^H \rightarrow \varepsilon_i^S \oplus \varepsilon_i^H$ & $a_j^L \rightarrow \varepsilon_i^S \oplus \varepsilon_i^L$
0	Sensor	$a_j^H \rightarrow \varepsilon_i^S$ & $a_j^L \rightarrow \varepsilon_i^S$
0	Parameter	$a_j^H \rightarrow \varepsilon_i^S$ & $a_j^L \rightarrow \varepsilon_i^S$

Before the diagnostic patterns can be created, just one more important issue needs to be examined. The parameter variables required to derive the Primary models are set to their normal and/or extreme values when those models are evaluated to generate their corresponding residuals. If it is truly an extreme value (in this example, leaks are assumed not to be occurring so their normal and one bounded extreme value would be 0.0), then the actual value of the parameter variable can deviate in one direction only (for a leak this would be only positive values if the system is pressurized and be only negative values if the system is under vacuum). For this example, the three possible leaks can deviate high and significant leaks would cause the two Primary models to be violated high (i.e., for sign($\partial \varepsilon_i / \partial a_j$) being—then $a_j^H \rightarrow \varepsilon_i^S \oplus \varepsilon_i^H$ and a_j^L does not occur and thus can be ignored in the following analysis).

As is obvious from this assignment logic, the various relationships are describing cause (assumption variable deviation) to effect (residual response). In fault diagnosis, we actually reason from observed effects to underlying causes, a reasoning mechanism known as abduction (Charniak and McDermott 1985). Succinctly, if A implies B, and B is true, we then conclude A is true. This is called causality and is not the same as logical implication.[14] Consequently, abduction is only **plausible inference**: there could be many other plausible explanations for B being true. A technique commonly used to reduce, as much as possible, the other plausible explanations is to make the closed world assumption (Genesereth and Nilsson 1987; Schoning 1994) (regarded as the most straightforward and simplest way to complete a theory). This reduces the possible set of all potentially plausible hypotheses to only those explicitly referred to in the causality logic.

---

[14]In logical implication, knowing A and knowing A implies B, we then know B. This rule of inference is usually referred to as a syllogism (Parsaye and Chignell 1988a).

Table 8.2   sign $(\partial \varepsilon_i / \partial a_j)$.

Model	P_1	P_2	S_1	S_2	S_3
Assumption variable					
a_1 (i.e., F1) sensor	+	+	+	0	+
a_2 (i.e., F2) sensor	+	+	0	−	−
a_3 (i.e., F3) sensor	−	−	−	+	0
a_4 (i.e., L1) parameter	−	−	−	0	−
a_5 (i.e., L2) parameter	−	−	0	+	+
a_6 (i.e., L3) parameter	−	−	−	+	0
a_7 (i.e., T1) sensor	0	+	+	+	+
a_8 (i.e., T2) sensor	0	+	+	+	+
a_9 (i.e., T3) sensor	0	−	−	−	−

The sign table (Table 8.2) along with the above assignment logic is the basis for performing pattern matching, an area widely studied in the field of artificial intelligence (Reggia et al. 1984). Pattern matching is a process of assigning values to unbound variables consistent with the constraints imposed by the relevant facts and rules (Parsaye and Chignell 1988b). The patterns useful for direct proactive process fault analysis are derived by the following set covering technique. Each model can fail either high or low or be satisfied. In single fault situations, the fact that a model is satisfied does not imply all its associated assumption deviations are definitely satisfied, only that the magnitude of the given assumption deviation may be insufficient to violate that model. The patterns of residual response identified by the above assignment logic can be superimposed on the model derivative sign table to create many potential patterns for identifying possible single fault situations (Table 8.3).

Unique model state patterns in the various rows in the above table constitute valid diagnostic rules that offer perfect resolution of their associated assumption variable deviations. A SV&PFA diagnostic rule may be found for each assumption variable deviation by choosing as many residual response deviations as possible from those appearing in that assumption variable deviation's row and adding satisfied responses for the other relevant residuals. If other assumption variable deviations are compatible with this pattern, then the associated diagnostic rule has to conclude a disjunction of all those assumption deviations, conjoined with the satisfied state of all the assumptions whose deviations are not compatible. When more than one assumption variable deviation is compatible with the pattern, the associated diagnosis has lower diagnostic resolution. Sometimes the pattern can be relaxed by replacing a residual response deviation with the disjunction of

Table 8.3 a_j **Deviation consequences.**

Model	P_1	P_2	S_1	S_2	S_3
Assumption deviation					
a_1^L (i.e., F1^L)	$P_1^S \oplus P_1^L$	$P_2^S \oplus P_2^L$	$S_1^S \oplus S_1^L$	S_2^S	$S_3^S \oplus S_3^L$
a_1^H (i.e., F1^H)	$P_1^S \oplus P_1^H$	$P_2^S \oplus P_2^H$	$S_1^S \oplus S_1^H$	S_2^S	$S_3^S \oplus S_3^H$
a_2^L (i.e., F2^L)	$P_1^S \oplus P_1^L$	$P_2^S \oplus P_2^L$	S_1^S	$S_2^S \oplus S_2^H$	$S_3^S \oplus S_3^H$
a_2^H (i.e., F2^H)	$P_1^S \oplus P_1^H$	$P_2^S \oplus P_2^H$	S_1^S	$S_2^S \oplus S_2^L$	$S_3^S \oplus S_3^L$
a_3^L (i.e., F3^L)	$P_1^S \oplus P_1^H$	$P_2^S \oplus P_2^H$	$S_1^S \oplus S_1^H$	$S_2^S \oplus S_2^L$	S_3^S
a_3^H (i.e., F3^H)	$P_1^S \oplus P_1^L$	$P_2^S \oplus P_2^L$	$S_1^S \oplus S_1^L$	$S_2^S \oplus S_2^H$	S_3^S
a_4^H (i.e., L1^H)	$P_1^S \oplus P_1^H$	$P_2^S \oplus P_2^H$	$S_1^S \oplus S_1^H$	S_2^S	$S_3^S \oplus S_3^H$
a_5^H (i.e., L2^H)	$P_1^S \oplus P_1^H$	$P_2^S \oplus P_2^H$	S_1^S	$S_2^S \oplus S_2^L$	$S_3^S \oplus S_3^L$
a_6^H (i.e., L3^H)	$P_1^S \oplus P_1^H$	$P_2^S \oplus P_2^H$	$S_1^S \oplus S_1^H$	$S_2^S \oplus S_2^L$	S_3^S
a_7^L (i.e., T1^L)	P_1^S	$P_2^S \oplus P_2^L$	$S_1^S \oplus S_1^L$	$S_2^S \oplus S_2^L$	$S_3^S \oplus S_3^L$
a_7^H (i.e., T1^H)	P_1^S	$P_2^S \oplus P_2^H$	$S_1^S \oplus S_1^H$	$S_2^S \oplus S_2^H$	$S_3^S \oplus S_3^H$
a_8^L (i.e., T2^L)	P_1^S	$P_2^S \oplus P_2^L$	$S_1^S \oplus S_1^L$	$S_2^S \oplus S_2^L$	$S_3^S \oplus S_3^L$
a_8^H (i.e., T2^H)	P_1^S	$P_2^S \oplus P_2^H$	$S_1^S \oplus S_1^H$	$S_2^S \oplus S_2^H$	$S_3^S \oplus S_3^H$
a_9^L (i.e., T3^L)	P_1^S	$P_2^S \oplus P_2^H$	$S_1^S \oplus S_1^H$	$S_2^S \oplus S_2^H$	$S_3^S \oplus S_3^H$
a_9^H (i.e., T3^H)	P_1^S	$P_2^S \oplus P_2^L$	$S_1^S \oplus S_1^L$	$S_2^S \oplus S_2^L$	$S_3^S \oplus S_3^L$

that deviation with that residual's satisfied response without changing the set of assumption variables the pattern is compatible with (i.e., diagnostic resolution is not diminished with this disjunction). Consequently, the following SV&PFA diagnostic rules can be directly derived from this table with this procedure to identify the various associated fault situations with the highest diagnostic resolution.

Sensor Validation and Proactive Fault Analysis Diagnostic Rules: At the highest possible levels of diagnostic resolution (i.e., discrimination possible between different potential faults).

$$P_1^L \wedge (P_2^L \oplus P_2^S) \wedge S_1^L \wedge S_2^S \wedge S_3^L$$
$$\rightarrow a_1^L \wedge a_2^S \wedge a_3^S \wedge a_4^S \wedge a_5^S \wedge a_6^S \wedge a_7^S \wedge a_8^S \wedge a_9^S.$$

$$P_1^L \wedge (P_2^L \oplus P_2^S) \wedge S_1^S \wedge S_2^H \wedge S_3^H$$
$$\rightarrow a_2^L \wedge a_1^S \wedge a_3^S \wedge a_4^S \wedge a_5^S \wedge a_6^S \wedge a_7^S \wedge a_8^S \wedge a_9^S.$$

$$P_1^S \wedge P_2^L \wedge S_1^S \wedge (S_2^S \oplus S_2^H) \wedge S_3^H$$
$$\rightarrow a_2^L \wedge a_1^S \wedge a_3^S \wedge a_4^S \wedge a_5^S \wedge a_6^S \wedge a_7^S \wedge a_8^S \wedge a_9^S.$$

$$P_1^L \wedge (P_2^L \oplus P_2^S) \wedge S_1^L \wedge S_2^H \wedge S_3^S$$
$$\rightarrow a_3^H \wedge a_1^S \wedge a_2^S \wedge a_4^S \wedge a_5^S \wedge a_6^S \wedge a_7^S \wedge a_8^S \wedge a_9^S.$$

$$P_1^S \wedge P_2^L \wedge S_1^L \wedge S_2^H \wedge S_3^S$$
$$\rightarrow a_3^H \wedge a_1^S \wedge a_2^S \wedge a_4^S \wedge a_5^S \wedge a_6^S \wedge a_7^S \wedge a_8^S \wedge a_9^S.$$

$$P_1^H \wedge (P_2^H \oplus P_2^S) \wedge S_1^H \wedge S_2^S \wedge S_3^H$$
$$\rightarrow (a_1^H \vee a_4^H) \wedge a_2^S \wedge a_3^S \wedge a_5^S \wedge a_6^S \wedge a_7^S \wedge a_8^S \wedge a_9^S.$$

$$P_1^H \wedge (P_2^H \oplus P_2^S) \wedge S_1^S \wedge S_2^L \wedge S_3^L$$
$$\rightarrow (a_2^H \vee a_5^H) \wedge a_1^S \wedge a_3^S \wedge a_4^S \wedge a_6^S \wedge a_7^S \wedge a_8^S \wedge a_9^S.$$

$$P_1^S \wedge P_2^H \wedge S_1^S \wedge (S_2^L \oplus S_2^S) \wedge S_3^L$$
$$\rightarrow (a_2^H \vee a_5^H) \wedge a_1^S \wedge a_3^S \wedge a_4^S \wedge a_6^S \wedge a_7^S \wedge a_8^S \wedge a_9^S.$$

$$P_1^H \wedge (P_2^H \oplus P_2^S) \wedge S_1^H \wedge S_2^L \wedge S_3^S$$
$$\rightarrow (a_3^L \vee a_6^H) \wedge a_1^S \wedge a_2^S \wedge a_4^S \wedge a_5^S \wedge a_7^S \wedge a_8^S \wedge a_9^S.$$

$$P_1^S \wedge P_2^H \wedge S_1^H \wedge S_2^L \wedge S_3^S$$
$$\rightarrow (a_3^L \vee a_6^H) \wedge a_1^S \wedge a_2^S \wedge a_4^S \wedge a_5^S \wedge a_7^S \wedge a_8^S \wedge a_9^S.$$

$$P_1^S \wedge P_2^L \wedge (S_1^L \oplus S_1^S) \wedge S_2^L \wedge (S_3^L \oplus S_3^S)$$
$$\rightarrow (a_7^L \vee a_8^L \vee a_9^H) \wedge a_1^S \wedge a_2^S \wedge a_3^S \wedge a_4^S \wedge a_5^S \wedge a_6^S.$$

$$P_1^S \wedge P_2^H \wedge (S_1^H \oplus S_1^S) \wedge S_2^H \wedge (S_3^H \oplus S_3^S)$$
$$\rightarrow (a_7^H \vee a_8^H \vee a_9^L) \wedge a_1^S \wedge a_2^S \wedge a_3^S \wedge a_4^S \wedge a_5^S \wedge a_6^S.$$

These complete patterns do not always develop if the magnitudes or rates of occurrence of the particular assumption variable deviations occur at levels that violate some but not all affected model residuals (i.e., as mentioned above, these theoretical minimum assumption deviations are normally different for each model). By relaxing these complete patterns further, new assumption variable deviations become compatible with them and have to be added as part of a disjunction of deviations in the conclusions of the rules. This leads to more ambiguous diagnoses (i.e., lower diagnostic resolution) as demonstrated by the following SV&PFA diagnostic rules[15]:

---

[15]The original FALCON System only included the SV&PFA diagnostic rules for diagnosing faults at the highest possible level of diagnostic resolution; FALCONEER and all subsequent FALCONEER™ IV applications contain SV&PFA diagnostic rules that handle all possible levels of diagnostic resolution.

Sensor Validation and Proactive Fault Analysis Diagnostic Rules: At lower possible levels of diagnostic resolution but higher levels of diagnostic sensitivity.

$$P_1^H \wedge (P_2^H \oplus P_2^S) \wedge S_1^H \wedge S_2^S \wedge S_3^S$$
$$\rightarrow (a_1^H \vee a_3^L \vee a_4^H \vee a_6^H) \wedge a_2^S \wedge a_5^S \wedge a_7^S \wedge a_8^S \wedge a_9^S.$$

$$P_1^H \wedge (P_2^H \oplus P_2^S) \wedge S_1^S \wedge S_2^L \wedge S_3^S$$
$$\rightarrow (a_2^H \vee a_3^L \vee a_5^H \vee a_6^H) \wedge a_1^S \wedge a_4^S \wedge a_7^S \wedge a_8^S \wedge a_9^S.$$

$$P_1^L \wedge (P_2^L \oplus P_2^S) \wedge S_1^L \wedge S_2^S \wedge S_3^S$$
$$\rightarrow (a_1^L \vee a_3^H) \wedge a_2^S \wedge a_4^S \wedge a_5^S \wedge a_6^S \wedge a_7^S \wedge a_8^S \wedge a_9^S.$$

$$P_1^L \wedge (P_2^L \oplus P_2^S) \wedge S_1^S \wedge S_2^H \wedge S_3^S$$
$$\rightarrow (a_2^L \vee a_3^H) \wedge a_1^S \wedge a_4^S \wedge a_5^S \wedge a_6^S \wedge a_7^S \wedge a_8^S \wedge a_9^S.$$

$$P_1^S \wedge P_2^H \wedge S_1^H \wedge S_2^S \wedge S_3^H$$
$$\rightarrow (a_1^H \vee a_4^H \vee a_7^H \vee a_8^H \vee a_9^L) \wedge a_2^S \wedge a_3^S \wedge a_5^S \wedge a_6^S.$$

$$P_1^S \wedge P_2^H \wedge S_1^S \wedge S_2^L \wedge S_3^S$$
$$\rightarrow (a_2^H \vee a_3^L \vee a_5^H \vee a_6^H) \wedge a_1^S \wedge a_4^S \wedge a_7^S \wedge a_8^S \wedge a_9^S.$$

$$P_1^S \wedge P_2^H \wedge S_1^H \wedge S_2^S \wedge S_3^S$$
$$\rightarrow (a_1^H \vee a_3^L \vee a_4^H \vee a_6^H \vee a_7^H \vee a_8^H \vee a_9^L) \wedge a_2^S \wedge a_5^S.$$

$$P_1^S \wedge P_2^H \wedge S_1^S \wedge S_2^S \wedge S_3^H$$
$$\rightarrow (a_1^H \vee a_4^H \vee a_7^H \vee a_8^H \vee a_9^L) \wedge a_2^S \wedge a_3^S \wedge a_5^S \wedge a_6^S.$$

$$P_1^S \wedge P_2^L \wedge S_1^L \wedge S_2^S \wedge S_3^L$$
$$\rightarrow (a_1^L \vee a_7^L \vee a_8^L \vee a_9^H) \wedge a_2^S \wedge a_3^S \wedge a_4^S \wedge a_5^S \wedge a_6^S.$$

$$P_1^S \wedge P_2^L \wedge S_1^S \wedge S_2^H \wedge S_3^S$$
$$\rightarrow (a_2^L \vee a_3^H) \wedge a_1^S \wedge a_4^S \wedge a_5^S \wedge a_6^S \wedge a_7^S \wedge a_8^S \wedge a_9^S.$$

$$P_1^S \wedge P_2^L \wedge S_1^L \wedge S_2^S \wedge S_3^S$$
$$\rightarrow (a_1^L \vee a_3^H \vee a_7^L \vee a_8^L \vee a_9^H) \wedge a_2^S \wedge a_4^S \wedge a_5^S \wedge a_6^S.$$

$$P_1^S \wedge P_2^L \wedge S_1^S \wedge S_2^S \wedge S_3^L$$
$$\rightarrow (a_1^L \vee a_7^L \vee a_8^L \vee a_9^H) \wedge a_2^S \wedge a_3^S \wedge a_4^S \wedge a_5^S \wedge a_6^S.$$

Early fault detection and diagnosis by MOME requires only that one Primary model be violated. This results in the most sensitive diagnosis possible but also in the lowest diagnostic resolution between fault hypotheses by the process fault analyzer. This directly trades diagnostic resolution for the most sensitive alert possible for this model-based analysis.

Sensor Validation and Proactive Fault Analysis Diagnostic Rules: At the lowest possible levels of diagnostic resolution but the highest possible levels of diagnostic sensitivity.

$$P_1^L \wedge (P_2^L \oplus P_2^S) \wedge S_1^S \wedge S_2^S \wedge S_3^S$$
$$\rightarrow (a_1^L \vee a_2^L \vee a_3^H) \wedge a_4^S \wedge a_5^S \wedge a_6^S \wedge a_7^S \wedge a_8^S \wedge a_9^S.$$

$$P_1^H \wedge (P_2^H \oplus P_2^S) \wedge S_1^S \wedge S_2^S \wedge S_3^S$$
$$\rightarrow (a_1^H \vee a_2^H \vee a_3^L \vee a_4^H \vee a_5^H \vee a_6^H) \wedge a_7^S \wedge a_8^S \wedge a_9^S.$$

$$P_1^S \wedge P_2^H \wedge S_1^S \wedge S_2^S \wedge S_3^S$$
$$\rightarrow (a_1^H \vee a_2^H \vee a_3^L \vee a_4^H \vee a_5^H \vee a_6^H \vee a_7^H \vee a_8^H \vee a_9^L).$$

$$P_1^S \wedge P_2^L \wedge S_1^S \wedge S_2^S \wedge S_3^S$$
$$\rightarrow (a_1^L \vee a_2^L \vee a_3^H \vee a_7^L \vee a_8^L \vee a_9^H) \wedge a_4^S \wedge a_5^S \wedge a_6^S.$$

As demonstrated in this example, perfect resolution between different process fault hypotheses is not always possible or is possible only at larger magnitudes or rates of occurrence of the specified fault (i.e., at a magnitude or rate sufficient enough to violate all so affected relevant Primary and Secondary model residuals). This is the classic tradeoff between timely process fault detection and correctly identifying the underlying fault(s). Trading lower diagnostic resolution for higher diagnostic sensitivity in the fashion demonstrated above allows the process fault analyzer to narrow down the potential process faults that could be currently occurring into a reasonable number of plausible explanations for the current process state. These plausible hypotheses can then be further checked out by the process operators to determine the actual fault present. This directly flags potential incipient fault situations sooner rather than waiting until those faults' magnitude or rate of occurrence are severe enough to allow unique classification (i.e., perfect resolution if possible). This is why the methodology is called the method of minimal evidence: all plausible fault situations are diagnosed whenever even only just one Primary model residual indicates abnormal process operation.

It is imperative to also emphasize that because of the extensive use of inclusive OR logic in these rules conclusions, whenever more than one fault hypothesis is announced, it could actually be that one of the others or all is the actual fault situation occurring in the target process system; that is, all possible hypotheses and all of their possible combinations are considered to be all equally plausible explanations of current abnormal process behavior. In those cases, additional reasoning (e.g., perhaps based on either Occam's Razor (Reggia et al. 1984) or so-called Meta-knowledge (Davis 1980)) is required to resolve such ambiguous situations. Fortunately, because the MOME diagnostic strategy cross checks the current status of the states of all possible relevant Primary and Secondary models' residuals for each potential assumption deviation, as many such inclusive OR logic ambiguities as possible are directly eliminated. A generalized SV&PFA Fuzzy logic diagnostic rule utilizing MOME as described here for single fault situations is currently implemented in FALCONEER™ IV. This generalized rule is described in detail below.

Finally, almost any other combination of violated and satisfied Primary and Secondary models not shown above would directly indicate that a multiple fault situation was occurring. There are 243 ($=3^5$) potentially unique SV&PFA diagnostic rules for this target process system. The above example shows only 49 of these possible patterns.[16] These 49 patterns and most of the remaining 194 possible patterns are likewise interpreted by another generalized SV&PFA Fuzzy logic diagnostic rule also currently implemented in FALCONEER™ IV that directly diagnoses multiple faults. That generalized diagnostic rule is also similarly described below.

It should now be highly evident that automating the creation of MOME's diagnostic logic is extremely advantageous. Even for a target process system as trivial as this mixing tee requires a substantial effort to analyze a mere handful (15 possible assumption variable deviations) of potential process faults. When there are scores of Primary models and hundreds of necessary assumption variables, there are literally tens of thousands of unique SV&PFA diagnostic rules required to competently analyze the current process operating state. MOME logically creates up to 3^N such unique SV&PFA diagnostic rules (where N is the number of relevant Primary and Secondary models for each fault situation). Furthermore, the SV&PFA diagnostic rules

---

[16]It is also possible to easily derive other patterns for diagnosing single fault situations in this example that do not rely upon either Primary model being violated, but these are currently suppressed in FALCONEER™ IV. We require at least one Primary model to be violated for any fault diagnosis. In actual applications this makes the results more meaningful although in some odd situations not as sensitive as theoretically possible.

derived in Example 8.2 are just for diagnosing single fault situations. Other examples of additional MOME diagnostic reasoning, including that required for diagnosing multiple fault situations and the effective application of Performance Equations[17] in fault diagnosis, are given elsewhere (Fickelscherer and Chester 2013). The required diagnostic logic to properly handle this expanded MOME reasoning is completely straightforward.

8.5 MOME SV&PFA DIAGNOSTIC RULES' LOGIC COMPILER MOTIVATIONS

A critical step in the development of a process fault analyzer is verifying that the underlying diagnostic knowledge base performs competently. Such verification ensures that the SV&PFA diagnostic rules will always perform correctly. During the development of the FALCON System, this was accomplished by nearly exhaustively testing that analyzer with both actual and simulated fault situations. Thoroughly doing this turned out to be the most computationally intensive undertaking of the FALCON Project, with more than 5500 h of actual plant data and about 260 simulated process fault situations being analyzed before the FALCON System was installed on-line in the actual plant control room. The various phases of this verification effort required a period of approximately 3 years to complete.

Obviously, if such extensive verification efforts were a prerequisite for developing competent process fault analyzers in general, these programs would never be widely used within the processing industries. Fortunately though, efforts similar to that expended in verifying the original FALCON System are not required.

It is possible to dramatically reduce the effort required to verify a process fault analyzer based upon MOME. If all of the Primary process models contained within the declarative knowledge base are well-formulated, then any subsequent misdiagnoses made by the fault analyzer directly indicates that one or more of the SV&PFA diagnostic rules is incorrect. However, since MOME is a very systematic and logically structured procedure for

---

[17]Performance Equations are quantitative engineering models which have one unknown parameter assumption variable, as opposed to Primary models which have no unknown parameter assumption variables. They are also commonly referred to as Soft-sensors. An example of a Performance Equation is the accurate calculation of an overall heat transfer coefficient. Its true value depends upon the unknown current variable level of fouling actually present in the associated heat exchanger. Performance Equations thus do not close to zero as Primary models always do so by definition. (See Eq. 8.1.)

creating diagnostic rules, it has been possible to codify a patented Fuzzy logic version (Chester et al. 2008) of it in a compiler program, called FALCONEER™ IV, which automatically creates diagnostic knowledge bases based upon it. This program ensures that the Method of Minimal Evidence is always applied correctly (Fickelscherer and Chester 2013; Fickelscherer et al. 2005). Concurrently to this development, a prototype SV&PFA compiler based on the alternative Model Assumption State Differences Algorithm version of MOME (Fickelscherer and Chester 2016) was created and tested in 2017.

Compilation refers to any process that transforms a representation of knowledge into another representation that can be used more efficiently. Such transformations can include optimization as well as the tailoring of representations for potential instruction sets (Stefik et al. 1983b). Compiling is thus a process of knowledge chunking; that is, meaningful portions of knowledge are stored and retrieved as functional units (Harmon and King 1985a). A compiler is said to create complete solutions if it is capable of producing every possible solution; it is said to be non-redundant if in the course of compiling it produces each solution only once (Stefik et al. 1983a). With the compiler, the most important criteria when choosing the knowledge representation scheme for automating process fault analysis becomes picking one that allows all the necessary knowledge be represented directly and facilitates its use in compiling the solution of the underlying problem (Rich 1983). Since the knowledge underlying our automated process fault analysis (i.e., Primary models and Performance Equations) is firm, fixed, and formulated, an algorithmic computer program is more appropriate than a heuristic one (Buchanan et al. 1983). Furthermore, since the problem of automated process fault analysis is characterized by a small search space, mostly reliable data and highly reliable knowledge (i.e., evaluated well-formulated Primary models and Performance Equations), it permits correspondingly simple system architectures; that is, systems of this sort can effectively employ exhaustive search (Hayes-Roth et al. 1983).

With these MOME-based SV&PFA compilers, if all of the Primary process models are indeed well-formulated, then the resulting fault analyzer is guaranteed to perform competently. Consequently, verifying the correctness of various diagnostic rules is not necessary: only the correctness of the set of Primary process models and Performance Equations used to create those rules needs to be verified. This in effect converts the much more difficult task of performing competent automated fault analysis into an easier problem of process modeling. The reduction in required development effort is very substantial.

Consequently, process engineers now only have to derive and maintain the declarative knowledge base containing the various well-formulated Primary models and Performance Equations of normal operation. This is advantageous for the following reason. As opposed to other knowledge representation schemes (such as production rules, frames, etc.), Primary models and Performance Equations are represented as mathematical equations, an engineer's dream. Furthermore, since all of the Primary models within the diagnostic knowledge base are linearly independent of each other, each can be independently added, modified, or removed as required. The altered declarative knowledge database is then merely recompiled to create the improved process fault analyzer. This simplifies the maintenance of the process fault analyzer immensely. With the MOME diagnostic logic compiler, the process fault analyzer can be incrementally improved with minimal effort and cost as the target process system's operating behavior becomes better understood or its topology is modified. This greatly improves these analyzers' maintainability as their associated target process systems relentlessly evolve.

Having FALCONEER™ IV also greatly increases the resulting **KBSs' transparency**[18] by separating the domain specific knowledge (i.e., the various Primary models and Performance Equations of normal operation, constituting the entirety of its declarative knowledge) from its general problem solving procedure (i.e., the MOME diagnostic strategy and exhaustive search of the resulting SV&PFA diagnostic rules). Again, this directly simplifies the much more complicated problem of performing competent automated process fault analysis into an effort of correctly modeling normal process behavior. FALCONEER™ IV thus allows anyone capable of doing such modeling the means of directly creating and maintaining both highly competent and affordable automated process fault analyzers.

8.6   MOME FUZZY LOGIC ALGORITHM OVERVIEW

This section describes the details of the patented Fuzzy logic algorithm currently implemented in our FALCONEER™ IV program for performing optimal automated process fault analysis. This implementation generalizes the underlying Boolean logic version of MOME described previously to a highly comprehensive Fuzzy logic algorithm (Chester et al. 2008). This generalization allows for a more compact treatment of each potential single and multiple fault situations, at all levels of possible diagnostic resolution. It does so with both elegant and efficient uniform Fuzzy logic SV&PFA diagnostic rules for diagnosing all significant occurrences of those

---

[18]Transparency is defined as the KBS's understandability to both that system's developer and targeted user (Stefik et al. 1983c).

situations. This Fuzzy logic algorithm of the MOME diagnostic strategy thus allows the automation of the diagnostic reasoning necessary to continuously perform optimal process fault analysis: only the underlying well-formulated Primary models and Performance Equations are required to achieve such performance. Using FALCONEER™ IV consequently directly simplifies the solution of the more complicated problem of automated process fault analysis into the much simpler problem of adequate modeling of the target process system's normal operating behavior.

The original FALCON system derived its conclusions exclusively via the use of Boolean logic. Residuals were determined to be high, low or satisfied by comparing them against predetermined high and low thresholds. Consequently, a residual could only be in a high, low or satisfied state at any given instant in time. The diagnostic rule formats described previously optimize the resulting fault analyzer's diagnostic knowledge base for fault analyzers based upon Boolean logic. However, Boolean logic places a practical limitation upon the process fault analyzer's competency. With Boolean logic, an incremental change in plant state can sometimes radically alter the process fault analyzer's conclusions. This occurs if that incremental change in plant state causes the values of one or more of the various models' residuals to deviate enough so as to change their Boolean assignment. Kramer (1987) refers to this problem as **diagnostic instability**. He further argues that it is an inherent problem in the performance of any process fault analyzer whose reasoning is based upon Boolean logic, regardless of the diagnostic strategy employed to create the SV&PFA diagnostic rules.

Rather than using Boolean logic, the version of MOME incorporated in FALCONEER™ IV uses the Fuzzy logic (Zadeh 1988) certainty factor calculation algorithm described below to make its judgments. This algorithm directly eliminates "diagnostic instability" and minimizes the possible ambiguity in the SV&PFA diagnostic rules. Since the resulting certainty factors are continuous functions, diagnostic instability does not occur. Furthermore, each possible fault hypothesis is diagnosed by considering all possible diagnostic evidence for it in just one diagnostic rule. Each SV&PFA diagnostic rule contains all the possible patterns of relevant Primary and Secondary model residual states (i.e., only those that are pertinent to that particular fault situation). Credence in the possible associated low, high or satisfied states of the various required modeling assumption variables are each assigned values from any and all relevant Primary and Secondary residual values. This directly optimizes the diagnostic sensitivity along with diagnostic resolution as was described for the various SV&PFA diagnostic rules shown in Example 8.2.

Performing automated process fault analysis according to the MOME diagnostic strategy is in effect solving what Newell defines as a well-structured problem (Davis and Lenat 1982b); i.e., it can be stated in terms of numerical variables (evaluated and interpreted Primary and Secondary models and Performance Equations), its goals can be specified in terms of a well-defined objective function (classifying a subset of potential process faults) and there exists an algorithmic solution (directly based on the MOME diagnostic strategy). Furthermore, since we strive to immediately recognize the implication of all current evidence and all plausible solutions are required, it is recommended that an exhaustive, forward chaining search be employed (Kline and Dolins 1985).

As implemented in our Fuzzy logic algorithm, each SV&PFA diagnostic rule encapsulates a chunk of domain specific information identifying a specific fault situation if the conditions specified in its premise are fulfilled. These rules are judgmental, that is they make inexact inferences. As such, every rule premise is a conjunction of clauses, containing arbitrarily complex conjunctions and disjunctions nested within each clause. The conclusions are drawn if the premises are satisfied, making the rules purely inferential. Each rule thus succinctly states a single independent chunk of knowledge and states all necessary information explicitly in the premise (Davis and Lenat 1982a). The need to combine certainties is thus limited to the single existence of a given SV&PFA diagnostic rule. This greatly simplifies the certainty calculations of the rules' conclusions.[19]

The only drawback to this approach is that evidence supporting a given fault hypothesis is combined directly with the evidence negating that hypothesis into a simple metric for overall support. This makes it difficult to directly identify which support is either lacking or conflicting the given fault hypothesis. In general, this drawback is normally only encountered if the KBS user wants an explanation of why particular fault hypotheses are not given.

The various possible clauses of our diagnostic rules are represented as Fuzzy sets. Fuzzy sets (Stefik 1995) are used to define concepts or categories that have inherent vagueness and degree. Fuzzy set theory is thus based on the idea of gradual membership in a set. A Fuzzy set is characterized by a membership function that relates a degree of membership on a scale of 0.0 to 1.0 to inherent properties of the objects. Fuzzy logic consequently is thus increasingly perceived as a way of handling continuous valued variables rather than uncertainty (Russell and Norvig 1995b).

---

[19]Operating in such a fashion allows the following three attributes to exist in the resulting certainty calculation (Russell and Norvig 1995a): (1) **locality**—each rule should be separate from all other rules; (2) **detachment**—once determined a certainty can be understood independently of how it was derived; and (3) **truth functionality**—the truth of complex sentences can be computed from the truth of their components.

A Fuzzy set proposition is a statement that asserts a value of a Fuzzy variable. The Fuzzy set proposition formulae allow conjunctions, disjunctions, and negation in the antecedents of the SV&PFA diagnostic rules. The relevant Certainty Factor (CF) formulae are (Zadeh 1988):

For conjunction: $CF(A \wedge B) = MIN(CF(A), CF(B))$.

For disjunction: $CF(A \vee B) = MAX(CF(A), CF(B))$.

For negation: $CF(\neg A)$ $= 1.0 - CF(A)$.

The intuitive appeal of these formulae is strong (Sell 1985b). The probability of two events cannot be better than that of the least likely and the probability of either of two events cannot be worse than the most likely. Furthermore, the formula for negation is the standard one used in probability theory; it is based on the observation that the probabilities of A and not A add up to 1.0.

8.6.1 MOME Fuzzy Logic Algorithm Details

As described, MOME is based on the evaluation of quantitative process models of normal operation. The pattern of their residuals that result from evaluating real-time values of process measurements is interpreted to determine any underlying modeling assumption variables that are currently deviating significantly. More generally, such a quantitative process model residual may be represented generically as follows:

$$\text{residual} = f\left(x_1, x_2, \ldots, x_n\right) \tag{8.6}$$

where x_1, x_2, ..., x_n are modeling assumption variables, that is, measured variables and unmeasured parameters that define the normal operating state of a process at any given moment, and f is a function of those variables that computes, for example, a balance of mass or energy in a control volume.

Because the sensors that measure process variables may not be 100% accurate or provide exact readings, because the process models may not be perfect models of the relationship between the assumption variables, and because random process operations' perturbations (e.g., normal process noise) may occur, it is empirically observed that the residuals are not always zero, though they are usually close to zero when the process is operating normally. The mathematical model describing normal process operation used by our program requires that all residuals be zero, on average, when a monitored target process system is operating normally. Therefore, a calculation is made from historical plant data of the average value of each Primary model residual (its *Beta*) and that *Beta* is subtracted from the corresponding process model residual calculation.

In practice, then, each function f, above, will behave like a statistical random variable having a mean value β and a standard deviation σ. The mean value $\beta = \beta_0 \rho$ and the standard deviation $\sigma = \sigma_0 \rho$, where β_0 and σ_0 are constants and ρ is either 1 or a process variable that is the definitive measure of the production level of the process being monitored, etc. Usually, β is just a constant value, but sometimes it is the product of a constant times a process variable whose value determines the level of production at which the process is currently operating, etc.

The generic process model residual above (Eq. 8.6) can be replaced with a Primary model residual r defined as:

$$r = f\left(x_1, x_2, \ldots, x_n\right) - \beta \tag{8.7}$$

which has a mean value of zero and a standard deviation of σ. The equation defining r is referred to herein as a Primary model residual. The program examines the values of all such (adjusted) process model residuals and, among other things, infers from the full pattern of real-time Primary and Secondary (defined below) residual deviations from zero which sensors or other parts of the process may be faulty.

In more generic terms, if a process engineer provides the formula $f(\cdots)$ as the formula for a process model residual under normal process operating conditions, and the formula *mean* for the average of $f(\cdots)$ over time based on historical process system data, and the formula *sigma* for the standard deviation of $f(\cdots)$ over time, the program generates the Primary model residual:

$$r = f\left(\cdots\right) - \text{mean} \tag{8.8}$$

which has the property that the average of r is expected to be zero. In reality, any formula that can be expressed in the mathematical language of FALCONEER™ IV is allowed. The formula *sigma* is not used in evaluating the process model residuals, but is used to calculate certainty factors of the residuals' possible three states (i.e., low, satisfied and high), as discussed below.

As discussed previously, Primary model residuals are distinguished from certain linearly dependent process model residuals automatically generated by our program. Such additional models are referred herein to as Secondary model residuals, and are computed as follows: Suppose that r_1 and r_2 are Primary model residuals and both contain a common assumption variable v. If both Primary models are linear functions of v, they may be combined algebraically to remove the terms containing v. An easier approach is to use Eq. (8.4) to directly determine the desired residual. The standard deviation for this Secondary model residual can also be directly computed from its two parent Primary models' standard deviations using Eq. (8.5).

With these standard deviations, certainty factors regarding the degree of each of the three possible model states can be directly calculated for all the various Primary and Secondary models' real-time residuals. Any number of different functions for computing certainty factors from residuals may be used. A commonly used function, the Gaussian function, is defined as:

$$Gauss(r, sigma) = \exp\left(-\left(r/sigma\right)^2/2\right) \tag{8.9}$$

where *sigma* is the standard deviation of the model residual *r*.

When this program monitors a given target process system, it reads real-time sensor data, computes the associated relevant Primary and Secondary model residual values and their standard deviations, and then calculates three certainty factors, one for each possible residual state, for each residual value, as needed. Let *r* be one of the residual values and let *sigma* be its standard deviation. Residual *r* is expected to be zero, but often it is not. If it is only a little off from zero, the program considers it to be satisfactory, but the farther away from zero it gets, there exists less confidence that it is satisfactory. Using the Gaussian function for example, the certainty factor for *r* being satisfactory is calculated as follows:

$$cf(r, sat) = Gauss(r, sigma) \tag{8.10}$$

If *r* is much greater than zero, it is considered to be high, that is, higher than it is supposed to be. The certainty factor for *r* being high is represented as[20]:

$$cf(r, high) = 1.0 - Gauss(r, sigma) \text{ if } r > 0.0 \text{ } 0.0 \text{ otherwise} \tag{8.11}$$

Similarly, if *r* is much less than zero, it is considered to be low. The certainty factor for *r* being low is represented as:

$$cf(r, low) = 1.0 - Gauss(r, sigma) \text{ if } r < 0.0 \text{ } 0.0 \text{ otherwise} \tag{8.12}$$

Use of these three equations as shown converts all the Primary and Secondary models' residuals into three separate certainty factors, one for each potential model state (low, satisfied, high). These three certainties can alternatively be determined from any number of other possible continuous functions.[21] FALCONEER™ IV allows users to use one other

---

[20]Eq. (8.11) is plotted in Figure 8.4.

[21]In actual process monitoring, certainty factors based on the Gaussian distribution, even with residual filtering, typically lead to poor process fault analyzer performance results when analyzing even moderately noisy sensor data.

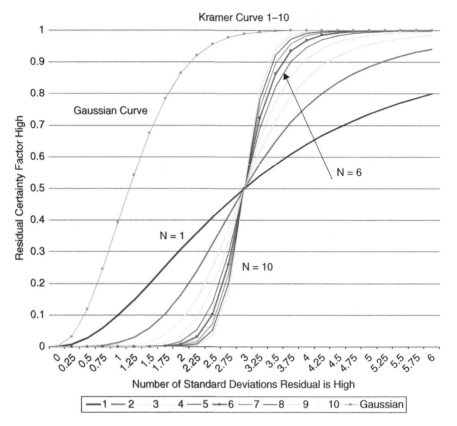

Figure 8.4   Gaussian and Kramer curve certainty factor high distributions.

such continuous function recommended by Kramer (1987). It is computed with the following formula:

$$Kramer(r, sigma) = 1.0 / \left(1.0 + \left(r / \left(3 * sigma\right)\right)\right)^{N} \qquad (8.13)$$

On the moderately to extremely noisy data normally encountered in most actual applications of FALCONEER™ IV, we recommend using Kramer's distribution[22] with an $N = 6$. (There is no unique or agreed upon method for dealing with uncertainties in knowledge-based systems (Sell 1985a)). Fortunately, as will be demonstrated, the details of the certainty factor calculations associated with the SV&PFA diagnostic rules matter less than the semantic and structural content of those rules themselves (Gashnig et al. 1983)).

---

[22]The Kramer curve formulas for various values of N using the high residual case are also plotted in Figure 8.4.

8.6.2 Single Fault Fuzzy Logic SV&PFA Diagnostic Rule

This implementation of the MOME diagnostic strategy computes certainty factors to identify faults and/or validate underlying modeling assumption variables. By definition, a "fault" is a pair consisting of an assumption variable v and a deviation direction d, and is designated as $\langle v, d \rangle$. The value of d can be either *high* or *low*, which are in turn defined by

$$high = 1 \qquad (8.14)$$

$$low = -1 \qquad (8.15)$$

For purposes of calculating CFs for faults, $\langle v, d \rangle$ signifies a fault; that is, v is an assumption variable and d is a direction, either *high* or *low*. To compute the certainty factor that a fault is present, the three certainty factors for all the relevant Primary and Secondary model possible states are examined to find evidence for the fault. If r is a residual, r provides evidence for fault $\langle v, d \rangle$ when it has deviated from zero in a direction that is consistent with variable v deviating in the direction d. For example, if $\partial r/\partial v$ is greater than zero, then both v and r can be expected to go high (or low) at the same time. If, however, $\partial r/\partial v$ is less than zero, then v and r can deviate in opposite directions. The certainty factor for r in the appropriate direction is then the strength to which r can provide evidence for the fault. One strong piece of evidence for the fault is enough to strongly conclude that the fault is present, unless there is also strong evidence that it is not present.

The evidence for fault $\langle v, d \rangle$ is this set of certainty factors for all relevant Primary model residuals:

$$evidence\text{-}for\text{-}fault\left(\langle v,d \rangle\right) = \left\{ \mathrm{cf}\left(r, sign\left(\partial r/\partial v\right)d\right) \mid \left(\partial r/\partial v\right)\right.$$
$$\left. \neq 0 \text{ and } r \text{ is a Primary residual} \right\} \qquad (8.16)$$

In some applications, the certainty factor for any such possible model residual, whether Primary or Secondary, may be included in this set.[23] The strength of the evidence for the fault is the maximum of the values in this set.

Similarly, if a residual deviates in the opposite direction from what is expected when the fault is present, that deviation is evidence against the fault being present. The evidence against fault $\langle v, d \rangle$ is this set of certainty factors for all relevant residuals:

$$evidence\text{-}against\text{-}fault\left(\langle v, d \rangle\right) = \left\{ \mathrm{cf}\left(r, -sign\left(\partial r/\partial v\right)d\right) \mid \right.$$
$$\left. \left(\partial r/\partial v\right) \neq 0 \right\} \qquad (8.17)$$

---

[23]By design, as discussed above, FALCONEER™ IV currently only relies upon Primary model residuals for calculating evidence-for-fault.

Certainty factors for both Primary and Secondary model residuals are included in this set. The strength of the evidence against the fault is the maximum of the values in this set. If that value is subtracted from one, the strength to which this evidence is consistent with the fault being present is determined.

An additional consideration is significant in evaluating a certainty factor for a fault. Some residuals are not functions of v and so are not expected to deviate from zero when the fault $\langle v, d \rangle$ is present. The Secondary model residual that was formed by eliminating v from its two parent Primary models generates such a residual. It is relevant to evaluating the presence of the fault, so this Secondary model residual is expected to have a high certainty factor of being satisfactory when the fault involves v. Also, if two Primary model residuals were combined to generate a Secondary model residual by eliminating some variable other than v, and one of these Primary model residuals is a function of v but the other is not, it is expected that the Primary model residual that is not a function of v is satisfactory. This Primary model residual is thus considered relevant to the fault as well. In addition, r may be a function of v, but at the moment, $(\partial r/\partial v) = 0$, making its satisfaction currently relevant. The neutral-evidence for fault $\langle v, d \rangle$ is this set of certainty factors for all these possible relevant residuals:

$$neutral\text{-}evidence\big(\langle v, d \rangle\big) = \big\{ \mathrm{cf}\big(r, sat\big) \big| r \text{ is relevant as } neutral$$
$$\text{-} evidence \text{ for } v \big\} \tag{8.18}$$

The strength of this evidence is the minimum of the set because if any one of the residuals that are supposed to be satisfactory is in fact high or low, that weakens the evidence for the fault $\langle v, d \rangle$.

Some Primary model residuals may not be functions of v and are not combined with any models that are. These and all possible Secondary model residuals formed by combining pairs of them to eliminate all other linear assumption variables are considered to be not relevant to the fault $\langle v, d \rangle$. Another fault can be present and cause them to deviate from zero, but this will not affect the assessment for fault $\langle v, d \rangle$. This allows a diagnosis of the presence of several single faults that happen not to interact with each other (i.e., a **multiple fault diagnosis**[24]).

---

[24]Such situations are referred to as **non-interactive multiple fault situations** (Fickelscherer 1990). These situations are defined as those for which the intersection of the sets of relevant Primary and Secondary models required to diagnose their member single fault situations results in the empty set. These multiple fault diagnoses are directly possible with Eq. (8.19) because independent inference pathways exist through the SV&PFA diagnostic rules based on MOME. This is an extremely important artifact of this methodology because combinations of fault and/or non-fault situations generating process operating events frequently occur and must be properly handled by the fault analyzer for it to be considered robust in actual applications.

The certainty factors in these three sets, *evidence-for-fault, neutral-evidence*, and *evidence-against-fault*, are considered as Fuzzy logic values (Zadeh 1988), and are combined using a common interpretation of fuzzy "AND" as the minimum function, fuzzy "OR" as the maximum function, and fuzzy "NOT" as the complement function (1.0 minus the value of its argument). For finite sets, the quantifier "SOME" is just the "OR" of the values in the set, so it is equivalent to taking the maximum of the set. Similarly, for finite sets, the quantifier "ALL" is just the "AND" of all the values in the set, so it is equivalent to taking the minimum of the set. Putting this all together, the certainty factor for fault $\langle v, d \rangle$ is defined in this Fuzzy logic implementation as the following single fault SV&PFA diagnostic rule:

$$\mathrm{cf}\left(\langle v, d \rangle\right) = SOME\left(evidence\text{-}for\text{-}fault\left(\langle v, d \rangle\right)\right) AND$$

$$ALL\left(neutral\text{-}evidence\left(\langle v, d \rangle\right)\right) AND$$

$$NOT\left(SOME\left(evidence\text{-}against\text{-}fault\left(\langle v, d \rangle\right)\right)\right) \quad (8.19)$$

If this value is above the alert threshold, the corresponding fault is displayed as a plausible fault hypothesis to the process operators.

Example 8.3   Example of a Fuzzy logic SV&PFA single fault diagnostic rule. Consider once again the mixing tee shown in Figure 8.1 and modeled in Examples 8.1 and 8.2. The following is our Fuzzy logic SV&PFA diagnostic rule for diagnosing flow sensor F2 failing low.
Equation (8.16) would equal:

$$evidence\text{-}for\text{-}fault\left(F2, low\right) = \left\{\mathrm{cf}\left(\varepsilon_{P1}{}^{L}\right), \mathrm{cf}\left(\varepsilon_{P2}{}^{L}\right)\right\}.$$

Equation (8.18) would equal:

$$\text{neutral-evidence}\left(F2, low\right) = \left\{\mathrm{cf}\left(\varepsilon_{S1}{}^{S}\right)\right\}.$$

Equation (8.17) would equal:

$$\text{evidence-against-fault}\left(F2, low\right)$$
$$= \left\{\mathrm{cf}\left(\varepsilon_{P1}{}^{H}\right), \mathrm{cf}\left(\varepsilon_{P2}{}^{H}\right), \mathrm{cf}\left(\varepsilon_{S2}{}^{L}\right), \mathrm{cf}\left(\varepsilon_{S3}{}^{L}\right)\right\}.$$

Equation (8.19) would thus equal:

$$\mathrm{cf}\left(F2, low\right) = MIN\left(MAX\left(\left\{\mathrm{cf}\left(\varepsilon_{P1}{}^{L}\right), \mathrm{cf}\left(\varepsilon_{P2}{}^{L}\right)\right\}\right), MIN\left(\left\{\mathrm{cf}\left(\varepsilon_{S1}{}^{S}\right)\right\}\right),$$
$$\left(1.0 - \left(MAX\left(\left\{\mathrm{cf}\left(\varepsilon_{P1}{}^{H}\right), \mathrm{cf}\left(\varepsilon_{P2}{}^{H}\right), \mathrm{cf}\left(\varepsilon_{S2}{}^{L}\right), \mathrm{cf}\left(\varepsilon_{S3}{}^{L}\right)\right\}\right)\right)\right)\right).$$

This single fault SV&PFA diagnostic rule can diagnose this fault at any significant level of that assumption deviation directly with the best diagnostic resolution possible for that level. There is one such rule corresponding to each 15 potential single fault situations for this target process system. More than one such rule can fire at any given time, reducing the possible diagnostic resolution of the diagnosis.

Regarding the display of a fault $\langle v, d \rangle$: If v is a sensor variable, its current real-time value for that sensor reading is substituted for the variable in computing all the Primary and Secondary model residual values. If $d = high$, a conclusion is drawn that the sensor reading is higher than the true value for that process variable. If $d = low$, a conclusion is drawn that the sensor reading is lower than the true value for that process variable. In either case, a conclusion is drawn that the sensor reading is faulty. If $cf(\langle v, d \rangle)$ is about zero for both cases, $d = high$ and $d = low$, then the sensor reading has been validated (i.e., the sensor is correctly measuring its associated process variable).

In the case of an unmeasured variable v (i.e., a parameter variable), such as a process leak, a high certainty factor for $\langle v, low \rangle$ means that the assumed value of v, which can be viewed as the reading from a virtual sensor, is low compared to the true value. In order to display a conclusion about the true value of the unmeasured variable, the program displays a message that v is high in this case (i.e., there is a positive leak out of the process). Similarly, if the certainty factor for $\langle v, high \rangle$ is high, it displays a message about v being low. If neither of these cases apply, a conclusion is drawn that the true value of v is about equal to its assumed normal and/or extreme value (i.e., the associated fault is not present and the parameter's true value is currently indeed its normal and/or extreme value).

The certainty factor calculation of $\langle v, d \rangle$ described above can be directly used to calculate certainty factors of satisfied assumption variables. The original FMC FALCONEER KBS actually contained explicit sensor validation diagnostic rules like those shown in the Example 8.2; FALCONEER™ IV KBS applications do not. Instead, they rely upon the following logically equivalent reasoning to make the identical conclusions.

In the case of the Boolean logic representation presented above, the following identity[25] always holds:

$$\left(a_i^L \oplus a_i^S \oplus a_i^H \right)$$

---

[25]Similarly, these two identities $(P_i^L \oplus P_i^S \oplus P_i^H)$ and $(S_i^L \oplus S_i^S \oplus S_i^H)$ also always hold. They have significance when logically specifying SV&PFA diagnostic rules for particular multiple fault situations. See Example 8.4.

This directly leads to the following statement:

$$a_i^S \leftrightarrow \left(\neg a_i^L \wedge \neg a_i^H\right)$$

Consequently, this equivalence statement for SV&PFA diagnostic reasoning about a given assumption variable's satisfaction can be written as the following Fuzzy logic calculation:

$$cf(v, sat) = MIN\left(\left(1.0 - cf(v, low)\right), \left(1.0 - cf(v, high)\right)\right) \qquad (8.20)$$

8.6.3 Multiple Fault Fuzzy Logic SV&PFA Diagnostic Rule

In the vast majority of actual process systems, multiple fault situations normally occur very much less frequently than do single fault situations. Nevertheless, the ability to diagnose multiple fault situations is important because: (1) some major process disasters have occurred as a consequence of a series of two or more concurrent process fault situations, and (2) the patterns of diagnostic evidence (i.e., the SV&PFA diagnostic rules) used to diagnose single fault situations may sometimes misdiagnose multiple fault situations. The latter reason is equally valid for SV&PFA diagnostic rules used by human troubleshooters, and in fact represents a contributing factor in some major process disasters (Kletz 1985). Furthermore, certain types of multiple fault situations may occur much more frequently than others. Kramer (1987) describes three classes of such multiple faults as follows: (1) faults causing other faults (so-called induced failures), (2) latent faults which are not detectable until additional faults occur, and (3) intentional operation in the presence of one or more faults, with the sudden occurrence of an additional fault.

The Fuzzy logic SV&PFA diagnostic rules described above (i.e., Eqs. 8.19 and 8.20) were derived to directly detect and distinguish the single cause of process operating events. This single cause may be either a fault or a non-fault situation. Thus, the methodology described above is inherently a strategy for diagnosing any and all single assumption variable deviations that cause detectable operating events, not just fault situations per se.

This is an important distinction because not all of the potential assumption variables required in the Primary models and Performance Equations for describing the normal operation of a typical process system correspond to the absence of actual process fault situations. For instance, unsteady state operation, unusually low production rates, changeover of a feed supply, normal process shutdown, etc. are all examples of assumption variable deviations which may generate diagnostic evidence. Moreover, such resulting process operating events may occur very frequently, much more so than those caused by the vast majority of single fault situations alone, and in many instances such non-fault disturbances may accompany single fault situations. Consequently, if the process fault analyzer does not properly account for the potential occurrences of multiple assumption variable deviations, the correctness of its diagnoses would always be questionable. Even if a given process fault analyzer cannot directly diagnose all such multiple assumption variable deviations (i.e., all possible combinations of process fault and non-fault situations), it still needs to ensure that its SV&PFA diagnostic rules will not misdiagnose those situations by excluding the correct fault hypothesis when making other plausible hypotheses. This is an extension of the conservative philosophy underlying MOME to the handling of multiple assumption variable deviations. It must always be remembered that all fault hypotheses generated by the process fault analyzer, whether single and/or multiple, are all equally plausible explanations of the current abnormal process operating behavior.

Throughout the following discussion, the term multiple fault situation will actually refer to any situation of multiple assumption variable deviations, whether or not all of those deviations correspond to associated actual process fault situations. This will not affect the generality of this discussion at all.

The Fuzzy logic SV&PFA diagnostic rule defined above (Eq. 8.19) may be generalized to sets of faults by redefining what counts as evidence for the set, evidence against the set, and what counts as neutral-evidence. An inference may be drawn that a set of faults is present when no subset of them may be inferred to be present. This means there must be at least one appropriate Primary model state value for each member fault in the multiple fault set deviating in the direction that those particular individual faults can cause.[26] This leads to the following generalized fuzzy multiple fault SV&PFA diagnostic rule of this methodology:

[26]These situations are referred to as **interactive multiple fault situations** (Fickelscherer 1990). These situations are defined as those for which the intersection of the sets of relevant Primary and Secondary models required to diagnose their member single fault situations does not result in the empty set. There are three different types of such situations further identified as being **fully discernible, partially discernible**, and **indiscernible**.

Fully discernible interactive multiple fault situations are those diagnosed directly by Eq. (8.22) and/or by lower resolution single fault situations diagnosed by Eq. (8.19). Furthermore, if identified by Eq. (8.22), it is possible that Eq. (8.19) may also identify other erroneous but equally plausible single fault hypotheses.

Indiscernible interactive multiple faults are those which cannot be diagnosed by either Eq. (8.19) or Eq. (8.22). This occurs because there is not at least one Primary model state

Let

$$fault - set = \left\{ \langle v_1, d_1 \rangle, \ldots, \langle v_n, d_n \rangle \right\} \qquad (8.21)$$

Then

$$\mathrm{cf}\left(fault\text{-}set \right) = SOME\left(evidence\text{-}for\text{-}fault\left(\langle v_1, d_1 \rangle \right) \right) AND$$

$$\cdot$$
$$\cdot$$
$$\cdot$$

$$SOME\left(evidence\text{-}for\text{-}fault\left(\langle v_1, d_1 \rangle \right) \right) AND$$
$$ALL\left(neutral\text{-}evidence\left(fault\text{-}set \right) \right) AND$$
$$NOT\left(SOME\left(evidence\text{-}against\left(fault\text{-}set \right) \right) \right) \qquad (8.22)$$

value that uniquely depends only upon each of the member assumptions violated by that multiple fault situation. The fault analyzer would report no fault diagnoses emerging from either Eqs. (8.19) and (8.22) in this situation. This is considered acceptable performance by it because to correctly classify a fault situation the fault analyzer must either get the correct diagnosis or remain silent. This behavior by the fault analyzer is considered conservative because it is only venturing diagnoses it is highly certain of. Remaining silent about the actual multiple fault situation in these cases thus does no harm, as long as the underlying abnormal process operating behavior is independently noticed and acted upon by the process operators in a timely fashion. Their training largely focuses on properly performing such diligent process monitoring. Consequently, such misdiagnoses are considered acceptable performance by the process fault analyzer.

Partially discernible interactive multiple fault situations are those that are not diagnosed directly by Eq. (8.22) but Eq. (8.19) identifies one or more plausible single fault situation hypotheses. Although they may or may not be partially incorrect, these single fault diagnoses are still all considered equally plausible explanations of the current process state. These types of misdiagnoses however represent a major potential problem because the unannounced actual fault situations could be extremely hazardous, requiring immediate corrective actions be taken by the process operators. It is also possible that the choice of the appropriate corrective actions will depend upon whether or not any of the unannounced fault situations are actually currently occurring. This is also true about indiscernible interactive multiple fault situations but with the following caveat: if the process fault analyzer only announces an incorrect (although completely logically plausible) single fault diagnosis, the process operators may be led to believe that the fault being announced is the only one that could be possibly occurring. This may inadvertently cause them to take inappropriate corrective actions, subsequently causing the actual fault situation to become more dangerous. This is the one Achille's heal of the MOME diagnostic strategy. It is something to always be aware of when using the program in actual process applications.

The component evidence sets are defined as follows:

- The set *evidence-for-fault*($\langle v_i, d_i \rangle$) is the set of cf(r, sign($\partial r/\partial v_i$) d_i) such that r is a Primary residual and ($\partial r/\partial v_i$) $\neq$ 0. It is the same set that was used for single faults.
- The set *neutral-evidence*(*fault-set*) is the set of cf(r, sat) such that r is relevant to one or more of the faults in *fault-set* as neutral-evidence and such that ($\partial r/\partial v$) = 0 for all the assumption variables v in *fault-set*.
- The set *evidence-against*(*fault-set*) is the set of cf(r, $-$sign($\partial r/\partial v$) d) such that ($\partial r/\partial v$) $\neq$ 0 for at least one variable v in *fault-set* provided that $-$sign($\partial r/\partial v$)d has the same value for all such assumption variables v in *fault-set*. (In other words, all the faults that can influence r must influence it in the same direction; if two faults can influence r to deviate in opposite directions, we can learn nothing about *fault-set* from that residual.)[27]

Example 8.4 Example of a Fuzzy logic SV&PFA multiple fault diagnostic rule.

Consider once more the mixing tee shown in Figure 8.1 and modeled in Examples 8.1 and 8.2. The following is the Fuzzy logic SV&PFA diagnostic rule for diagnosing both flow sensor F2 failing low and thermocouple T3 failing high.

Equation (8.16) would equal for these two faults:

$$evidence\text{-}for\text{-}fault\left(F2, low\right) = \left\{\text{cf}\left(\varepsilon_{P1}{}^L\right), \text{cf}\left(\varepsilon_{P2}{}^L\right)\right\}$$

$$evidence\text{-}for\text{-}fault\left(T3, high\right) = \left\{\text{cf}\left(\varepsilon_{P2}{}^L\right)\right\}$$

Adjusted Eq. (8.18) would equal:

$$neutral\text{-}evidence\left(F2, low; T3, high\right) = \left\{empty\right\}$$

Adjusted Eq. (8.17) would equal:

$$evidence\text{-}against\left(F2, low; T3, high\right) = \left\{\text{cf}\left(\varepsilon_{P1}{}^H\right), \text{cf}\left(\varepsilon_{P2}{}^H\right), \text{cf}\left(\varepsilon_{S1}{}^H\right)\right\}.$$

---

[27]Consequently, because the particular multiple fault situation influences the given Primary or Secondary model both high and low, that model can be in any of its three possible states whenever that multiple fault situation occurs, the actual state depending upon the actual relative magnitudes or rates of occurrence of the underlying member faults. Fortunately, the following equivalence is always a true statement: (A ∧ B) ↔ A—if and only if B is valid, as in this case; both ($P_i^L \oplus P_i^S \oplus P_i^H$) and ($S_i^L \oplus S_i^S \oplus S_i^H$) are always valid statements. Such validities can thus be left out of the SV&PFA multiple fault diagnostic rules.

Equation (8.22) would thus equal:

$$cf\left(F2,low;T3,high\right) = MIN\left(\left(MAX\left(\left\{cf\left(\varepsilon_{P1}{}^{L}\right),cf\left(\varepsilon_{P2}{}^{L}\right)\right\}\right)\right),\right.$$

$$\left(MAX\left(\left\{cf\left(\varepsilon_{P2}{}^{L}\right)\right\}\right)\right)$$

$$\left(MIN\left(\left\{1.0\right\}\right)\right),\left(1.0 - \left(MAX\left(\left\{cf\left(\varepsilon_{P1}{}^{H}\right),cf\left(\varepsilon_{P2}{}^{H}\right),cf\left(\varepsilon_{S1}{}^{H}\right)\right\}\right)\right)\right)$$

This multiple fault SV&PFA diagnostic rule can diagnose these two faults at any significant levels of those assumption deviations directly with the best diagnostic resolution possible for those levels. There is one such rule corresponding to each potentially realistic interactive multiple fault situation for this target process system.

The generalized rule for multiple fault situations (Eq. 8.22) is currently limited in FALCONEER™ IV to compute only certainty factors for potential pairs of interactive multiple fault situations. Typically, the diagnostic resolution by the process fault analyzer of such situations is almost always considerably much lower than the typical diagnostic resolution of potential single fault situations: there are normally almost always many more possible combinations of potential multiple fault effects on the current process operating state to consider.

8.7 SUMMARY OF THE MOME DIAGNOSTIC STRATEGY

The chief advantage of the MOME diagnostic strategy is that it provides a uniform framework for examining quantitative process models and their associated required modeling assumption variables. In this framework, each Primary model represents the relationship that exists between these modeling assumption variables during normal process operation. Any significant residuals resulting from the evaluation of these models with real-time sensor data directly indicate that one or more of the associated suspected modeling assumption variables are currently deviating significantly. The MOME diagnostic strategy directly permits the derivation of standardized SV&PFA diagnostic rule formats to directly identify these deviations, thus allowing the process fault analyzer development to be completely systematic once all the possible Primary models and Performance Equations are derived and deemed well-formulated.

As discussed, the key logical feature of the MOME diagnostic strategy centers about the SV&PFA diagnostic rule formats it uses. These formats have been chosen to ensure that the diagnostic knowledge bases will always perform competently; that is, it will only make all plausible fault diagnoses or not any diagnoses. These formats also ensure that the resulting diagnostic knowledge bases can diagnose those faults with the best diagnostic sensitivity and diagnostic resolution possible for the current magnitude of the current process operating event. Each diagnostic rule requires only the minimal amount of diagnostic evidence necessary to uniquely discriminate its associated fault situation(s) from all of the other implausible modeling assumption deviations. Requiring just the minimal amount of diagnostic evidence in each SV&PFA diagnostic rule in order to diagnose a given fault in this manner also allows many possible multiple fault situations to be diagnosed directly. This diagnostic methodology can also be directly used to determine the strategic sensor placement to better perform process fault analysis, improving diagnostic sensitivity and/or resolution for diagnosing particular process operating problems. It can also further be used to shrewdly distribute independent process fault analyzers throughout a large, highly integrated processing plant complex.

8.8 ACTUAL PROCESS SYSTEM KBS APPLICATION PERFORMANCE RESULTS

The development of the original FMC **FALCONEER System** (Fickelscherer et al. 2003; Skotte et al. 2001) proved to be a systematic application of MOME to their Electrolytic Sodium Persulfate (ESP) process. Developing the 35 Primary models and 5 Performance Equations that describe normal process operation, evaluating them with sufficient process data to determine their normal variances and *Betas*, and then hand compiling the 15,000+ SV&PFA diagnostic rules consumed the vast majority of the development effort. It took an order of magnitude less development time and effort to create this program compared to the original FALCON Project program (1 person year versus approximately 15 person years) although the FMC ESP process (Figure 8.3) was more than twice the scope of DuPont's adipic acid process (Figure 8.2A and B).

Even this impressive improvement in the required development effort for the original FALCONEER System was still an order or two of magnitude higher in time requirements than those now necessary. This development time has been remarkably shortened by the automation of MOME in **FALCONEER™ IV** (Fickelscherer and Chester 2013; Fickelscherer et al. 2005). Specifically, the entire development effort for FMC's Liquid Ammonium Persulfate process required deriving and evaluating the required statistics of about 30 Primary models and 5 Performance Equations (requiring about 2 weeks effort). All the requisite diagnostic logic required to evaluate our two Fuzzy logic diagnostic rules is now directly derived from the underlying quantitative models of normal process operation. Consequently, all of the development and maintenance effort required to create such process fault analyzers can be directed at the derivation of these models of normal process operation (the declarative knowledge) in the configuration database. This constitutes the data structure input into FALCONEER™ IV's compiler. Since the set of all the Primary models are linearly independent of each other, they can be added, improved or deleted as need be and then the entire application recompiled to include these changes. Consequently, with this compiler, the process fault analyzer can now be incrementally improved with minimal effort as the target process system's operating behavior becomes better understood or the target process system topology changes, allowing the process fault analyzer to easily evolve along with its associated target process system. This in turn substantially reduces both the development and maintenance effort and subsequent overall costs of such programs. It effectively converts the much more difficult problem of automated process fault analysis into the much simpler and more directly tractable (and incrementally solvable) problem of process modeling.

To date, FALCONEER™ IV has competently monitored a diverse assortment of highly complex target process systems for possible process fault situations and other non-fault causes of process operating events as a diligent and relentless watchdog; that is, it has exhibited high utility. A summary of the direct real-world benefits derived from such continuous, timely process fault analysis was independently reported by FMC (Lymburner et al. 2006).

8.9 CONCLUSIONS

Performing automated process fault analysis as detailed in this chapter requires the ability to create the underlying well-formulated quantitative models describing normal process behavior. Where this is possible, directly using these models to then perform SV&PFA in real-time proves the assertion that "Models are the means by which data can be converted to meaningful information" (Kramer and Mah 1994). These models are based on the most fundamental understanding of normal operating behavior of the given process system. This fundamental knowledge thus constitutes an unimpeachable source for directly generating relevant information concerning normal and abnormal process behavior. Such information is immensely useful for logically inferring highly insightful conclusions about the target process system's current operating state.

Automatically performing this inference based on MOME with FALCONEER™ IV after each update of real-time process sensor data allows these process fault analyzers to continuously perform highly effective "intelligent process supervision" of the daily operations of their associated target process systems. Doing such supervision continuously in real-time mirrors an ideal concept of intelligence called rationality; that is, a given knowledge-based system is considered rational if it does the right thing (Russell and Norvig 1995c). Thus, there currently exists a simple and easy to use tool available to the process industries for proactively examining live, real-time process information on-line for immediate diagnosis of any underlying process faults. This directly allows continuous monitoring of a given target process system for safer operation, better performance, and ultimately higher efficiency and/or production levels. It has consequently now become a very straightforward and cost-effective proposition to develop and maintain extremely competent automated process fault analyzers throughout the processing industries.

ACKNOWLEDGMENTS

The authors would like to thank Professor David J. Courtemanche at the State University of New York at Buffalo for reviewing this treatment. His comments helped improve it immensely.

REFERENCES

Buchanan, B.G., Barstow, D., Bechtal, R. et al. (1983). Constructing an expert system. In: *Building Expert Systems* (ed. F. Hayes-Roth, D.A. Waterman, and D.B. Lenat), 127. Reading, MA, USA: Addison-Wesley Publishing Co.

Charniak, E. and McDermott, D. (1985). *Introduction to Artificial Intelligence*, 453–455. Reading, MA: Addison-Wesley Publishing Co.

Chester, D. L., L. Daniels, R. J. Fickelscherer, and D. H. Lenz, "Method and System of Monitoring, Sensor Validation and Predictive Fault Analysis", 2008. United States Patent No.: US 7,451,003

Das, A., Maiti, J., and Barnerjee, R.N. (2012). Process monitoring and fault detection strategies. *International Journal of Quality & Reliability Management* 29 (7): 720–752.

Davis, R. (1980). Meta-rules: reasoning about control. *Artificial Intelligence* 15: 179–222.

Davis, R. and Lenat, D.B. (1982a). *Knowledge-Based Systems in Artificial Intelligence*, 244–245. New York: McGraw-Hill, Inc.

Davis, R. and Lenat, D.B. (1982b). *Knowledge-Based Systems in Artificial Intelligence*, 412. New York: McGraw-Hill, Inc.

Fickelscherer, R.J. (1990). *Automated Process Fault Analysis*, Ph.D. Thesis. Newark, DE: Department of Chemical Engineering, University of Delaware.

Fickelscherer, R.J. (1994). A generalized approach to model-based process fault analysis. In: *Proceedings of the 2nd International Conference on Foundations of Computer-Aided Process Operations* (ed. D.W.T. Rippin, J.C. Hale, and J.F. Davis), 451–456. Austin, TX: CACHE.

Fickelscherer, R.J. and Chester, D.L. (2013). *Optimal Automated Process Fault Analysis*. New York: AIChE/John Wiley and Sons, Inc.

Fickelscherer, R.J. and Chester, D.L. (2016). Automated quantitative model-based diagnostic protocol via assumption state differences. *Computers and Chemical Engineering* 90: 94–110.

Fickelscherer, R.J., Lenz, D.H., and Chester, D.L. (2003). Intelligent process supervision via automated data validation and fault analysis: results of actual CPI applications. In: *Paper 115d, AIChE Spring National Meeting, New Orleans, LA*.

Fickelscherer, R.J., Lenz, D.H., and Chester, D.L. (2005). *Fuzzy Logic Clarifies Operations*, 53–57. InTech.

Gashnig, J., Klahr, P., Pople, H. et al. (1983). Evaluations of systems: issues and case studies. In: *Building Expert Systems* (ed. F. Hayes-Roth, D.A. Waterman, and D.B. Lenat), 265. Reading, MA, USA: Addison-Wesley Publishing Co.

Genesereth, M.R. and Nilsson, N.J. (1987). *Logical Foundations of Artificial Intelligence*, 117–121. Los Altos, CA: Morgan Kaufmann Publishers, Inc.

Harmon, P. and King, D. (1985a). *Expert Systems: Artificial Intelligence in Business*, 30. New York: John Wiley & Sons, Inc.

Harmon, P. and King, D. (1985b). *Expert Systems: Artificial Intelligence in Business*, 2. New York: John Wiley & Sons, Inc.

Harmon, P. and King, D. (1985c). *Expert Systems: Artificial Intelligence in Business*, 212. New York: John Wiley & Sons, Inc.

Hayes-Roth, F., Waterman, D.A., and Lenat, D.B. (1983). An overview of expert systems. In: *Building Expert Systems* (ed. F. Hayes-Roth, D.A. Waterman, and D.B. Lenat), 20. Reading, MA, USA: Addison-Wesley Publishing Co.

Kennedy, J.P. (1994). Data treatment. In: *Proceedings of the 2nd International Conference on Foundations of Computer-Aided Process Operations* (ed. D.W.T. Rippin, J.C. Hale, and J.F. Davis), 21–44. Austin, TX: CACHE.

Kletz, T.A. (1985). *What Went Wrong? Case Histories of Process Plant Disasters*. Houston, TX: Gulf Publishing Co.

Kline, P.J. and Dolins, S.B. (1985). *Choosing Architectures for Expert Systems*," CCSC Technical Report #85-01-001, 66. Dallas, TX: Texas Instruments Inc.

Kramer, M.A. (1987). Malfunction diagnosis using quantitative models and non-Boolean reasoning in expert Systems. *AICHE Journal* 33: 130–147.

Kramer, M.A. (1990). Letter to the editor in regard to Petti et al. *AICHE Journal* 36: 1121.

Kramer, M.A. and Mah, R.S.H. (1994). Model-based monitoring. In: *Foundations of Computer-Aided Process Operations II* (ed. D.W.T. Rippin, J.C. Hale, and J.F. Davis), 45–68. Austin, TX: CACHE, Inc.

Lymburner, C., Rovison, J., and An, W. (2006). Battling information overload. *Control* 95–99.

Ma, J. and Jiang, J. (2011). Applications of fault detection and diagnosis methods in nuclear power plants: a review. *Progress in Nuclear Energy* 53 (3): 255–266.

Parsaye, K. and Chignell, M. (1988a). *Expert Systems for Experts*, 73. New York: John Wiley & Sons, Inc.

Parsaye, K. and Chignell, M. (1988b). *Expert Systems for Experts*, 116–117. New York: John Wiley & Sons, Inc.

Petti, T.F., Klein, J., and Dhurjati, P.S. (1990). Diagnostic model processor: using deep knowledge for process fault diagnosis. *AICHE Journal* 36 (4): 565–575.

Reggia, J.A., Nau, D.S., and Wang, P.Y. (1984). Diagnostic expert systems based on a set covering model. In: *Developments in Expert Systems* (ed. M.J. Combs), 35–58. London: Academic Press.

Rich, E. (1983). *Artificial Intelligence*, 242. New York: McGraw-Hill, Inc.

Russell, S.J. and Norvig, P. (1995a). *Artificial Intelligence: A Modern Approach*, 415. New York: Prentice-Hall International, Inc.

Russell, S.J. and Norvig, P. (1995b). *Artificial Intelligence: A Modern Approach*, 467. New York: Prentice-Hall International, Inc.

Russell, S.J. and Norvig, P. (1995c). *Artificial Intelligence: A Modern Approach*, 845–847. New York: Prentice-Hall International, Inc.

Schoning, U. (1994). *Logic for Computer Scientists*, 2nde, 151–152. Boston, MA: Birkhauser, Inc.

Sell, P.S. (1985a). *Expert Systems—A Practical Introduction*, 86. New York: John Wiley & Sons, Inc.

Sell, P.S. (1985b). *Expert Systems—A Practical Introduction*, 88–89. New York: John Wiley & Sons, Inc.

Skotte, R., Lenz, D., Fickelscherer, R. et al. (2001). Advanced process control with innovation for an integrated electrochemical process. In: *AIChE Spring National Meeting, Houston, TX*.

Stefik, M. (1995). *Introduction to Knowledge Systems*, 531. San Francisco, CA: Morgan Kaufmann Publishers, Inc.

Stefik, M., Aikins, J., Balzer, R. et al. (1983a). Basic concepts for building expert systems. In: *Building Expert Systems* (ed. F. Hayes-Roth, D.A. Waterman, and D.B. Lenat), 71. Reading, MA, USA: Addison-Wesley Publishing Co.

Stefik, M., Aikins, J., Balzer, R. et al. (1983b). The architecture of expert systems. In: *Building Expert Systems* (ed. F. Hayes-Roth, D.A. Waterman, and D.B. Lenat), 121. Reading, MA, USA: Addison-Wesley Publishing Co.

Stefik, M., Aikins, J., Balzer, R. et al. (1983c). The architecture of expert systems. In: *Building Expert Systems* (ed. F. Hayes-Roth, D.A. Waterman, and D.B. Lenat), 122. Reading, MA, USA: Addison-Wesley Publishing Co.

Venkatasubramanian, V., Rengaswamy, R., and Kavuri, S.N. (2003a). A review of process fault detection and diagnosis part 2: qualitative models and search strategies. *Computers and Chemical Engineering* 27: 313–326.

Venkatasubramanian, V., Rengaswamy, R., Kavuri, S.N., and Yin, K. (2003b). A review of process fault detection and diagnosis part 3: process history based methods. *Computers and Chemical Engineering* 27: 327–346.

Venkatasubramanian, V., Rengaswamy, R., Yin, K., and Kavuri, S.N. (2003c). A review of process fault detection and diagnosis part 1: quantitative model based methods. *Computers and Chemical Engineering* 27: 293–311.

Zadeh, L.A. (1988). Fuzzy logic. *Computer* 21 (4): 83–93.

8.A

FALCONEER™ IV FUZZY LOGIC ALGORITHM PSEUDO-CODE

8.A.1 INTRODUCTION

This appendix describes in detail how the FALCONEER™ IV fault analyzer actually works. When the analyzer detects new data, it reads it, computes the Primary model residuals (**pmrs**), Secondary model residuals (**smrs**), and their certainty factors and partial derivatives. Then it computes the current certainty factors for all single faults, also automatically considering all possible non-interactive multiple faults. If two or more modeling assumption variable deviations have evidence for them, the certainty factors for interacting pairs of faults are also computed. (Identifying such interacting multiple faults is currently limited to only pairs of such faults). Finally, the results of all these computations are displayed.

8.A.2 SINGLE AND NON-INTERACTIVE MULTIPLE FAULTS

The critical things to understand are how the certainty factors for single and interacting pairs of faults are computed. Below is the actual code for these two computations. We will make a few comments about the code,

Artificial Intelligence in Process Fault Diagnosis: Methods for Plant Surveillance,
First Edition. Edited by Richard J. Fickelscherer.
© 2024 John Wiley & Sons, Inc. Published 2024 by John Wiley & Sons, Inc.

but we will also give a more readable explanation of how the computations are done.

Here is the function that computes the certainty factor for the nth modeling assumption variable, which we will denote by Vn. If d is True, it computes cfHigh(Vn); if d is False, it computes cfLow(Vn). The function contains two FOR loops, one for evaluating the pmrs and one for evaluating the smrs. They are the same except for which set of residuals they operate on, so we'll only explain the first loop. The lines are numbered on the right, and corresponding footnotes are below.

There are two auxiliary arrays being used here, pmrMask and pmrPattern. (There are corresponding arrays for the smrs too.) Array pmrMask holds some strings of 0s and 1s to indicate which residuals are relevant to which variables. Array pmrPattern indicates whether the variable and the residual change in the same direction, the opposite direction, or that the residual does not change when the variable changes. These conditions correspond to the pattern values 1, −1, and 0, respectively. The pattern value is 0 if it is known that the formula for the residual does not contain the variable Vn or if the derivative of the residual with respect to Vn is 0. More accurately, the pattern value for residual i with respect to variable Vn is the sign of the derivative of pmr(i) with respect to Vn if Vn is a measured (sensor) variable, and it is minus the sign of the derivative of pmr(i) with respect to Vn if Vn is an unmeasured variable (set valued parameter (e.g., a process leak)). (The difference in treatment is due to the fact that for a measured (sensor) variable, we are drawing conclusions about the measurement of the process variable, and for an unmeasured variable, we are drawing conclusions about the current constant value of the process variable (parameter) itself.)

Relevant residuals can be divided into classes: the neutral residuals (pattern value = 0) and the affected residuals. Depending on the direction in which an affected residual has changed, it might be evidence for Vn being a fault (in the direction indicated by d) or it might be evidence against Vn being a fault. The local variable minSoFar keeps track of the minimum cfSatisfactory value for the neutral residuals. For the affected residuals, maxFor keeps track of the maximum certainty factor for those residuals that are evidence for the Vn fault, and maxAgainst keeps track of the maximum certainty factor for those residuals that are evidence against the fault. The certainty factor of the Vn fault is then the minimum of minSoFar, maxFor, and (1−maxAgainst).

Note that if the Vn fault can cause pmr(i) to rise, the cfHigh value for pmr i is the certainty level to which pmr i is evidence for the fault, and the cfLow value for pmr i is the certainty level to which pmr i is evidence against the fault. (The opposite holds true if the fault can cause pmr(i) to fall.) If maxAgainst is the certainty factor of the strongest evidence against the fault, then the certainty factor for the fault can't be any higher than 1−maxAgainst. (Remember, 1−X is the fuzzy version of NOT(X).) So basically, the rule being used is the fuzzy version of the Boolean rule

```
FAULT-IS-PRESENT  =        SOME(EVIDENCE-FOR-FAULT)
                           AND ALL (NEUTRAL-EVIDENCE-IS-OK)
                           AND NOT SOME(EVIDENCE-AGAINST-FAULT)
```

(The quantifier SOME acts like an OR expression. SOME(X) means X1 OR X2 OR ... OR Xn. That is why CF(SOME(X)) = max{CF(Xi) | i = 1,...,n}.)

Single Fault Pseudo-code

```
Function computedVarCF(n As Integer, d As Boolean) As Double
    Dim cfMax As Double
    Dim minSoFar As Double
    Dim maxFor As Double
    Dim maxAgainst As Double
    Dim i As Integer
    Dim cf As Double
    minSoFar = 1
    maxFor = 0
    maxAgainst = 0
    For i = 1 To m
        If Mid(pmrMask(n), i, 1) = "1" Then                      [1]
            If pmrPattern(n * m + i) = 0 Then                    [2]
                minSoFar = min(minSoFar, cfSatisfactory(pmrCF(i)))
            ElseIf pmrPattern(n * m + i) = 1 Then                [3]
                If d Then                                        [4]
                    maxFor = max(maxFor, cfHigh(pmrCF(i)))
                    maxAgainst = max(maxAgainst, cfLow(pmrCF(i)))
                Else
                    maxFor = max(maxFor, cfLow(pmrCF(i)))
                    maxAgainst = max(maxAgainst, cfHigh(pmrCF(i)))
                End If
            ElseIf pmrPattern(n * m + i) = &minus;1 Then         [5]
                If d Then                                        [4]
                    maxFor = max(maxFor, cfLow(pmrCF(i)))
                    maxAgainst = max(maxAgainst, cfHigh(pmrCF(i)))
                Else
                    maxFor = max(maxFor, cfHigh(pmrCF(i)))
                    maxAgainst = max(maxAgainst, cfLow(pmrCF(i)))
                End If
            End If
        End If
    Next
    For i = 1 To dm                                              [6]
        If Mid(smrMask(n), i, 1) = "1" Then
            If smrPattern(n * dm + i) = 0 Then
                minSoFar = min(minSoFar, cfSatisfactory(smrCF(i)))
```

```
        ElseIf smrPattern(n * dm + i) = 1 Then
            If d Then
                maxFor = max(maxFor, cfHigh(smrCF(i)))
                maxAgainst = max(maxAgainst, cfLow(smrCF(i)))
            Else
                maxFor = max(maxFor, cfLow(smrCF(i)))
                maxAgainst = max(maxAgainst, cfHigh(smrCF(i)))
            End If
        ElseIf smrPattern(n * dm + i) = &minus;1 Then
            If d Then
                maxFor = max(maxFor, cfLow(smrCF(i)))
                maxAgainst = max(maxAgainst, cfHigh(smrCF(i)))
            Else
                maxFor = max(maxFor, cfHigh(smrCF(i)))
                maxAgainst = max(maxAgainst, cfLow(smrCF(i)))
            End If
        End If
    End If
Next
If d Then                                                    [7]
    var2High(n) = maxFor
Else
    var2Low(n) = maxFor
End If
computedVarCF = min(minSoFar, min(maxFor, 1 - maxAgainst))
End Function
```

[1] If Mid(pmrMask(n), i, 1) = "1" then pmr i is relevant to Vn.

[2] If pmrPattern(n*m+i) = 0 then pmr i is neutral for Vn.

[3] If pmrPattern(n*m+i) = 1 then a change in Vn causes pmr(i) to change in the same direction.

[4] If d then we are computing cfHigh(Vn), else we are computing cfLow(Vn).

[5] If pmrPattern(n*m+i) = −1 then a change in Vn causes pmr(i) to change in the opposite direction.

[6] The smrs are handled exactly the same way as the pmrs.

[7] This If statement just saves the maxFor value to indicate when the fault should be considered for double faults. See below.

8.A.3 PAIRS OF INTERACTIVE MULTIPLE FAULTS

Pairs of interactive faults are computed by the following subroutine. It computes the certainty factor for the pair of modeling assumption variables Vi and Vj. If b is True, the fault with Vi is that it is high; if b is False, the fault is that Vi is low. Similarly, if c is True, we are computing for the case where Vj is high, and if c is False, we are computing for the case where Vj is low.

Again, the strategy is to find the certainty factor of the evidence for the pair of faults, the certainty factor of there not being evidence against the pair of faults, and then take the minimum. For the pair of faults involving Vi and Vj, there must be SOME evidence for Vi that is not explainable by Vj, AND SOME evidence for Vj that is not explainable by Vi. Local variable maxISo-Far is set to the maximum certainty factor for a residual that Vi can change but that Vj cannot change, and local variable maxJSoFar is set to the maximum certainty factor for a residual that Vj can change but that Vi can't change. Local variable minSoFar is set to the minimum certainty factor of a residual that might count as evidence against the pair (the certainty factor used is the certainty that it is NOT in fact evidence against the pair of faults). The certainty factor for the pair of faults is the minimum of minSoFar, max-ISoFar and maxJSoFar.

Again, the code contains two FOR loops, one for the pmrs and one for the smrs. Otherwise, they are the same.

Basically, the rule being used can be thought of as the fuzzy version of the Boolean rule

```
FAULT-PAIR-IS-PRESENT = SOME(EVIDENCE-EXCLUSIVELY-FOR-FIRST-FAULT)
AND SOME(EVIDENCE-EXCLUSIVELY-FOR-SECOND-FAULT)
AND ALL( NEUTRAL-EVIDENCE-IS-OK)
AND NOT SOME(EVIDENCE-AGAINST-THE-PAIR)
```

The tricky part of interactive fault pairs is determining what counts as neutral evidence and what counts as evidence against the pair. This is determined by the nextMin function used to update minSoFar. What counts as exclusive evidence for a fault is determined by function next-Max, which is used to update both maxISoFar and maxJSoFar. See both of those functions below.

Multiple Fault Pseudo-code

```
Sub pairCF(i As Integer, b As Boolean, j As Integer, c As Boolean)
    Dim minSoFar As Double
    Dim maxISoFar As Double
    Dim maxJSoFar As Double
    Dim finalCF As Double
    Dim iP As Integer
    Dim jP As Integer
    Dim k As Integer
    minSoFar = 1
    maxISoFar = 0
    maxJSoFar = 0
```

```
For k = 1 To m
    If Mid(pmrMask(i), k, 1) = "1" Or Mid(pmrMask(j),
    k, 1) = "1" Then                                            [8]
        If b Then                                               [9]
            iP = pmrPattern(i * m + k)
        Else
            iP = -pmrPattern(i * m + k)
        End If
        If c Then                                               [10]
            jP = pmrPattern(j * m + k)
        Else
            jP = -pmrPattern(j * m + k)
        End If
        minSoFar = nextMin(minSoFar, iP, jP, k, True)           [11]
        maxISoFar = nextMax(maxISoFar, iP, jP, k, True)         [12]
        maxJSoFar = nextMax(maxJSoFar, jP, iP, k, True)         [13]
    End If
Next
For k = 1 To dm
    If Mid(smrMask(i), k, 1) = "1" Or Mid(smrMask(j),
    k, 1) = "1" Then
        If b Then
            iP = smrPattern(i * dm + k)
        Else
            iP = -smrPattern(i * dm + k)
        End If
        If c Then
            jP = smrPattern(j * dm + k)
        Else
            jP = -smrPattern(j * dm + k)
        End If
        minSoFar = nextMin(minSoFar, iP, jP, k, False)
        maxISoFar = nextMax(maxISoFar, iP, jP, k, False)
        maxJSoFar = nextMax(maxJSoFar, jP, iP, k, False)
    End If
Next
finalCF = min(minSoFar, min(maxISoFar, maxJSoFar))
If finalCF > displayThreshold Then                             [14]
    pairCount = pairCount + 1
    pairList = Array(i, b, j, c, finalCF, pairList)
End If
End Sub
```

[8] If Mid(pmrMask(i), k, 1) = "1" Or Mid(pmrMask(j), k, 1) = "1" then pmr k is relevant to either Vi or Vj (or both).

[9] If b then iP is set to the pattern value for pmr k with respect to variable Vi; otherwise iP is set to minus the pattern value for pmr k with respect to variable Vi. (If b is True, Vi is failing high, and if b is False, Vi is failing low, so the direction of influence on pmr k changes.)

[10] If c then jP is set the same way iP was as explained in [9].

[11] Update minSoFar according to neutral and possibly negative evidence.

[12] Update maxISoFar with evidence exclusively for Vi.

[13] Update maxJSoFar with evicence exclusively for Vj.

[14] If finalCF > displayThreshold then add the pair to the list of fault pairs.

The NextMin function is used to update minSoFar whenever it sees neutral evidence or possible evidence against the fault pair. Variable msf is the previous value of minSoFar. Variable iP is the pattern value for pmr k (if b is True) or smr k (if b is False) with respect to Vi, and variable jP is the pattern value for pmr k (if b is True) or smr k (if b is False) with respect to Vj. The residual k is classified as neutral or as evidence against the pair according to the following table:

iP	jP	Neutral	Evidence against pair
1	1	No	If residual is low
1	0	No	If residual is low
1	−1	No	No
0	1	No	If residual is low
0	0	Yes	No
0	−1	No	If residual is high
−1	1	No	No
−1	0	No	If residual is high
−1	−1	No	If residual is high

If residual k is evidence against the pair because it is low (one or both modeling assumption variables Vi and Vj should be causing it to go high), the certainty that it is NOT in fact evidence for the fault pair is (1−the certainty factor of residual k being low). If residual k is evidence against the pair because it is high, the certainty that it is NOT in fact evidence for the fault pair is (1−the certainty factor of residual k being high). If the certainty factor for residual k as new evidence is lower than the previous minimum, that is what is returned to be the new value of minSoFar.

```
Function nextMin(msf As Double, iP As Integer, jP As Integer, k As
Integer, b As Boolean)
    Dim scf As Double
    If b Then
        scf = pmrCF(k)
    Else
        scf = smrCF(k)
    End If
```

```
If iP = 0 And jP = 0 Then
    nextMin = min(msf, cfSatisfactory(scf))
ElseIf (iP = 0 Or jP = 0 Or iP = jP) Then
    If iP = 1 Or jP = 1 Then
        nextMin = min(msf, 1 - cfLow(scf))
    ElseIf (iP = &minus;1 Or jP = &minus;1) Then
        nextMin = min(msf, 1 - cfHigh(scf))
    End If
Else
    nextMin = msf
End If
End Function
```

The function nextMax is used to update maxISoFar and maxJSoFar whenever it sees exclusive evidence for one of the faults. Internally, msf is the previous value of whichever program modeling assumption variable is being updated, iP is the pattern value for the fault that the evidence is exclusively for, and jP is the pattern value for the other fault. Residual k is seen as exclusive evidence if iP is not 0 and jP is 0. If iP is 1, the residual k should be high to be counted as evidence for the fault. IF iP is −1, the residual k should be low to be counted as evidence for the fault. Since we are looking for SOME exclusive evidence for the fault, we take the maximum of the certainty factors for such evidence. The variable b simply indicates whether the residual is a primary residual or a secondary residual.

```
Function nextMax(msf As Double, iP As Integer, jP As Integer, k As
Integer, b As Boolean)
    Dim scf As Double
    If b Then
        scf = pmrCF(k)
    Else
        scf = smrCF(k)
    End If
    nextMax = msf
    If jP = 0 Then
        If iP = 1 Then
            nextMax = max(msf, cfHigh(scf))
        ElseIf iP = &minus;1 Then
            nextMax = max(msf, cfLow(scf))
        End If
    End If
End Function
```

8.A.4 SUMMARY

The Fuzzy logic pseudo-code above implementing the method of minimal evidence (**MOME**) competently diagnoses both all single and non-interactive multiple faults and most pairs of interactive multiple faults. This algorithm requires just the underlying quantitative models describing normal process operation and their associated residual means and variances: all the required diagnostic logic required in the corresponding fault analysis is incorporated into the compiler program, separate from this domain knowledge. This greatly reduces both the original development and on-going maintenance efforts required for initially creating and then evolving competent fault analyzers as process operations' changes dictate. FALCONEER™ IV has to date been successfully implemented in many highly complex industrial processing plants.

8.B

MOME CONCLUSIONS

8.B.1 OVERVIEW

This appendix summarizes our model-based diagnostic strategy, the MOME, and its various advantages for optimally automating process fault analysis. It also describes a procedure for developing automated process fault analyzers using our program, called FALCONEER™ IV, which is directly based upon the Fuzzy logic implementation of MOME described previously. This program allows fault analyzer development and maintenance to be highly cost effective in actual process system applications, allowing process engineers to create and maintain such programs themselves for their targeted process systems. The benefits directly derived from using FALCONEER™ IV on two real-world, highly complex electrolytic process systems formally owned and operated by FMC Corporation in Tonawanda, New York were discussed in Chapter 8.

8.B.2 SUMMARY OF THE MOME DIAGNOSTIC STRATEGY

The chief advantage of MOME is that it provides a uniform framework for examining process models and their associated modeling assumptions. In this framework, each Primary model represents the relationship that exists between its required modeling assumption variables during normal process

Artificial Intelligence in Process Fault Diagnosis: Methods for Plant Surveillance,
First Edition. Edited by Richard J. Fickelscherer.
© 2024 John Wiley & Sons, Inc. Published 2024 by John Wiley & Sons, Inc.

operation. Any significant residuals resulting from the evaluation of these models directly indicate that one or more of the associated modeling assumption variables are deviating significantly. All modeling assumption variable deviations can be classified into one of three possible categories. These three categories are distinguished from one another by their effect on the residuals of the Primary models they violate. This directly permits the derivation of standardized SV&PFA diagnostic rule formats, thus allowing the fault analyzer development to be completely systematic once the Primary models are derived and deemed well-formulated.

As discussed, the major advantage of the MOME diagnostic strategy centers about the diagnostic rule format it uses. The format has been chosen to ensure that the diagnostic knowledge bases will always perform competently; that is, it will only make the correct fault diagnoses or not any diagnoses. The format also ensures that the resulting diagnostic knowledge bases can diagnose those faults with the best diagnostic sensitivity and diagnostic resolution possible for the current magnitude of the process operating event. Each diagnostic rule contains only the minimal amount of diagnostic evidence required to uniquely discriminate its associated fault situations from all of the other likely process operating events. As discussed, using just the minimal amount of diagnostic evidence in each diagnostic rule in this manner also allows many possible multiple fault situations to be diagnosed directly. Also as discussed, this diagnostic methodology can be directly used to determine the strategic sensor placement for performing process fault analysis that further maximizes diagnostic sensitivity and/or resolution for particular process operating events. It can further be used to shrewdly distribute fault analyzers throughout a large process system.

8.B.3 FALCONEER™ IV KBS APPLICATION PROJECT PROCEDURE

The FALCONEER™ IV Process Performance Suite program is a real-time, on-line process performance monitoring software tool that assists process operators and engineers by auditing current process operations to help ensure they are optimal and to identify root causes of problems when those operations become abnormal. It accomplishes this by continuously monitoring specific sensor measurements, performing advanced calculations and analysis with them, and then immediately giving advisory alerts when the process is not performing optimally or actual processing problems occur. These alerts go beyond typical distributed control system alarms because FALCONEER™ IV directly uses engineering models of normal process operation and advanced statistical calculations in its analysis. From this analysis, FALCONEER™ IV is able to determine whether the monitored sensor measurements are correct, whether they are currently in control, whether any other process faults are occurring, and whether the current overall process performance is optimal. The resulting timely alerts should help users to more effectively and optimally control and operate their process systems.

The advanced and systematic process auditing performed by FALCONEER™ IV while performing its associated State ID, SV&PFA, and Virtual Statistical Process Control (SPC) analysis extracts additional, highly useful information about current process operations continuously from the current sensor measurements. The results of this analysis are continuously available in a variety of user-friendly formats to automatically and immediately report those results to the appropriate personnel (e.g., immediate intelligent alerts to operators, systematic e-mails to process engineers, etc.). This should enable corrective actions for improving process operations to be taken in a more timely, proactive manner, both reducing operating costs and improving process safety. These benefits continuously occur from more effectively extracting all of the relevant information currently available in the process data already being collected by the client's control system.

In general, the following series of steps are followed during a FALCONEER™ IV application project:

(1) Identify a target process system for analysis.

(2) Segment that target process logically according to its major unit operations or equipment, identifying them with unique process area names (or identifiers).

(3) Identify all of the measured and unmeasured variables associated with each process area and determine their engineering measurement units.

(4) Identify key measured variables that directly indicate the current process operating state and their associated limits. These are used directly by the State ID module to determine if further analysis of the measured variables collected during the current process state is warranted by FALCONEER™ IV.

(5) Create as many Primary models and Performance Equations (also known as, key performance indicators (KPIs) or soft sensors) as possible with the sensor measurements available in the target process.

(6) Evaluate these Primary models with sufficient normal process data (i.e., at least 3 to 6 months' worth) to determine typical model residual standard deviations and offsets as functions of production. These models and associated statistics are used directly by the fault analysis module to perform SV&PFA.

(7) Evaluate the normal standard deviation and mean of each measured variable for which Virtual SPC will be performed on. These statistics are used directly by the Virtual SPC module to do its associated analysis.

(8) Determine upper and lower alert limits (if both are applicable) on Performance Equations (also Known as, KPIs or soft sensors). These limits are used directly by the Virtual SPC module to do its associated analysis.

(9) Configure the FALCONEER™ IV application database (input Primary models, Performance Equations, and associated calculated statistics) according to the suggested configuration hierarchy.

(10) Choose the various FALCONEER™ IV time intervals to help ensure its analysis is performed consistently and correctly.

(11) Run the FALCONEER™ IV application with actual process data to perform the SV&PFA sensitivity analysis; make adjustments to FALCONEER™ IV's various SV&PFA tuning parameters as necessary.

(12) Run the FALCONEER™ IV application with actual process data to perform the Virtual SPC sensitivity analysis; make adjustments to FALCONEER™ IV's various Virtual SPC tuning parameters as necessary.

(13) Start using the FALCONEER™ IV application on-line in real time to help improve process operations.

For a moderately complicated process system it typically requires less than a person-month of effort to completely configure and fully validate a FALCONEER™ IV application.

8.B.4 OPTIMAL AUTOMATED PROCESS FAULT ANALYSIS CONCLUSIONS

Process data becomes valuable only when it is at the right place at the right time and there are mechanisms to properly interpret and use it (Harmon and King 1985b). Process data thus cannot be considered an asset unless correctly presented, analyzed, and converted into information (Kennedy 1994). "Moreover, we really don't want information; we want knowledge ... we want information analyzed, converted, and organized in a useful way" (Harmon and King 1985c). Using models to do this analysis and conversion and then performing SV&PFA based on those results in real time proves the assertion that "Models are the means by which data can be converted to meaningful information" (Kramer and Mah 1994). These models are based on a fundamental understanding of normal operating behavior of the given process system. This fundamental knowledge thus constitutes an unimpeachable source for directly generating relevant information concerning normal and abnormal process behavior. Such information is immensely useful for logically inferring conclusions about the current process operating state. Automatically performing this inference based on MOME with FALCONEER™ IV after each update of process sensor data allows such fault analyzers to continuously perform highly effective "intelligent supervision" of the daily operations of their associated process systems. "Thus, the whole business environment of process operations should become more rational. More information will be gathered, synthesized, and put into useful form more rapidly ... information organized in a useful way, after knowledgeable inferencing about its content" (Harmon and King 1985c). FALCONEER™ IV is thus a simple and easy to use tool now available to the process industries for proactively examining live process information for any faults so as to allow continuous monitoring of a given process for safer operation, better performance, and ultimately higher efficiency and/or production levels. It has now become a very straightforward and cost-effective proposition to develop and maintain extremely competent automated process fault analyzers throughout the processing industries.

9

FAULT DETECTION USING ARTIFICIAL INTELLIGENCE AND MACHINE LEARNING

Philippe Mack[1,2] and Lieven Dubois[3]

[1]*PPITE SA, Liège, Belgium*
[2]*Université de Liège, Liège, Belgium*
[3]*ISA 18, 88 and 101, Working Group 8 of ISA 18.2, Dept. of Alarm Management and Training, Manage 4U, Leiden, Netherlands*

CHAPTER HIGHLIGHTS

- Applicable to process systems in which traditional process modeling is not possible.
- Enumerates all potential such AI-based algorithmic methods currently employed in PFD.

ABBREVIATIONS USED

AI   artificial intelligence
ANN   artificial neural network
BN   Bayesian network, Bayes network

Artificial Intelligence in Process Fault Diagnosis: Methods for Plant Surveillance, First Edition. Edited by Richard J. Fickelscherer.
© 2024 John Wiley & Sons, Inc. Published 2024 by John Wiley & Sons, Inc.

CPU   central processing unit
GPU   graphical processing unit
KBS   knowledge-based system
LLM   large language models
PCA   principal component analysis

9.1 INTRODUCTION

In this chapter, we distinguish possible diagnosis by a subdivision of AI. It is based on algorithms which are basically hidden for the user or engineer. This in contrast to knowledge-based systems, where the goal is to make the knowledge contained in the system as explicit as possible.

9.2 ARTIFICIAL INTELLIGENCE

Wikipedia (2022a) defines artificial intelligence as the theory and development of computer systems able to perform tasks that normally require human intelligence, such as visual perception, speech recognition, decision-making, and translation between languages. The main concepts of artificial intelligence were invented by Alan Turing (and described in paper published in 1948 in a report entitled "Intelligent Machinery").

There are basically two main subdomains in artificial intelligence. First one is knowledge-based systems, also known as *expert systems*. Industrial applications of such systems are already in use for a few years now. Gensysm (Gensym Corp 2023) has marketed a product called G2 since 1986. The main drawback of such systems is that they rely on a knowledge base to infer decisions. The quality of the decision and the reasoning will depend on the quality and exhaustiveness of the knowledge base designed by domain experts.

The other main subdomain of AI is *machine learning*, where data are used to infer decision models. In this case, the advantage is that we do not depend so much on domain experts to build the decision models but more on available data. As models are learned from examples (the data), the robustness of models will mostly depend on the quality and quantity of data available.

In this chapter, we will only focus on machine learning as this has probably the widest spectrum of application in advanced diagnostics and fault detection.

9.3 MACHINE LEARNING

Machine learning (Wikipedia 2022b) is the study of computer algorithms that can be built (or learnt) automatically through experiences, using data. Machine learning is a field of artificial intelligence that has gained recently a lot of attention and success with the emergence of big data and increased CPU power. Some CPUs are even now specifically designed for the purpose of artificial intelligence (e.g., GPU; Wikipedia 2023).

The purpose of machine learning is to train decision models with data. In industry, data can be found in different locations and structured in different ways (log files, images, historian, spreadsheets, etc.). At the end, all these data should fit in a normalized flat data table, where rows depict the cases or the examples and the columns the features of these cases. Examples of a data table for varnish process in automotive industry are shown in Figure 9.1.

There are three main categories of machine learning: supervised machine learning, unsupervised machine learning, and reinforcement learning (Figure 9.2).

With supervised machine learning, the goal is to model output(s) based on a set of features. This can be used, for example, to predict the amount of energy generated by a solar panels based on weather conditions (Figure 9.3).

In unsupervised learning, the goal is to discover relationships or similarities between features. It can be used to detect abnormal situation or detect various regimes of operations. If a new situation occurs outside known clusters, this can be considered as an anomaly and a proper alarms can be generated (Figure 9.4).

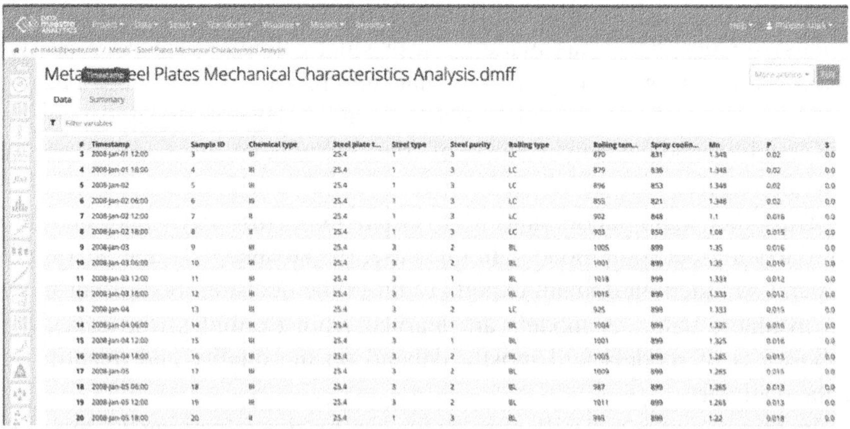

Figure 9.1 A data table where all examples are the rows and columns are the variables is the key input to use machine learning.

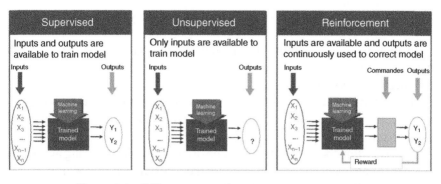

Figure 9.2 Different types of machine learning models.

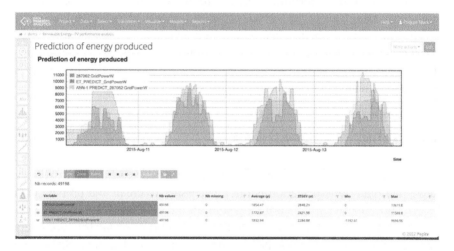

Figure 9.3 Example of prediction energy production on PV solar panels based on weather conditions.

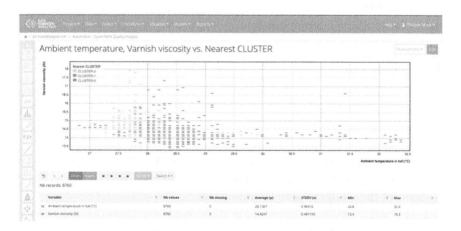

Figure 9.4 Example of clusters build from K-Means clustering machine learning algorithm.

In reinforcement learning, the model is constantly updated with fresh data to optimize an output. In industry, this can be typically used for advanced process control.

It must be reminded that data in itself will change depending on modifications in the applied infrastructure (see Chapter 10 on Knowledge Based Systems).

To stay within the field of fault diagnosis, any machine learning algorithm that cannot be made aware of changes in the instrumentation will fail (partly or substantially) when instruments are swapped or replaced by newer technology without the algorithm being aware of such change and recalibrated with fresh data.

To "fix" this problem, instrument and machine suppliers now often provide an xml description of the functions and features of the applied device. But it is up to the user (the customer) to integrate this information and make machine learning applications "aware" of such information.

9.4 ENGINEERED FEATURES

The variety and velocity of data generated in a process can be extremely large, and data transformation is often necessary to clean and prepare data before applying machine learning per se. In the context of anomaly detection, sound, vibration, and videos are typical sources of unstructured data that need data transformation before applying machine learning. Some data transformation can be basic, and some others might be more complex such as standard signal processing. Standards signal processing such as fast Fourier transform and wavelets creates new features that can be used as inputs for machine learning models.

9.4.1 Fast Fourier Transformation and Signal Processing

Fourier transformation is a mathematical transformation that decomposes a function depending on space or time into functions depending on spatial or temporary frequency. The term refers to both the frequency domain representation and the mathematical operation.

Process data often include certain frequencies. A rolling mill runs at a set frequency, and all the involved engines turn at set speeds. The currents and voltages of these engines can be monitored.

Fast Fourier transform allows to perform *real-time* transformation of the minor variations in the measured values and transform these time series into the frequency domain where a certain spectrum is generated. A normal operation provides a given spectrum after transformation, deviations in this spectrum can be detected and rules established for informing maintenance, operations, or service providers if such deviations occur or persist. Basically, fast Fourier transform can be applied as an event generator for Fault Trees.

9.4.2 Principal Component Analysis

Principal component analysis (PCA) is a data transformation technique used to reduce a complex system of correlated variables into a smaller number of orthogonal dimensions (the principal components). These features can then be used as inputs to machine learning.

9.5 MACHINE LEARNING ALGORITHMS

There are many types of algorithms in machine learning. The purpose of this section is not to depict all of them but illustrate how some of them that can be successfully applied to multiple situations to solve problems in manufacturing and process industries.

9.5.1 Decision Trees and Ensemble Trees

Decision trees are supervised machine learning that can be extremely useful as they can be easily interpreted by a domain expert. Drawback of decision tree is their poor robustness regarding variance. This means that small perturbations in data can lead to quite different models. To deal with this weakness, several models can be generated with different learning data sets. This process is known as ensemble models where the output is calculated as the average outputs of all individual models.

An example of decision tree is shown in Figure 9.5. The principle of the decision tree is to identify conditions leading to a potential risk of severe degradation in the process. In this example, the decision tree identifies conditions leading to a high risk of producing bad products. In other words, it identifies a sequence of events that are known to be prone to poor performance. In this case, if all these events are simultaneously detected: (i) thermalization temperature < 59.25, (ii) rinsing temperature < 36.1, and (iii) rinsing temperature > 33.35, then you have almost 50% chance to produce a bad product.

The disadvantage of this method is its poor reliability if process conditions have not been yet recorded in data used to train the decision tree model. However, when the data workflow is well designed, recalibration of the

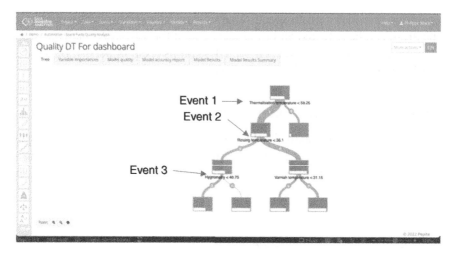

Figure 9.5 Example of a decision tree AI models to detect sequences of critical events.

models with newer data would obviously capture automatically new problematic conditions. This model could also be used with other anomaly detection models to identify conditions never seen in past data.

9.5.2 Artificial Neural Networks

Artificial neural networks is probably one of the oldest and most popular machine learning technologies that has been originated in 1943 by Warren McCulloch and Walter Pitts. Initially, their research was at the frontier of computational theory and biology. Its goal was to model the way brain works.

The first artificial neural networks, the perceptron, as we know it today, was invented by Frank Rosenblatt. Artificial neural networks (ANNs) are computing systems emulating the way brains work. A neural network is based on a collection of connected units, the neurons. The connections, called the synaptic connections, transmit signals from one neuron to another neuron. Once a neuron received a signal, this signal is processed by an activation function and sent through the synaptic connections to other neurons and so on. This process ends when a final unit is reached.

The architecture of neural networks is usually based on one input layer, several hidden layers, and one output(s) layer. The machine learning of an artificial neural network consists of adjusting the weight of the synaptic connection to optimize the errors on the learning set (Figure 9.6).

The latest significant development in algorithms, CPUs, and memory allowed to learn neural networks with a very large number of neurons and

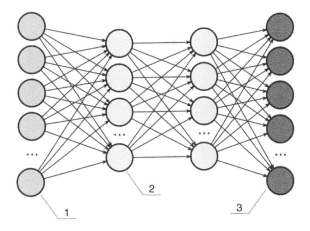

Figure 9.6   Example of an artificial neural net structure.

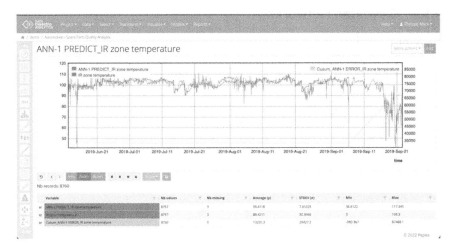

Figure 9.7   Example of a neural networks to predict a critical temperature in a chemical process.

layers. Deep learning is an extremely active area in the AI community where significant improvements have been demonstrated on multiple applications, especially in pattern and image recognitions (Figure 9.7).

Auto-encoders are particular artificial neural networks where the number of inputs neurons is similar to the number of outputs. This can be particularly useful to detect anomalies in input data. Hidden layers encode the "normal behavior" of the system learnt from historical data. Once new inputs are shown, then the auto-encoder predicts the most probable "normal behavior." Differences between input values and output values could then be interpreted as an anomaly in input data (Figure 9.8).

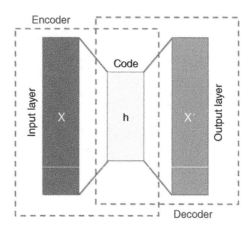

Figure 9.8 Autoencoders illustration.

9.5.3 Bayesian Networks

Depending on how it is used, Bayesian networks can be used as both supervised and unsupervised machine learning. A Bayesian network is acyclic graph that models the causality or the relation between features. The "weight" of the relation is then determined by the conditional probability table of all features values.

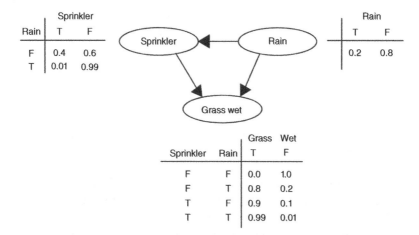

The structure of the graph itself can be modeled by a domain expert or trained by machine learning, while the conditional probability table is learned from historical data.

The advantage of using BN is that it can easily mix the knowledge of experts (relationship between parameters and causality model with the graph structure), while machine learning calculates the actual distribution probability.

9.5.4   High-Density Clustering

The main key aspect of supervised machine learning is that you need enough representative cases to learn the relationships between inputs and outputs. In case of anomaly detections, the challenge is that anomalies are often rare events. Rare events create extremely unbalanced data sets which are difficult if not impossible to predict directly with supervised machine learning.

One way to deal with this is to create a physical, first principles, model and apply Monte-Carlo simulations where would create artificially more abnormal situations. This has been applied in electrical transmission system where blackouts (major outages) are simulated using "first principles" digital twins of the electrical grid. However, it is not always possible to have an accurate first-principles model of an equipment and simulate all possible failures and generate synthetic data. In such a case, pure data-driven models could be cumbersome to detect anomalies (Figure 9.9).

Another way to deal with this issue is to use unsupervised machine learning models and calculate a distance to normal operations. This approach is extremely interesting as the learning process learns a digital twin from healthy conditions of the system. When deployed, the system compares in real time the actual behavior with its digital counterpart.

Clustering is a type of unsupervised machine learning algorithm that is used to identify similar patterns or behaviors within a dataset. When clustering is performed on the normal, healthy behaviors of a system, any new data points that fall outside of these established clusters can be considered as outliers or anomalies. By calculating the distance between these anomalous data points and the "normal" cluster, it is possible to determine how far they deviate from expected behavior. This distance metric can then be used as a criterion for detecting anomalies and identifying the parameters that are responsible for the deviation.

9.5.4.1 *AI Opportunities and Limitations*   Based on the information above, one could draw the conclusion that machine learning can be a faster method for developing a mathematical model of system behavior.

A mathematical model that has been validated and performs well is known as a white box model if it is constructed using well-known mathematical functions by a domain expert who is guided by first principles. These models are considered white box models because their behavior is well-understood.

Most machine learning models are typically classified as black-box models because the underlying mathematical equations are often difficult or

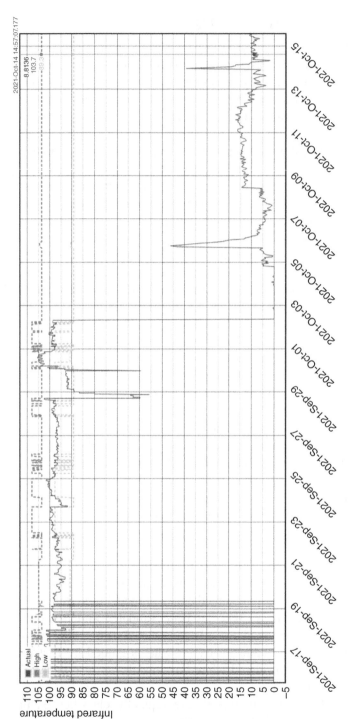

Figure 9.9 Example of a health monitoring systems showing dynamic limits calculated in real-time. Source: Courtesy of Pepite.

impossible to interpret, which can make it challenging to gain trust in their operational use, especially in process industries. However, there are several approaches that can be used to improve the interpretability and trust of black-box models and turn them into what is called gray models.

One such approach is to develop hybrid models that combine physical models with machine learning models. One way to do this is to incorporate physical models during the learning stage. One can construct features that are based on raw measurements, which can then be used as inputs or outputs to the machine learning model. For example, if you want to diagnose the drift of a boiler performance, you will calculate with thermodynamics the efficiency of the boiler. This would be used as an output of black-box models that would detect and diagnose drifts of boiler efficiency in real time.

Another approach is to visualize the behavior of the model by simulating changes of its inputs and inspect its behavior on historical data. This can help understand the behavior of the models especially on values that are outside of the training set.

By employing such techniques, it is possible to make black-box models more explainable and increase their acceptance and trustworthiness in industrial applications (Figure 9.10).

Explainable AI is a booming research area and will probably provide some more answers in the coming decades allowing more trusted AI to be deployed in factories.

9.5.5 Large Language Models and the Future AI-Driven Factories

The latest developments in artificial intelligence have also significantly contributed to the development of large language models. This has gained

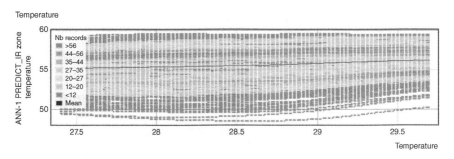

Figure 9.10 Partial dependency plots showing the change of model outputs based on one input variation.

significant popularity in 2022 and has already started to impact many businesses.

Large language models are trained with deep learning on large databases of text corpus to generate human-like texts. Based on a question provided to the LLM in natural language, the LLM will be able to formulate a human-like answer.

In diagnosis and alarms management, this can be a game changer at different levels:

- Create more explicit and human understandable alarms available in different languages.
- Solve more quickly problems as the LLM could also generate answers to support the diagnosis.

The current drawbacks of the systems available is their relative inaccuracy for specialized topics such as diagnosis for plant operations. Most of this models have been designed for rather generic purposes. However, these systems can be improved and specialized with feedback from end-users or/and trained with more specific data.

As a simple illustration, we can, for example, ask ChatGPT why the temperature of a heat exchanger is getting too high (Figure 9.11).

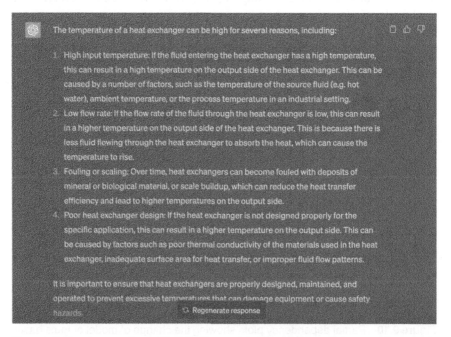

Figure 9.11 Answer from ChatGPT on why temperature of heat exchanger is too high.

Although it may seem like a simple solution, the system could provide a valuable checklist of responses for operators and engineers to use. Looking ahead, it is possible that software in factories will incorporate more advanced technologies that generate clear and detailed alarms in a variety of languages. This could significantly improve operational safety and reduce the risk of human errors.

As AI technologies continue to evolve, we can anticipate that voice recognition and large language models (LLMs) will enhance the capabilities of plant operators by enabling more efficient and user-friendly interactions with the technical operations of the factory. For instance, a chatbot could be linked to plant operations, making it easier to detect and diagnose critical events in real-time.

REFERENCES

Gensym Corp. (2023). http://dev.gensym.com/platforms. Retrieved from: http://dev.gensym.com/platforms/g2-standard.

Wikipedia. (2022a, November 20). Artificial Intelligence. Retrieved from: https://en.wikipedia.org/wiki/Artificial_intelligence

Wikipedia. (2022b January 20). Machine Learning. Retrieved from https://en.wikipedia.org/wiki/Machine_learning.

Wikipedia. (2023). Graphical Processing Unit. Retrieved from https://en.wikipedia.org/wiki/Graphics_processing_unit

10

KNOWLEDGE-BASED SYSTEMS

Lieven Dubois[1] and Philippe Mack[2,3]

[1]ISA 18, 88 and 101, Working Group 8 of ISA 18.2, Department of Alarm Management and Training, Manage 4U, Leiden, Netherlands
[2]PPITE SA, Liège, Belgium
[3]Université de Liège, Liège, Belgium

CHAPTER HIGHLIGHTS

- Discusses primary distinctions between knowledge and information.
- Describes details of various possible logical reasoning strategies for searches of PFD solutions.
- Outlines possible methods to handle temporal state transitions.

ABBREVIATIONS USED

CAD computer-aided design
CAE computer-aided engineering
DCS distributed control system
HMI human–machine interface

Artificial Intelligence in Process Fault Diagnosis: Methods for Plant Surveillance,
First Edition. Edited by Richard J. Fickelscherer.
© 2024 John Wiley & Sons, Inc. Published 2024 by John Wiley & Sons, Inc.

KB   knowledge base
KBS   knowledge-based system
PLC   programmable logic controller
SCADA   supervisory control and data acquisition
TCP/IP   transmission control protocol/Internet protocol

10.1 INTRODUCTION

Knowledge-Based Systems (KBSs) are computer applications containing "knowledge." In this chapter, we will discuss the following terms:

- Knowledge
  - Different forms of knowledge
  - Knowledge representation
  - Maintaining and updating knowledge
  - Learning and machine learning
- Expert systems
  - Some history
  - Typical characteristics
  - Benefits and drawbacks
- Application of knowledge-based systems on fault diagnosis

10.2 KNOWLEDGE

10.2.1 Definition

According to the Cambridge Dictionary (2023), "knowledge" is defined as "understanding of or information about a subject that you get by experience or study, either known by one person or by people generally."

Merriam-Webster (2023) defines "knowledge" as "information, understanding, or skill that you get from experience or education" and as "awareness of something: the state of being aware of something." Merriam-Webster defines knowledge also as "the sum of what is known."

For the purpose of this chapter, we will limit "knowledge" as "understanding of or information about a subject, gained by experience or study" and for the purpose of knowledge-based systems, to "understanding of and information on about a subject, as being stored in a software application."

Is Google a knowledge-based system? No, it is not, it is an information-based system. It just displays information on a subject, based on "personalized" search engines (or, "paid-for" algorithms, think of Google ads). So, what is the difference between information and knowledge?

Information are facts about a situation, person, event, etc. (Cambridge Dictionary 2023). How this information is stored and transmitted is not discussed. Basically, a person obtains information through his/her senses, the ability to see, hear, smell, taste, and feel. The most common information transmission of repository to person is visual, by text, images, or film. The second most important transmission is auditory, either by language, sound, or music. Sommeliers will use smell and taste, and pilots (racers) will certainly use feel.

Information is not knowledge. In fact, you can have contractionary information (think of news and fake news). So, what is the difference? Knowledge is information put in a context and complemented with some form a reasoning, calculus, or association. For fault diagnosis, this is of utmost importance. The reading of a sensor, for example, a temperature, is information. Whether or not this information is of importance or not, correct or false is based on a process of contextualization.

10.2.2   Mathematical – Physical Contextualization

Over time, researchers have tried to define mathematical formulas to calculate relationships between (process) variables, such as thermodynamic relationships, flow rate relationships, gravity, gas laws, chemical reactions, and electrical. Depending on the number of variables in an equation (and/ or the number of equations), if you know one or more values, you can calculate the other(s) (Figure 10.1).

Equations can assist in contextualizing information. If the result of the calculation does not agree with the observation or measurement, then

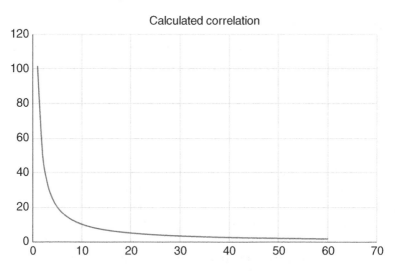

Figure 10.1   Example of a calculated correlation.

something may be wrong. Equations are helpful when the variable we want to monitor cannot be measured.

Ergo, one important pillar of a knowledge-based system is its ability to compute with data. That is exactly why computers (analog and digital) were invented for.

10.2.3   Procedural Knowledge

An important skill of a human being is its ability to perform several tasks in a prescribed sequence. If you want to know or to do something, you probably need to perform a few operations in a set sequence. For example, to validate a measurement, the value of the transmitter must be transformed in a physical value (e.g., from milliamps to temperature), it must be compared with its preceding readings, a trend might be calculated, it must be compared with expected values or be put in a range, and only then it is trusted measurement.

Hence, the way information is processed will define its value and importance. Digital computers (and PLCs) are extremely well fit to perform a fixed and repetitive set of operations and tasks. Usually, these sequences are called programs. We will discuss some of the disadvantages of programs in one of the next sections in this chapter.

10.2.4   Heuristic Knowledge

Heuristics are a set of (self-learned) rules human beings apply to information or situations to assess them. For example, when filling a bottle, if the bottle is nearly full, then decrease the inflow (the outflow of the valve used). From a neurological point of view, heuristics come from the mammal part of the brain which, evolutionary, was developed after the reptile and mammal part of the brain. Most psychological biases are heuristics. The pre-frontal cortex, which is, evolutionary, the part of the brain which makes humans different from other mammals, allows human beings to express these heuristics in language. Expert Systems shells were developed in the 1970s and 1980s to capture these rules as part of the attempt to store and re-use the neurological capabilities of subject matter experts.

We will discuss these Expert Systems shells in more detail in the next section.

10.2.5   Topological Knowledge

According to Cambridge Dictionary (2023), topology is "the way the parts of something are organized or connected." This is of importance for calculations (mathematical knowledge), sequence of tasks or operations (procedural knowledge), and rules (heuristic knowledge).

A topology is something that is not temporal and not easy to change, think of infrastructure (highways, railroad tracks), circuits (electric, pneumatic), industrial installations (connected valves, tubes, vessels, etc.), and so on.

Over time, we identified the following types of topologies in industrial plants:

- The logistic flows: How the raw materials and semi-products make it to packed production?
- The energy flows: How energy in its different forms (steam, hydraulics, pneumatics, electrical, etc.) is applied to the logistic flow?
- The information and control flows: How information and control is provided to and retrieved from the logistic flow?

The logistic flow is established by the infrastructure of an industrial plant: its tubes, vessels, cranes, robots, etc. Mostly, mechanical and its topology is often stored in the CAD/CAE system of the plant.

The energy flows are established by a mixture of tubes, wires, axes, drives, etc. Mostly, electro-mechanical and its topology is stored not only in the CAD/CAE system of the plant but also in the electrical circuitry diagrams.

The information flow – a new kid on the block – defines how information and control is provided to and retrieved from the logistic flow. In a totally manual operated system (think of a kitchen 200 years ago), this information flow was entirely human. With the arrival of automated control, a topology was established, either by wires (4–20 mA) or even wireless (Wi-Fi, 4G, Mesh networks, …). Instruments are often connected to bus systems these days. Wireless access points are connected with a TCP/IP network, and remote sensing might depend on external suppliers of transmission capacity.

For automated fault diagnosis, this is an often-forgotten component.

10.3   INFORMATION REQUIRED FOR DIAGNOSIS

10.3.1   Introduction

In order to perform automated diagnosis, software requires knowledge and information. While knowledge is rather static, information is volatile. We learned that information is data placed in a context. Let us examine what contextual information is available.

10.3.2   Association

There is no good single word that describes this aspect of information for this subject matter. Cambridge Dictionary describes association as "the fact of being involved with or connected to someone or something."

The ingredients to bake bread are not part of any topology but are essential to the process. The container in which lots of wheat are pored or blown into is part of the logistic topology. A lot of wheat used to bake a bread is not connected with the bread, but it is associated.

One might distinguish associated items from connected items by their temporal nature. There is no topological connection between the bread and the wheat used. One item can be associated with one or more other items. There are four types of relationships:

- One to one
- One to many
- Many to one
- Many to many

The car being produced is associated with an order and the bill-of-materials is also associated with the car. When the car body moves through the assembly plant when it is being built, it will have temporal associations with the working stations along the assembly line.

The batch being produced in a process cell is temporarily associated with that cell and an associated recipe. Association is of less importance in continuous processes, but of more importance in batch production and (discrete) manufacturing.

10.3.3 State and State Transition

Any item in any topology can be in a certain state. Think of a pump being off or on. Some processes have several states, and their associated items (the car body, the electrical current, and the product flow) transit from one state to another. Usually, the time it takes to go from one state to another defines the number of intermediate states.

State information is of importance for fault diagnosis, as the modern facilities deal more and more with so-called intelligent instruments, which can transmit their state (healthy, faulty, …) to the automation level.

State diagrams illustrate the states a process or a process unit can be in Figures 10.2 and 10.3.

10.4 KNOWLEDGE REPRESENTATION

10.4.1 Introduction

Human beings disseminate knowledge in various ways. The tribal way (up to the Celts in Europe, the aboriginals, and tribes in rural isolated areas) was oral, in poems, chants, and songs. Oral tradition led to myths and sagas,

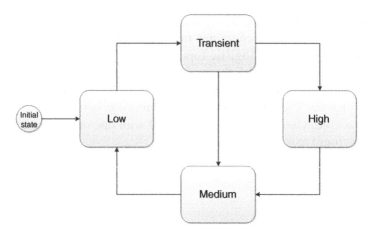

Figure 10.2 Example of a simple state diagram.

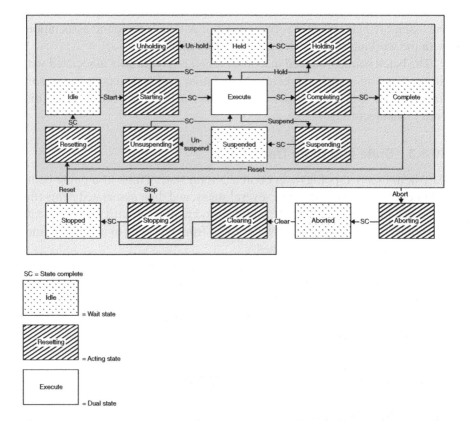

Figure 10.3 Base state model of batch production. Source: By courtesy of ISA–TR88.00.02 (ISA 88 Committee 2015).

without writing. Today, we do not use our auditive memory anymore to pass on or express knowledge. We do use music to express and pass on emotions and spiritual messages.

History (historiography), according to historians, begins with the recording of knowledge in written form. We can distinguish several written (and therefore visually captured) knowledge expressions.

10.4.2 Mathematical

The mathematical "language" is said to be the one and only global language. It uses "universal" symbols and expressions. It is developed to "calculate," which is no more or less than to predict "results" or "estimations" based on a set of data and known (investigated and widely accepted) relationships, such as physical, chemical, economical, probabilistically relationships.

Since these relationships are one of the cornerstones of general knowledge, it is a requirement that a knowledge base system can use the same symbols and equations to express and work with such knowledge. This is, however, not always the case.

Mathematical expressions are also context sensitive. The same expression in one context can have a significant different meaning in another.

Example:

$P = U\,I$. This formula depends on what variables are assigned to P, U, and I. Electrotechnicians might immediately associate P with power, U with tension, and I with current. But for a computer program, this association is meaningless, it is just a multiplication. It could be the surface of an item, whereby P is the surface, U is the length, and I is the width. Mathematics and consequently any equation start with an agreement on terms and symbols. Mathematical equations applied in a domain such as engineering or chemistry require on top of that an agreement on units.

Mathematical expressions in a knowledge base system are used for:

- Validating input data
- Calculating derived data
- Calculating probabilities
- Simulation

In the context of this book, it is only of importance to acknowledge that mathematical expressions are used for various purposes in fault diagnosis.

10.4.3 Linguistic

Natural language (in contradiction to mathematical language) is used to express heuristic knowledge. Typically, for knowledge base purpose, general expressions can be transformed in so-called rules.

Example:

I want climate control in my office. When the temperature is high, I want cold air. If it is too cold, I would like to have warm air.

Can be transformed as follows:

If the temperature in my office is high, then cold air must be blown into my office.

If the temperature in my office is low, then warm air must be blown into my office.

Since we deal with language, each word in such expression has a meaning. A dictionary will assist in defining the meaning of each word used.

In a KBS, these words need to be recognized by the software used. Items like "office" must match a computer memory position to derive what is meant. Also, adjectives like "my" must be understood by the KBS.

In fault diagnosis, the steps to follow to find the (root) are often prescriptive, i.e., in natural language.

10.4.4 Graphical

In many cases, a linguistic or mathematical description will not suffice to describe the entire subject matter. Describing the entire layout of a house just in words or mathematical expressions would require many words. A drawing or sketch of the house tells more than 10,000 words (is a widely used expression).

Human beings have used the capability to draw extensively. Drawings are widely used in building and construction, but also in other engineering practices such as electric schematics, maps, piping and instrumentation diagrams (P&ID), process flow diagrams, but also in areas like risk estimation, fault diagnosis, flow-charts, software modeling, PLC programming, etc.

A graphical representation often expresses a topological model, i.e., how things are connected to each other or what to do or check first.

A graphical language requires an agreement on the use of graphical elements in a given domain (subject matter). The same symbols could be used, but could mean a totally different thing in a different context. For example, a rectangle would mean an operation in a flow chart but would mean a room in an architectural drawing.

It would be beneficial if a knowledge-based system could use that type of knowledge to reason with or to express conclusions.

10.4.5 Conclusion

A knowledge base system should be capable of capturing and using all types of knowledge and be able to express this "captured or applied knowledge" in the most appropriate way.

10.5 MAINTAINING, UPDATING, AND EXTENDING KNOWLEDGE

10.5.1 Introduction

When writing this section title, it is in the context of maintaining, updating, and extending the knowledge in a knowledge-based system. As discussed, there are various ways to represent knowledge.

10.5.2 Progressing Insight

Progressing insight is the main driver for updating and extending a knowledge base. Every day, new discoveries are made on a global scale, but also on a micro-scale, for example, in your plant, your infrastructure, your organization, and the applied legislation to your operations.

In a knowledge base, it should be easy to add this progressing insight, hopefully only once and by preference with the lightest impact on your organization, i.e., in a comprehensible way for all stakeholders (from operators to plant manager).

10.5.3 Software as a Knowledge-Based System

One could argue that any program (software) is a knowledge-based system, as the knowledge on the functions performed (automated by the program) is represented in the programming language.

If the set of input data is in the range of the expected values (and a big chunk of a software application consists of testing and validating that these data are within that range), the knowledge stored in the application can perform its functions, usually a sequence of operations or a more complex algorithms, involving mathematical functions and calculations, logical and sequential operations, topological reasoning, or probabilistic estimations.

The downside of any knowledge stored in a software application is obvious: if you want to update or extend this knowledge, you need a person who can add functions or logic to the existing application in the programming language used.

Typically, in an expert system, the knowledge representation and the code to activate and execute the knowledge are separated.

10.5.4   A Database as a Knowledge-Based System

There are many databases created in the eighties and nineties of the previous century. A database turned out to be an excellent place holder of *information*, and the relations between the data were usually defined by the designer of the database (and thus fixed). Updating the database was/is performed by entering changes in the data manually, by scanning or other means of identification, and by deleting or archiving data after a while.

However, changing the structure of the database (the relationships between the tables) required a well-prepared and well-conceived effort from the database administrator, which turns a database, a good store of information, into a difficult carrier of knowledge.

10.5.5   Search Engines

With the venue of the Internet and HTML, a lot of data (information) from databases and other sources became published and thus publicly available. Internet Explorer programs started using search engines to find relevant (published) generic information. Web search engines became popular in the early 2000s, Google being the most widely used. Search engines do not add information or knowledge; they use clever algorithms to find and link information, based on (hidden) user profiles or (paid) advertisements.

10.5.6   Online Encyclopedia

Before computers came along, information – and to a large extend knowledge – was stored in so-called encyclopedia. The positive side on a published encyclopedia is that it was reviewed by subject matter experts. The downside is that once published it could quickly become obsolete. For example, a bio of a living person would become obsolete as soon as that person would die or change functions or alike.

The success of on-line encyclopedia is that published articles are reviewed by bots (software robots that review contributions automatically) and live persons (like the review by subject matter experts in the printed versions).

On-line encyclopedia like Wikipedia are probably the best available, be it static, knowledge-based systems of this time. They do include mathematical knowledge (formulas), heuristics (often in the form of schema and drawings), topological knowledge (maps and figures), linguistic information, etc.

Downside is that this knowledge is static and generic.

Static in the sense that the knowledge is not applied to real-life data from your plant, for example, that you often cannot feed formulas with data to calculate something, to diagnose something, or alike.

Generic in the sense that they do not contain the topology of your plant, the equipment you use, the recipes applied, etc.

10.5.7   Learning

According to the Cambridge Dictionary, learning is described as:

- To get knowledge or skill in a new subject or activity.
- To make yourself remember a piece of writing by reading it or repeating it many times.
- To understand that you must change the way you behave.
- To be told facts or information that you did not know.

So, learning in a human context is gathering information, digesting it, analyzing it, combining it, and applying it to the subject matter of your interest.

Without any algorithms for learning, a knowledge-based system will be static and become obsolete when not maintained properly. Digitalization is probably the solution to keep knowledge-based systems synchronized with actual reality (as built). One cannot categorize automatic updates of a knowledge-based system as "learning." Machine learning is discussed in "Diagnosis with AI."

10.6   EXPERT SYSTEMS

10.6.1   Introduction

Expert systems (Wikipedia 2022b) are a form of artificial intelligence, where a computer system emulated the decision-making ability of a human expert. They are designed to solve complex problems, such as fault diagnosis, by reasoning through bodies of knowledge, usually and mainly represented as if-then rules rather than through conventional procedural code.

An expert system is divided into two subsystems: the inference engine and the knowledge base. The knowledge base represents facts and rules. The inference engine applies the rules to the known facts and deduces conclusions (or new facts).

10.6.2 Inference Engine

The inference engine is an automated reasoning system that evaluates the current state of the related knowledge base, applies relevant rules, and then asserts new knowledge into the knowledge base.

The inference engine may also include abilities to explain these conclusions, tracing back over the applied rules.

There are two main modes for an inference engine: forward chaining and backward chaining.

10.6.2.1 *Forward Chaining*   A new fact or new data entered would trigger a rule, and the inference engine would try to find other rules that use the obtained conclusion (the that part of an if-then rule) to trigger (fire) other rules.

For example:

If the temperature is over 23 °C, then conclude that it is warm.

If the humidity is greater than 80%, then conclude that it is humid.

If it is warm and humid, then conclude to switch on the air conditioning.

If the air conditioning is switched on, then broadcast an email to close doors and windows.

A new temperature reading would trigger the first rule, which in turn would trigger the third rule. The conclusion of the third rule would fail, until the humidity is high enough. Once both conditions are fulfilled, the third rule would trigger the fourth rule.

10.6.2.2 *Backward Chaining*   Backward chaining is a bit less straightforward. In backward chaining, the system looks at possible conclusions and works backward to see if they might be true. Applied to the above example, one could ask the expert system if it would be good to broadcast an email to close doors and windows. That would look for evidence to support this (potential) action (the if part of the rule). So, the inference engine would look for rules to see if the air condition should be on. The inference engine would find the third rule, which would cause the inference engine to look for the first and second rules.

10.6.3 Generic – Specific

It is definitely an advantage if you can separate the reasoning from the subject matter, you are reasoning about. But, it soon turns out that the same reasoning is nevertheless highly dependent on the subject. To go back to the example: we probably do not want the temperature to be set at the same level in all rooms. If we do, we need to add something to such a rule so that it either becomes specific or starts referring to something specific from the subject, the object, which is being reasoned about.

Suppose there is a cold room in the building that you want the expert system to reason about. Then, you will have to modify the rule, for example, as follows:

If the temperature in the cold room is over 4 °C, then conclude that it is warm.

Suppose you have different types of rooms, or worse, that you want to allow the person in charge of the room to set the temperature themselves, then you will have to add something to both the rule and the knowledge base.

Back to specific rules for a moment. Suppose you have a building with 100 rooms, do you want to set up a rule for each room? Do you want to have that rule maintained by the person in charge of the room, or do you want to just give the person access to the temperature and humidity setting?

In such expert system, quickly hundreds or thousands of rules would need to be created, maintained, and executed, which often led to unmaintainable systems.

In the 1980s, a software revolution began that has carried over into the twenty-first century, namely, the classification of the real world through software classes and instances of those classes, objects. In the knowledge base, one could construct a class "room" and give it attributes, such as the name of the room, the desired temperature, the measured temperature, the desired humidity, and the measured humidity. Maybe also a status to indicate the airco heating, cooling, or doing nothing.

The rules of the above example could then be made generic:

If the measured temperature in any room Room > the desired temperature of the Room, then conclude that the temperature state of the Room is warm.

If the measured humidity in any room Room > the desired humidity of the Room, then conclude that the humidity state of the Room is humid.

If the temperature state of any room Room is warm and the humidity state of any room Room is humid, then conclude to set the airco-state of the Room to on.

If the airco state of any room Room is on, then broadcast an email to the residents of the Room to "close doors and windows of the [the name of the Room]."

Note that deliberately the word "room" is written in both lower case as with a capital. The inference engine would trigger on any update of any measured temperature of any instance of the class "room" in the knowledge base, detect which instances would be inferred, pass on those instances (objects) to the next rule, and so forth. The word Room with a capital R is the subset of the instances of the class "room" in the knowledge base. The rule would also update the attribute "measured temperature" of the inferred instance (object). The third rule would be triggered (fired) by such update and check if the other attributes comply to the conditions of the rule and so forth.

10.6.3.1 Truth Maintenance   One of the problems one must deal with when conceiving an expert system (or, as a matter of fact, any software program, and by consequence any automated fault diagnosis) is the validity of the conclusions. Once a conclusion is reached, it should not be valid forever. A high temperature does not remain high forever. There are two options. One is to iterate the rule (or the program) over the data with a given frequency (a scan frequency). This frequency will depend on the processing power, the number of rules affected, etc. Most PLC programs work this way. The other option is to attribute a validity interval at the data source level. A slow changing variable from a sensor could have a long validity interval, a fast changing a short, independent of the rules applied. As soon as the validity expires, all dependent conclusions could be withdrawn.

Usually, this a better way of doing than to attribute a scan time to a rule.

10.6.3.2 Uncertainty   With the venue of computers and PLCs, so-called crisp login was introduced. For example, if the temperature in a room was over 18 °C, crisp logic would derive that it was warm in the room. If it is less than 18 °C, crisp logic would conclude that it was cold and the logic would start the heating, which would then switch off when the temperature would be over 18°. To avoid switching too often, a so-called hysteresis is usually built in (older) temperature controllers: heating would start (at full force) when the temperature would drop below say 17°, thus the hysteresis would be 1 C. In reality, the human being does not consider 17.9° as cold. Some human beings would consider 18 °C as cold, others as warm, also depending on the type of labor the human being is performing (Figure 10.4).

Similar for an expression like "on time" (Figure 10.5).

If a train or other vehicle is "on-time," that moment lasts exactly 1 second. In all other cases, it is too early or too late.

Fuzzy sets are often defined or represented as triangle or trapezoid-shaped curves, as each value on the horizontal axis will have a slope where

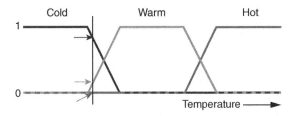

Figure 10.4   Trapezoid function. Source: By courtesy of Wikipedia.

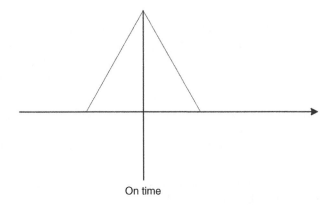

Figure 10.5   Triangular function.

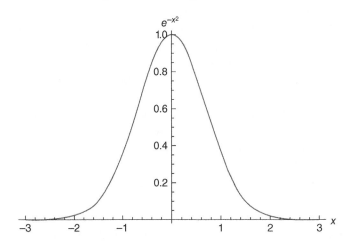

Figure 10.6   Gaussian function.

the value on the vertical axis is increasing, peaking at a value equal to 1, and a slope where the value is decreasing (Figure 10.6).

When using human language, such as in heuristic expressions, the human brain "translates" a given variable (temperature, timeliness, accuracy, quality, etc.) into a truth value. Very often, this is related to perception, and to what the majority of a (given) population perceives as "warm temperature," "on time," "accurate," "good quality," etc.

Fuzzy logic addresses this translation from crisp values into linguistic expressions (Zadeh 1965), but also vice versa, what is meant by defuzzification, i.e., translating a linguistic expression such as "fan moderate" when it is getting hot into an output value to control the fan.

Likewise, any further conclusions in a rule base system should carry forward (propagate) the uncertainty or probability of the original value, especially, when combining uncertainties and probabilities in a reasoning. Usually, this is achieved by multiplying the uncertainties. When the said temperature in a room is reasonably hot (say 50%) and the humidity quite high (say 80%), the outcome of the rule to switch the air conditioning on is 0.5×0.8 or 40%. If there is no means of using the air conditioning installation in a moderate way (on or off), one needs to decide at which calculated value something can be switched on.

Fuzzy "and" logic multiplies likelihoods, so the resulting likelihood becomes smaller.

Suppose your car engine will fail when the fuel is low. The lower the fuel level, the greater the likelihood of failing.

At the same time, the car engine will also fail (or automatically shut off) when the cooling system temperature increases.

When both things happen at the same time, the likelihood of the engine shutting down will become bigger and it might be safer to stop the engine manually, before it is shut down, consequently.

Fuzzy "or" logic adds likelihoods, so the resulting likelihood becomes bigger.

10.6.3.3 Completeness
An expert system shall never be complete. Practitioners of fault tree analysis, failure mode, and effects analysis will agree that no analysis is 100% complete. The more a system is used, the more complete such analysis shall be complete, on condition that the knowledge base is maintained and any non-explained incident properly analyzed.

10.6.3.4 Reproducibility
For a given set of conditions, an expert system should always reach the same conclusion. When this cannot be achieved, the confidence in such expert system will decrease rapidly. Which brings up the next point.

10.6.3.5 Ability to Explain the Inferred Conclusions
When using complex reasoning, applying fuzzy logic, topology reasoning, any conclusion presented to a human operator should be accompanied by an explanation

how such conclusion is achieved and/or the reliability of the conclusion. Typically, weather forecasts include a probability of precipitation. When combining, especially in fault diagnosis, multiple parameters with fuzzy logic, an indication of the reliability of the conclusion is welcome.

If an expert system proposed to replace a sensor because it is broken, it should be able to explain this conclusion, either in language (the sensor is broken because the measured values are not in line with the expected values and all other involved systems are ok) or in a diagram, highlighting the path in the diagram that contributed to this conclusion.

10.6.4 Interactive – Real-Time

Expert systems and knowledge bases can be used in two modes:

- Interactive, on demand
- Real-time, continuously

10.6.4.1 Interactive Typically, using Google or Wikipedia is in interactive mode. A question or problem is formulated in natural language. The knowledge base (or knowledge bases) is queried, and several results presented maybe ranked according to probability of likelihood. An expert system connected to (live) plant data can be much more precise, as it can rule out unlikely results.

As for fault diagnosis, interactive mode could be applied in situations where a user would like to find failures or root causes on a given (alarmed) situation.

10.6.4.2 Real-Time One of the benefits of using software in real-time mode, i.e., connecting a fault diagnosis knowledge base or expert system to real-time plant data is that it can perform real-time diagnosis, 24 hours per day, 7 days a week.

For this purpose, the reasoning schematics or rules need to be provided with real-time data from the plant (e.g., using OPC), historical data (e.g., from a data historian or event logging), or database data (using dynamic queries).

When using graphical objects to represent a fault diagnosis, one can design blocks that serve as real-time data sources, for example, an object that gets updated whenever a new value arrives.

Although very often not perceived as "real-time," updates on selected records in a database or entries added at the end of a log file can also be considered as "real-time" for reasoning purposes.

Such real-time updates can trigger the reasoning schematics the data entry object is used in. This way, fault diagnosis can be performed in near real-time, all the time. The advantage is that a diagnosis can be performed before an unwanted event (alarm) or be included in a warning (alert or other notification).

As stated above, the reasoning used (rules, schematics, …) needs to take into account the validity of such retrieved data from seconds in fast-changing process variables to minutes or hours for data in databases. It needs to take into account the quality of such data, the combined probability of such data, and the persistence of the combined data. Conclusions might need to be withdrawn if data changes, call for action requires a high probability, and a defined persistence of any conclusions.

10.6.5 Polling

In contrast to real-time (forward driven) reasoning, one could also design a system that would query or poll for data only when needed.

In a graphical representation, blocks (objects) can be designed to perform a given query in a database, retrieve selected data from OPC, a data historian, or read an entire or part of a file.

The reasoning schematic is triggered by an event (e.g., an alarm) or upon user (HMI) request.

Such backward reasoning has less trouble with validity consistency (unless the reasoning time exceeds the validity of the polled data) and creates less CPU load.

The disadvantage is that in such mode, the diagnosis comes after the (unwanted) event.

10.6.6 Conclusions

Expert System shells provide the ability to create and test many forms of hypothetical reasoning. An environment that enables development of graphical reasoning schematics, that enables creation and maintenance of (plant or machine) topologies, that can keep track of validities and probabilities, and that is able to explain and demonstrate conclusions or call for actions presented is well fit to perform fault diagnosis.

During the development of such systems, one must be aware that during the life cycle of such diagnosis, many aspects can change, from replacement of instrumentation to changes in the layout of the plant or machine.

As mentioned, fault diagnosis might never be 100% complete, so there exists the challenge to extend and/or modify the diagnostic reasoning.

Incident analysis and change management should, therefore, include a work process of updating the automated fault diagnosis and/or related knowledge base(s).

Automated fault diagnosis which is not maintained will lose credibility over time and become obsolete. It is also of importance to include and involve the end-users (operators, maintenance technicians, contractors, and service providers) in the design and maintenance of such systems (applications). Without their support and involvement, automated fault diagnosis will also become obsolete.

It has been proven around the world that graphical representation of the reasoning is the best way to explain conclusions, maintain the reasoning structure, and facilitate maintenance. Graphical representation is quite universal, is language independent, and has fast neural response from the human brain, especially when the conclusions are supported by highlighting the reasoning path for a given diagnosis.

10.6.7   Documentation

A well-structured knowledge base and/or expert system requires little to no documentation. When reasoning schematics are represented graphically and when the plant, machine, or infrastructure topology is graphically available, all these schematics together establish the documentation.

When no programming code is used, or at least hidden for the common user, the established application can be understood by almost every involved stakeholder.

10.7   DIGITIZATION, DIGITALIZATION, DIGITAL TRANSFORMATION, AND DIGITAL TWINS

10.7.1   Digitizing Information

Digitizing information is what has been done since the invention of the analog-digital converter (ADC) and the use of electronic tools like excel and word.

Process computers of the 1970s and 1980s used analog-to-digital boards to transform measured values to numeric ones for calculation purposes and digital-to-analog cards to provide analog signals to actuators.

Documents were written in word and converted to pdf, so they could be transferred (uploaded, downloaded) in an electronic (digital) way instead of being delivered with paper post. Scanners enabled digitizing paper documents in digital pictures. OCR readers digitize paper information into text.

On the plant design side, the introduction of CAD/CAE tools replaced human drawing by computer drawing, but the base was still a drawing. This transition was a copy of the paperwork processes at the time. Documents needed to be read and reviewed. The information in the document was transferred from the document to the head of the reader and the reader used this information for its right purpose. This required a human intervention, which might be error prone. And the biggest danger was that information would get lost in the transition processes between design, engineering, operation, and maintenance. Many companies struggled with documents (also the electronic copies thereof) not be "as built."

With the arrival of the World Wide Web, it became quickly clear that content (information) and presentation (display of the content) needed to be separated. If web designers could only access the most recent information for their web pages automatically …

10.7.2 Digitalization

A first step in accessing the information in digitized data is to know where the information is located, what form it has (numeric, text, language, …), and what purpose the information serves. Furthermore, information stands not in itself, it needs a context. The value "40" will have a different meaning in process control, accounting, or engineering. A human being could not "ask" an engineering document to find the maximum pressure in a pipe. The human could do a search in the document, but they would not look for the value but for a term like "pressure." If the word pressure is used at several places in the document, it depends on the ability of the human being to assess if this was the value they were looking for. If a human being cannot do it, then a machine cannot either.

In the 1990s, large organizations started to organize knowledge on information and standardizing on a modeling language. This resulted in several ISO standards (Wikipedia 2022a) and the development of eXtensible Mark-up Languages for a given subject. The representation, formal naming and definitions of the categories, properties and relationships between concepts, and information and entities of a given subject matter is called an ontology (Wikipedia 2022c). The World Wide Web consortium specified XML, a markup language and file format for storing, transmitting, and reconstructing arbitrary data (Wikipedia 2022e). The purpose of such language was that it would be both human readable and machine readable.

ISA 88 on Batch control started with a number of definitions on the subject matter and explaining the concepts (ISA 88 Committee 2010), while in Part 2, the required and desired information is listed (as a standard) (ISA 88

Committee 2001). This resulted in a technical report on machines and machine states (ISA 88 Committee 2015) and a mark-up language called PackML. PackML is an automation standard developed by the OMAC that makes it easier to transfer and retrieve consistent machine data (OMAC 2022).

The second step in digitalization is creating a repository to store instances of the subject matter. This can be done as software objects, a database, or other structures. What technology is applied is kept away from the different users. Human beings interact with the information stored through, for example, web pages, often with a track record on modifications made. Software applications can use methods like SOAP to exchange structured information in the implementation of web services in computer networks (Wikipedia 2022d).

10.7.3 Digital Transformation

Rather than moving documents around, work processes such as updating diagnostic applications can be automated. A major problem with diagnostic applications is changes made in the plant. Very often, after a near miss or an incident, a safety function will be added, and alarm settings adjusted. When the diagnostic application is not updated accordingly, it will use the wrong information, probably at the wrong time resulting in faulty diagnosis.

Digital transformation is about automating such workflows or work processes such that updates happen automatically or after approval. This reduces the risk of forgetting something in the automation chain.

10.7.4 Digital Twins

A new but very interesting development these last years is the creation of so-called digital twins. They emerged from the possibility to generate virtual images from digitalized CAD/CAE information. But even in the 1990s, it was envisioned to build a simulation model of the plant that could be used for model predictive control and diagnosis by feeding the model with real-time data and let it calculate data ahead in time. The same models were also used in operator training stations (OTS) to learn operators monitor the plant, operate the plant, and to learn them to react on abnormal situations.

Combining OTS with virtual 3D models resulted in so-called digital twins. The major problem of keeping the digital twin up-to-date has been solved by digitalizing all the required information from CAD, CAE, control logic, and other data sources. Digital twins are both an engineering exchange platform in the EPC phase and for continuous operations, e.g., maintenance, modifications, and optimization purposes.

Digital twins are by consequence an interesting environment to test and validate diagnostic logic.

10.8 FAULT DIAGNOSIS WITH KNOWLEDGE-BASED SYSTEMS

10.8.1 Introduction

Trouble shooting is a logical, systematic search for the source of a problem in a process, product, machine, or system, in order to solve it and make the process, product, machine, or system operational again.

10.8.2 Elimination

Determining the most likely cause is usually a process of elimination. Elimination is a logical method to identify an entity of interest among several options, by excluding all other entities. In other words, deleting options of which the probability (or likelihood) of the option is close to zero or significantly lower compared to other options.

For example, when there exist several events that can cause a failure, it can be represented by an OR diagram. However, the positive output of a crips logic OR gate does not indicate which input became true. So, the applied logic needs to test each input to determine which one is the cause.

For a fuzzy OR gate, the probability (likelihood) of each input is added. Suppose there is a likelihood for input 1 of 40% and a likelihood of another input of 40%, then the likelihood of the output is 80%

Elimination is required when or logic is used to determine the causing event, measurement, or observation. As pictured above, it can come up with more than one candidate for the failure or state.

10.8.3 Inductive (Forward) Reasoning

Inductive reasoning is a method of reasoning in which a body of observations is synthesized to come up with a general principle. The truth of a conclusion of an inductive argument is probable, based on the evidence given.

Types of inductive reasoning include the following:

- Generalization: A number of samples or observations lead to a conclusion. For example, several readings of process variables fail. If the instruments providing such readings are connected to the same bus, it is likely that there is a bus failure. If the instruments are all the same type and from the same supplier, it might be likely that the fault is related to the instruments.

○ *Statistical generalization*: A statistical generalization is a type of inductive argument in which a conclusion about a population is inferred using a statistically representative sample.

Example: Statistically speaking the majority if the Icelandic population is of light skin, so it can be concluded Icelandic inhabitants are of Caucasian origin.

○ *Anecdotal generalization*: An anecdotal generalization is a type of inductive argument in which a conclusion about a population is inferred using a non-statistical sample. This inference is less reliable (and thus more likely to commit the fallacy of hasty generalization) than a statistical generalization, first, because the sample events are non-random and second because it is not reducible to mathematical expression. Statistically speaking, there is simply no way to know, measure, and calculate as to the circumstances affecting performance that will obtain in the future.

Example: This animal has a trunk, big ears, tusks, weighs more than two tons. So, it is concluded that it is an elephant.

• Prediction: An inductive prediction draws a conclusion about a future instance from a past and current sample. Like an inductive generalization, an inductive prediction typically relies on a data set consisting of specific instances of a phenomenon. But rather than conclude with a general statement, the inductive prediction concludes with a specific statement about the probability that the next instance will (or will not) have an attribute shared (or not shared) by the previous and current instances.

• Inference regarding past events is similar to prediction in that one draws a conclusion about a past instance from the current and past samples. Like an inductive generalization, an inductive inference regarding past events typically relies on a data set consisting of specific instances of a phenomenon. But rather than conclude with a general statement, the inference regarding past events concludes with a specific statement about the probability that the next instance will (or will not) have an attribute shared (or not shared) by the previous and current instances.

• Inference regarding current events is similar to an inference regarding past events in that one draws a conclusion about a current instance from the current and past samples. Like an inductive generalization, an inductive inference regarding current events typically relies on a data set consisting of specific instances of a phenomenon. But rather than conclude with a general statement, the inference regarding current events concludes with a specific statement about the probability that the next instance will (or will not) have an attribute shared (or not shared) by the previous and current instances.

Typically, inductive reasoning seeks to formulate a probability. Inductive reasoning is also called bottom-up logic.

10.8.4   Deductive (Backward) Reasoning

Deductive reasoning is the process of reasoning from one or more statements (premises) to reach a logical conclusion. If all premises are true, the terms are clear, and the rules of deductive logic are followed, then the conclusion reached is necessarily true.

- Modus ponens: The first premise is a conditional statement. The second premise is the antecedent. The conclusion deduced is the consequent. Example:
  If the room temperature is higher than 18 C and lower than 24 C, then it is warm in the room.
  The room temperature is 20 C.
  Conclusion: The room is warm.
- Modus tollens: Also known as the law of the contrapositive. It validates an argument that has as premises a conditional statement or formula and the negation of the consequent and as conclusion the negation of the antecedent. Example
  When the car engine is running, there is fuel available.
  There is no fuel.
  Thus, the car engine is not running.
- Syllogism: Takes two conditional statements and forms a conclusion by combining the hypothesis of one statement with the conclusion of another. Example:
  If a device contains a CPU, it is a computer.
  A smartphone contains a CPU.
  Hence, a smartphone is a computer.

10.8.5   Conclusions

Different types of reasoning can be used. When designing a system, it must be taken into account that there are different ways to reason and that some reasoning mechanisms provide a probability rather than a 100% trustworthy conclusion. The rules used or formulated do not always cover the full spectrum of all possible cases or combinations of events or occurrences. Outside that spectrum/classification, the drawn conclusions are not valid and should not be presented to a user (they can be presented to the designer of the system).

10.9 GRAPHICAL REPRESENTATION OF FAULT DIAGNOSIS

10.9.1 Introduction

As described, the best way to represent a diagnosis is graphical. How to develop such diagnosis is described elsewhere in this book. In this section, several graphical representations are explained.

10.9.2 Fault Trees

A fault tree diagram is a key feature of fault tree analysis (FTA) and is best used to explain the relationship between events. Though the FTA diagram is not the intended result or purpose of fault tree analysis, it helps safety and reliability engineers spot and address issues that could lead to multiple undesired events.

10.9.2.1 Key Elements
Top event　The top event in a fault tree is the fault, the undesired event (Figure 10.7).

Intermediate event　The intermediate event is something that can be derived from a combination of other events (Figure 10.8).

Figure 10.7　Top event.

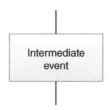

Figure 10.8　Intermediate event.

Basic event The basic event represents a root cause for an intermediate or top event. To create a basic event, one might need other blocks to convert data point(s) or events to a basic event (Figure 10.9).

Gates OR gates are used to combine different events (Figures 10.10–10.12).

Figure 10.9 Basic event.

Figure 10.10 Different OR gates.

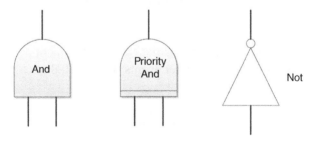

Figure 10.11 Different AND gates and NOT gate.

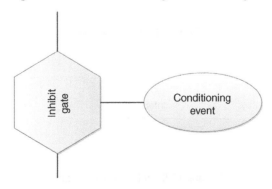

Figure 10.12 Inhibit gate.

Connection to data from the plant   See Figure 10.13.

Parsing block   Parsing blocks are typically used to isolate some relevant data from an incoming string (sent by a device). Typically, different string functions can be used to get the data out. Depending on how such block is implemented, it might require some (basic) programming (Figure 10.14).

Decision block   A decision block is used to compare a value from a data entry point, with a given value, and convert it to a truth value to typically establish a basic event or root cause event. It can also be used to compare a part of string from an event entry block with a target value, e.g., to check the quality of an OPC data point (should be "OK") (Figure 10.15).

Example of a root cause   In this example, not only the pump current is compared to be zero, it also includes a test on the quality of the OPC data.

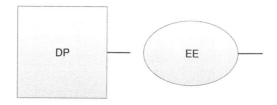

Figure 10.13   Connection blocks.

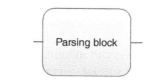

Figure 10.14   Parsing block.

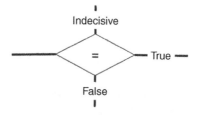

Figure 10.15   Decision block.

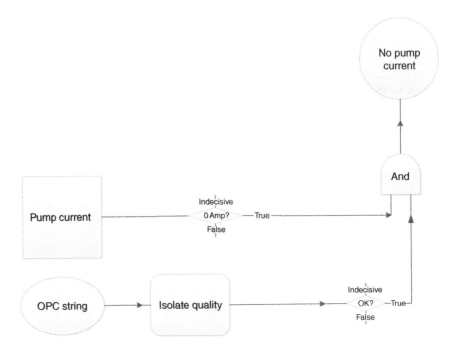

Figure 10.16   Example of a root cause generation diagram.

If both are OK, then it is reasonably certain to conclude that there is no current for the pump (Figure 10.16).

Such diagram can be extended with the presence of the supplying voltage. If there is voltage and no current, that is different from no voltage and no current.

10.9.2.2 *More Blocks*   One can imagine more (mathematical, statistical, time series, ...) blocks to validate input values, in order to filter out unwanted data, false data (dead signal), fleeting data (spikes and dips), and/or correlate data with other data to determine more root causes.

Examples   In the above example, there is no direction in the graphs. The purpose is to predict a trip when all three conditions occur at the same time. If real-time data would be provided to generate basic events, it will be data-driven. If, on the other hand, it is used when a compressor trip occurs, it would be used top-down. It could also be used polling every minute from the top event down to basic events (Figure 10.17).

The basic "events" can be established by (complex) logic.

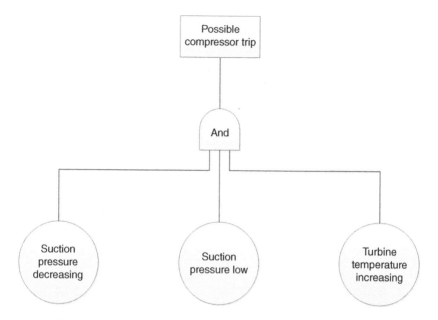

Figure 10.17   Example of a fault tree without direction graphs.

10.9.3   FMEA

Failure modes and effects analysis (FMEA, see Appendix 2A) is a systematic, proactive method for evaluating a process to identify where and how it might fail and to assess the relative impact of different failures, in order to identify the parts of the process that are most in need of change. FMEA includes review of the following:

Steps in the process:

- Failure modes (What could go wrong?)
- Failure causes (Why would the failure happen?)
- Failure effects (What would be the consequences of each failure?)

In a knowledge-based system – with a graphical representation – this is very often organized in schematics for a set of instances (e.g., pump defect FMEA tree for all pumps, i.e., all instances of the class pump in the application) or for specific object (e.g., the main batch reactor of the plant).

The method usually requires brainstorming sessions on what could hypothetically go wrong.

FMEA can combine fault trees with further effects. So basically, the same building blocks can be used (Figure 10.18).

As one can see, FMEA extends FTA. Very often spreadsheets are used to capture all possible causes for a given effect. But a textual representation is

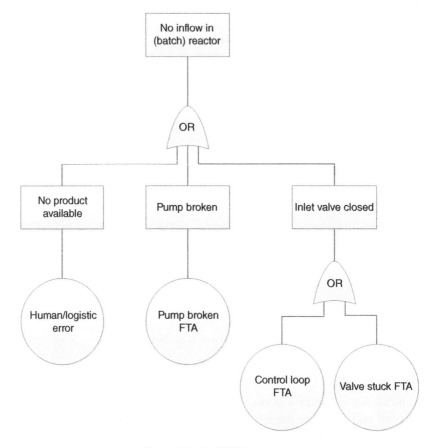

Figure 10.18 FMEA example.

less comprehensible than a single drawing. A good knowledge-based system will enable such drawings and usually offers the possibility to zoom in, e.g., clicking on a basic event will show even more detail on how such conclusion reached.

FMEA is also used for risk assessment. This leads us to yet another representation, described in the next section.

10.9.4 Bowtie Analysis

The Bowtie method is a qualitative risk evaluation method that allows an organization to determine the risks present in an organization and the preventative and control measures available. It can be used to analyze and demonstrate causal relationships in high-risk scenarios. The method takes its name from the shape of the diagram that you create, which looks like a man's Bowtie. The center of the Bowtie is the undesirable event or hazard. A Bowtie diagram does two things. First of all, a Bowtie gives a visual summary of all plausible accident scenarios or chains of events that could exist around a certain Hazard. The Bowtie diagram provides a powerful graphical representation of the risk assessment process which is readily understood by the "non-specialist."

The way it does that is by combining three different concepts. The first method is the fault tree method, which covers the left side of the Bowtie in a different form. Second method is the event tree that can be seen on the right side of the Bowtie top event. Lastly, the causal factors charting method, which is most likely the origin of Escalation factors.

The second thing the Bowtie does is to identify control measures that an organization has in place. This type of thinking is more easily explained with the famous "Swiss Cheese model" which was first described in the early nineties (Reason 2008).

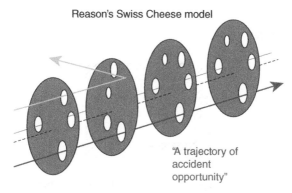

Reason's Swiss Cheese model

"A trajectory of accident opportunity"

Reason proposed the Swiss Cheese metaphor as an accident causation model. He hypothesized that hazards are prevented from causing losses by a series of controls, known as controls in the Bowtie method. Reason states that these controls, however, are never 100% effective. Each control has unintended or time varying weaknesses and when these so-called holes line up, a hazard can be released and consequences be triggered, e.g., explosion, flaring, and spills.

Also according to Reason, the common causes of the weaknesses in controls can often be found in the organization (latent failures). In the Bowtie method, these weaknesses are defined as escalation factors and are important features to fight the illusion of control that organizations sometime tend to have.

10.9.4.1 *Example*   The threats that will cause the trigger to escape are identified and listed to the left. The consequences of the top event have been identified and displayed to the right. The level of risk associated with each consequence can be assigned according to the organization's risk matrix and displayed on the diagram. Now that the hazard has been defined, preventive controls which will stop the event from occurring and recovery controls which will prevent the outcome can be put in place. Controls can be linked to specific sections or procedures of the organization's safety management system, can have responsibilities and inspection frequency assigned, and be ranked and color-coded for effectiveness (Figure 10.19).

The diagram above represents a typical Bowtie diagram. On each control, there is the ability to identify:

- The task
- Responsibilities for the task
- Classification of the control as appropriate
- Any document that may be applicable

10.9.4.2 Blocks
Hazard

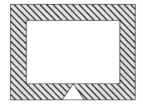

Hazard: the source of the risk to be managed. It suggests that it is unwanted. However, in Bowtie diagrams, it is inherent to business, e.g., making ethylene, moving oil or gas, or a flight guidance system.

The hazard needs to be managed, and as long as it is under control, it is of no harm.

Top event

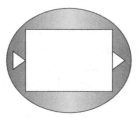

Certain events can cause the hazard to be released. The top event is not yet a catastrophe. The dangerous characteristics of the event are now in the

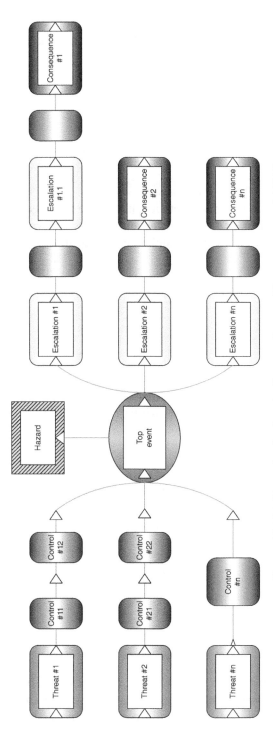

Figure 10.19 Example of a Bowtie diagram. Source: Blaauwgeers et al. (2013).

open. Example: too high pressure in an ethylene unit, oil outside the pipeline, two airplanes approaching each other. If not mitigated correctly, it can result in more unwanted events – consequences.

Consequence

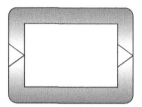

A consequence is a potential event resulting from the release of the hazard. Results in direct loss or damage. Unwanted events are events which "by all means" need to be avoided. Example: people get injured or killed, loss of image.

Threat

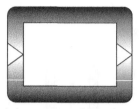

A threat is a factor which can cause a top event. Every threat itself can cause the top event.
Examples:

- Corrosion of a pipeline
- Lack of cooling in an exothermic reaction
- Radar failure in an aircraft

Control or barrier

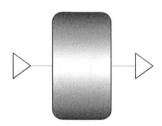

A control is a barrier to prevent certain events from happening. Control = any measure that acts against an undesirable force or intention. Control = maintain a desired state. Pro-active controls on left side of the diagram (prevention). Reactive controls on right side of diagram (recovery or mitigation).

Escalation factor

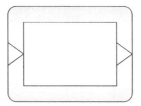

An escalation factor is a condition which makes a control fail or a condition resulting in increased risks.

Examples:

- Blow out preventer defect
- Inspection of pressure relieve valve
- Software error in emergency shutdown system

10.9.5 Complex Event Processing

Complex event processing (CEP) is a set of techniques for capturing and analyzing streams of data as they arrive to identify opportunities or threats in real time. CEP enables systems and applications to respond to events, trends, and patterns in the data as they happen. CEP is considered largely synonymous with event stream processing, though CEP is typically more associated with searching for complex patterns and dependencies in the incoming data.

Sensors in manufacturing facilities and large objects such as windmills and airplanes continuously collect data to determine whether there are any patterns indicating it is necessary to shut down equipment for predictive maintenance.

CEP usually looks at patterns of events to identify a specific event. Both the event detection and the underlying processes to get there could be quite complex. For example, a CEP application may recognize the sound of drums, the cheers and clapping of a crowd, the sounds of hounds, and the change in frequency of the instruments to put together a pattern that lets it determine that a specific type of event was occurring – a marching band was playing.

Algorithms can be configured to find patterns up to a length n within a timeframe T. The algorithm will then find patterns like A-B, A-B-C, A-B-C-D, and A-B-C-D-E within its runs, given there is occurrence (minimal

Timestamp	Event
15-Aug-2009 13:45:20	A
15-Aug-2009 13:45:25	B
15-Aug-2009 13:45:40	C
15-Aug-2009 13:45:50	D
15-Aug-2009 13:46:05	E
15-Aug-2009 13:46:10	E
15-Aug-2009 13:46:30	A
15-Aug-2009 13:46:40	A
15-Aug-2009 13:47:05	E
15-Aug-2009 13:47:10	D
15-Aug-2009 13:47:30	E
15-Aug-2009 13:47:40	D
15-Aug-2009 13:48:30	A
15-Aug-2009 13:48:35	E
15-Aug-2009 13:48:45	A
15-Aug-2009 13:48:50	B
15-Aug-2009 13:48:55	C
15-Aug-2009 13:49:05	B
15-Aug-2009 13:56:20	B

In this example a time window of 90 seconds has been used to produce the results below:

Pattern	# occurrences	Average time
EA	5	00:40.000
AE	4	00:27.500
ED	4	00:35.000
EE	4	00:37.500
EDE	4	00:63.750

Figure 10.20 Example of sequence of events mining algorithm. Source: Dubois et al. (2010).

occurrence of two) of such patterns in the database or log file under investigation, whereby the last event in the sequence is the unwanted one (Figure 10.20).

Patterns found by the above algorithm can be directly imported into fault trees to predict unwanted events (Figure 10.21)

Event pattern mining has two important pitfalls:

1. The event list can be corrupted by so-called chattering events, i.e., events recorded several times in each time frame, for example, due to bad control loop tuning, defect instruments, retries, etc. Those events will correlate with all other events.
2. Event sequences of faults rarely occur in today's automation. An algorithm as proposed above must occur at least twice in the event log scanned, to be recognized as a pattern.

For fault prediction, event analysis is rarely used. However, for incident analysis, event analysis is of utmost importance. Good software tools should be available for engineers to analyze this enormous stream of events. Software tools for

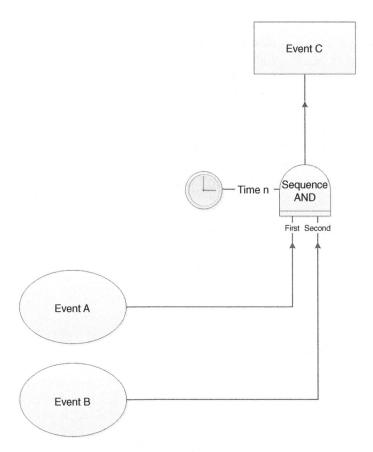

Figure 10.21   Example of a sequence of events fault tree.

complex event processing (like alarm and event historians) will probably force the organization to harmonize (e.g., by parsing into the same time stamp format) and synchronize events from different (sometimes remote) sources (e.g., by synchronizing clocks or taking into account different time zones).

As described in FTA, combinations of events with process data are common in fault diagnosis.

10.10   CONCLUSIONS

Knowledge-based systems need careful design. Rule-based engines are usually difficult to maintain as they rely on a set of rules produced by a domain expert. This makes it difficult to maintain as experienced domain experts are always needed when the system is changed or evolves.

A graphical representation of the (expert) knowledge is far better than a set of rules that are backward or forward chained. It offers the advantage to find the path that was followed when a conclusion (diagnosis) was reached.

Digital transformation will improve the maintenance of knowledge-based systems as updates on plant layout, applied instrumentation, etc. could be done semi-automatically, reducing the maintenance cost for diagnostic applications in such environment.

Expert system technology, extended with capabilities for artificial neural networks (see chapter), topology reasoning, graphical representation of fault trees, failure mode and effect analysis, Fourier analysis, and pattern recognition algorithms, is quite fit for advanced and automated fault diagnosis.

Automated fault diagnosis will reduce the time to repair, downtime, and searching for root causes.

In the section on alarm management (see Chapter 3), it is explained how these capabilities can be linked to alarms or alarm displays to support operators in industrial control rooms.

REFERENCES

Blaauwgeers, E., Dubois, L., and Ryckaert, L. (2013). Real-time risk estimation for better situational awareness. In: *IFAC HMS*, 6. Las Vegas: ResearchGate Retrieved from https://www.researchgate.net/publication/258846650_Real-Time_Risk_Estimation_for_Better_Situational_Awareness.

Cambridge Dictionary (2023). *Knowledge*. C. U. Press Retrieved from https://dictionary.cambridge.org/dictionary/english/knowledge.

Dubois, L., Forêt, J.-M., Mack, P., and Ryckaert, L. (2010). Advanced logic for alarm and event processing: methods to reduce cognitive load for control room operators. In: *IFAC HMS*, 5–6. Valenciennes: ResearchGate.

ISA 88 Committee (2001). *ANSI/ISA 88.00.02-2001 Batch Control Part 2: Data Structures and Guidelines for Languages*. Research Triangle Park, NC: ISA.

ISA 88 Committee (2010, December 6). *ANSI/ISA 88.00.01-2010 Batch Control Part 1: Models and Terminology. ANSI/ISA-88.00.01-2010*. Research Triangle Park, NC: ISA.

ISA 88 Committee (2015). *ISA-TR88.00.02-2015 Machine and Unit States: An implementation example of ANSI/ISA-88.00.01*. Research Triangle Park, NC: ISA.

Merriam-Webster. (2023). Merriam-Webster Dictionary. USA. Retrieved from https://dictionary.cambridge.org/dictionary/english/knowledge

OMAC. (2022). PackML. Retrieved from: https://www.omac.org/packml

Reason, J. (2008). *The Human Contribution*. Farnham, Surrey: Ashgate Publishing Company Limited.

Wikipedia. (2022a, 09). *ISO 15926*. Retrieved from: https://en.wikipedia.org/wiki/ISO_15926.

Wikipedia. (2022b, January 20). *Expert System*. Retrieved from https://en.wikipedia.org/wiki/Expert_system.

Wikipedia. (2022c, October 28). Ontology (Information Science). Retrieved from: https://en.wikipedia.org/wiki/Ontology_(information_science)

Wikipedia. (2022d, August 31). SOAP. Retrieved from: https://en.wikipedia.org/wiki/SOAP

Wikipedia. (2022e, October 26). XML. Retrieved from: https://en.wikipedia.org/wiki/XML

Zadeh, L. (1965). Fuzzy sets. *Information and Control* 338–353.

10.A

COMPRESSOR TRIP PREDICTION

Lieven Dubois

ISA 18 and 101, Working Group 8 of ISA 18.2, Department of Alarm Management and Training, Manage 4U, Leiden, Netherlands

When I was working for an A.I. company, active in alarm management, I was invited to present the technology at a major compressor plant, one of those where a pipeline from oil & gas fields came ashore.

The reason for my visit was they were suffering from too many alarms during compressor trips, exceeding the numbers imposed by the regulator. I proposed to analyze their alarm data to investigate how the technology could assist in reducing that number.

They provided us with one year of data on three compressor stations. Within that amount of alarm and event data we found 28 trips. Our consultant rang the plant and indicated we would not investigate 28 trips as a (free) presales effort. They answered we could pick two. We also wanted to have the P&IDs of the stations to make sense of the alarms and events. Topological knowledge is of importance to distinguish consequential alarm from possible causal events. Providing P&IDs required a NDA and approvals. Upon reception of a pile of drawings, our consultants could make sense out of the tag numbers in the alarm and event list.

Artificial Intelligence in Process Fault Diagnosis: Methods for Plant Surveillance,
First Edition. Edited by Richard J. Fickelscherer.
© 2024 John Wiley & Sons, Inc. Published 2024 by John Wiley & Sons, Inc.

340

The A.I. technology provided the possibility to play-back alarms and events in fast mode. Consequently, we could demonstrate in just a couple of minutes which alarms were annunciated during a compressor trip.

With this facility on board, the consultant could make a demo on how alarm reduction rules would reduce the number of alarms during a trip, by hiding redundant alarms, consequential alarms, and non-relevant alarms.

The consultant also discovered that the entire trip could be predicted by defining states and combining those with trends of some key process variables.

Armed with these two demos, I went back on site and demonstrated how the technology could play-back alarms, then activated the alarm reduction rules and demonstrated how some of these alarms were hidden or associated. The alarm rate during a trip was reduced from 86 alarms to 12, which was an acceptable level.

Next, I activated the prediction rules and demonstrated how the trip could have been announced minutes before it happened. The people attending the demonstration were astonished. One of the two trips analyzed was caused by running out of fuel while the compressor was running on fuel and not on gas. Somebody had removed or suppressed the alarm that fuel was low. Probably because it was a stale alarm. The other trip could be predicted by a combination of state detection and trend analysis (also known as fault diagnosis).

The potential customer agreed to build a prototype and bring this on-line. A budget was provided. The consultants succeeded in retrieving process data from one compressor train only because the OPC DA server would bring down the Local Control Network when subscribing to all process variables. It was then agreed to do the proof of concept on only one compressor train. In just a couple of days, the technology allowed to construct a small topology (based on the P&IDs provided), to create state detection diagrams and establish fault diagnosis rules to predict compressor trips. The DCS engineers were amazed that with such technology, state diagrams could be put on-line so quickly, as they had been trying to establish those within the DCS for more than a year.

When the demo was ready, the software was installed in the control room by the engineering department and our consultant. To avoid operators having to watch yet another monitor, the software application annunciated the compressor prediction by playing a mp3 file. The demo ran for three months and not a single trip occurred on the monitored train.

When being interviewed, the operators informed management they did not use the application, but just monitored the train better.

The visiting consultant discovered (from reading the announcements in the hallways) that the operators would receive a bonus if they could keep the flaring (usually caused by a trip) under the amount of a quarter million per month.

Talking to the control room operators, our consultants found out they did indeed not use the application (by logging in, something which could be found back in the event log of the application), but of course they heard the announcement (the mp3 file played) and then did everything on DCS level to prevent the trip from happening.

After silencing the application, sure a compressor trip happened within six weeks.

The engineering department made the (psychological) mistake not involving the operators in this prototype. The operators feared the application was designed to steal their bonusses, while it could have been presented as an aid to help them to achieve bonusses. After all, the bonusses were just a fraction of the cost each compressor trip caused.

Note: Because of non-disclosure agreements, the said companies, consultants and site cannot be mentioned/quoted.

11

THE FALCON PROJECT

Richard J. Fickelscherer, PE[1] and Daniel L. Chester[2]

[1]*Department of Chemical and Biological Engineering, State University of New York at Buffalo, Buffalo, NY, USA*
[2]*Department of Computer and Information Sciences (Retired), University of Delaware, Newark, DE, USA*

OVERVIEW

The FALCON Project investigated developing real-time, online knowledge base systems for actual advanced process control applications, specifically automated process fault analysis. The resulting FALCON System was a knowledge-based system capable of detecting and diagnosing process faults in a commercial adipic acid plant then currently owned and operated by DuPont. This chapter briefly describes: (i) the operation of DuPont's adipic acid process; (ii) the FALCON Project and its methods and procedures followed for studying and successfully solving the problem of actually automating process fault analysis; and (iii) the four main software modules of the resulting fielded FALCON System. It concludes by discussing the advantages of using the knowledge-based system paradigm for studying and solving ill-formed, real-world problems.

Artificial Intelligence in Process Fault Diagnosis: Methods for Plant Surveillance,
First Edition. Edited by Richard J. Fickelscherer.
© 2024 John Wiley & Sons, Inc. Published 2024 by John Wiley & Sons, Inc.

CHAPTER HIGHLIGHTS

- Detailed description of an actual expert system program development project.
- Emphasis on its particular fault diagnostic strategy selection and further methodology refinement through a trial and error approach.
- Overview of resulting Method of Minimal Evidence (MOME) quantitative process model diagnostic strategy.
- Summary of the advantages of employing the knowledge-based system paradigm in solving real-world problems.

11.1 INTRODUCTION

The *FALCON* (Fault AnaLysis CONsultant) Project was a joint venture between the University of Delaware, DuPont, and the Foxboro Company. Formally initiated in January 1984, its main objective was to develop a knowledge-based system (KBS) capable of performing continuous, real-time process fault analysis online in a commercial scale adipic acid plant operated by DuPont in Victoria, Texas. Officially commissioned at the plant on January 20, 1988, process operators used it online to analyze abnormal operating conditions in an adipic acid process. However, due to problems with the transmission of plant data to the FALCON System, it was decommissioned on April 15, 1988.

The major motivation for the FALCON Project was to identify general issues involved with developing knowledge-based systems for actual advanced process control applications, specifically automating process fault analysis. Its ultimate goal was to develop a generalized approach for doing so. Such a generalized approach would allow process fault analyzers to be rapidly and inexpensively developed and more easily maintained. This effort directly led to the development of the model-based diagnostic strategy known as the Method of Minimal Evidence (MOME) (Fickelscherer 1990,1994). This inevitably led to the development of a software package known as FALCONEER™ IV (Fickelscherer et al. 2005, Chester et al. 2008, Fickelscherer and Chester 2013), which automates the MOME algorithm in a Fuzzy logic implementation. FALCONEER IV enables the creation of competent and robust process fault analyzers automatically merely from quantitative engineering models describing normal process operation. This directly converts the solution of the problem of fault analysis into solving the much simpler problem of process modeling.

The overall scope and objectives of the original FALCON Project are described in greater detail by Lamb et al. (1985). See Rowan (1986, 1987, 1992) and Rowan and Taylor (1989) for DuPont's perspective of the FALCON Project, and for the Foxboro perspective, see Shirley and Fortin (1985) and Shirley (1986, 1987a, b).

11.2 THE DIAGNOSTIC PHILOSOPHY UNDERLYING THE FALCON SYSTEM

Tradeoffs exist in attempting to have a fault analyzer always correctly diagnose every possible process fault situation. These tradeoffs occur, as explained in Chapter 8, because there exists a spectrum of possible diagnostic sensitivities for the potential process fault situations at each level of diagnostic resolution sought.

One extreme of this spectrum represents very sensitive fault analyzers at the highest level of diagnostic resolution. These analyzers sometimes venture diagnoses based on patterns of evidence that are inadequate for the diagnostic discrimination desired. The major drawback of these systems is their tendency to misdiagnose faults either by identifying the wrong fault when a process fault situation exists or by concluding the presence of a fault when none exists. These types of misdiagnoses potentially create very confusing situations for the process operators that directly could easily lead to inappropriate operator actions. Consequently, fault analyzers performing in such a manner have the potential for making dangerous process operating situations even more dangerous. DuPont emphasized clearly at the beginning of the FALCON Project that such behavior by the FALCON System would rapidly diminish its credibility with the process operators and its further continued usage.

The other extreme of possible fault analyzer sensitivity at the highest level of diagnostic resolution represents conservative systems that only venture diagnoses based on definite and complete patterns of evidence. These systems tend to misdiagnose by concluding that no faults exist when a fault situation actually does. This type of misdiagnosis may lull the process operators into believing that the process system is operating normally, which in turn may lead to a serious time delay before appropriate operator actions are taken. Given their preference, DuPont considered this type of misdiagnosis to be more tolerable. This is because the process operators are already trained to closely monitor process operating conditions and to follow specific safety precautions whenever these operating conditions become abnormal. Consequently, the diagnostic knowledge base of the fielded FALCON System was designed to try to ensure that it would only misdiagnose in this manner. In this capacity, the FALCON System acted as an intelligent operator assistant, giving competent advice if it could, and prudently remaining silent if it could not.

11.3 TARGET PROCESS SYSTEM

The major motivation for the FALCON Project was to study the general issues involved with developing knowledge-based systems for automated fault analysis. The selection of the target process system was consequently considered the most critical factor in the project's eventual outcome. The main characteristics that the target process system should have were, therefore, defined a priori. These characteristics are fully described by Lamb et al. (1985) and are summarized here.

The target process system needed to be complicated enough to make the project's results meaningful. However, it should not be so complicated that it would be difficult to develop an operable fault analyzer within the time constraints of the project. The target process system would also need to have an automated data collection system already in place. A final requirement was that since the results of this project would eventually be published, the candidate target process system had to be a relatively low proprietary to facilitate publication of the results.

DuPont selected the adipic acid process as the target process system for the FALCON Project mainly because it was an established, highly developed, and widely known technology. Adipic acid is the monomer reactant in the polymerization of nylon-6,6. Since DuPont was firmly established as one of dominant producers and users of adipic acid worldwide, the threat posed by proprietary information inadvertently being released as a result of the project was minimal. Moreover, a wide variety of serious process fault situations, including explosions, had previously occurred in these plants. Consequently, developing an operable fault analyzer for the adipic acid process system could potentially have significant direct safety benefits. It was decided to focus the initial efforts on one particular process subsystem, the adipic acid process's low temperature converter (LTC) recycle loop. This subsystem is the main reactor of the process and met all of the specified requirements for the project's target process.

A schematic of the process side of the LTC recycle loop is given in Figure 11.1A. The main process unit in this system is the LTC itself. The LTC is a water cooled, plug flow reactor. Its main function is to remove the heat of reaction evolved from the oxidation via nitric acid (referred to as NAFM) of a mixture of cyclohexanone and cyclohexanol (referred to as TWKA) into

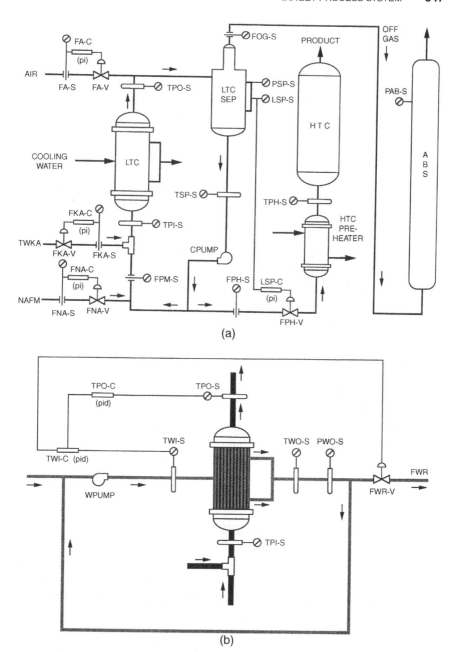

Figure 11.1 (A) DuPont adipic acid plant LTC recycle loop. (B) Figure DuPont adipic acid plant LTC cooling system.

adipic acid. The heat of reaction is removed from this highly exothermic reaction by cooling water flowing through the shell-side of the LTC. This reaction also produces a relatively large quantity of process off-gas. Air is injected into the process stream at the LTC's outlet to dilute the process off-gas in order to reduce the chances that it can ignite. The resulting gas/liquid mixture is separated in a unit called the LTC separator. The gas stream is sent to an absorber where the nitric oxides are recovered and recycled. The process liquid is pumped out of the separator, with the bulk of that liquid being recycled back to the LTC. The residual reactant in the product stream is oxidized in a unit known as the high temperature converter (HTC). The product stream is then sent to crystallizers where the adipic acid is removed.

The normal operation of the LTC occurs within a very narrow range of process temperatures. If the process temperature gets too high, the process liquid will degas, causing the process pressure to increase to the point where it could potentially rupture the process equipment. If the process temperature gets too low, the dissolved adipic acid will reach its saturation point and crystallize out of solution. When this occurs, the LTC's tubes become plugged and only a partial oxidation takes place in LTC. This operating problem is known as "frosting" and leads to incomplete reaction within the LTC. The remaining reactants are then converted in the downstream units not designed to remove the heat of reaction. This situation could also cause the process liquid to degas and consequently rupture the process equipment. To minimize the chances of such faults, the process temperature is maintained within the narrow range of temperatures with an elaborate temperature control system. A schematic of the LTC's cooling system is given in Figure 11.1B.

11.4 THE FIELDED FALCON SYSTEM

The fielded FALCON System consisted of two hardware components and five major software modules. The two hardware components were a DEC MicroVax II computer with 9 megabytes of RAM memory and a HP 2397A Color terminal with touchscreen capability. The five software modules were (i) the inference engine, (ii) the human/machine interface, (iii) the dynamic simulation model of the adipic acid process, (iv) the diagnostic knowledge base, and (v) the process data collection system. The process data collection system was developed by DuPont to interface the FALCON System with the plant's data collection computer system. The other four modules were developed primarily by the University of Delaware. Figure 11.2 illustrates

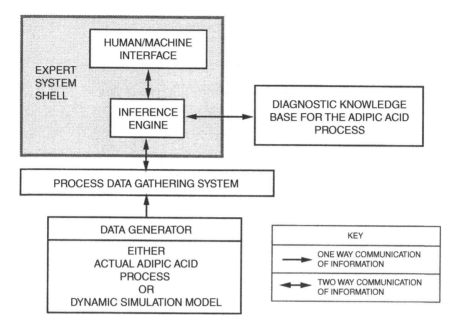

Figure 11.2 FALCON system structure.

how all of these modules are interconnected. The dynamic simulation was the only software module that was not used by the FALCON System when it was operating online in the plant: it was used mainly in the offline development and verification of the diagnostic knowledge base. Each of the four modules developed by the University of Delaware is briefly described below.

11.4.1 The Inference Engine

The purpose of the inference engine was to analyze real-time process data with the diagnostic knowledge base for the adipic acid process. The inference engine was written in Vax Common Lisp (over 1,600 lines of Lisp code) and ran under a VMS operating system on the DEC MicroVax II computer. Briefly, the inference engine was designed to: (i) accept a time-stamped data vector containing 31 process variables every 15 seconds from the process data collection system, (ii) analyze these data with the diagnostic knowledge base, and (iii) then report its conclusions to the human/machine interface. Each such inference cycle took approximately three CPU seconds to complete. Some additional features were included in the inference engine for checking the integrity of the process data obtained

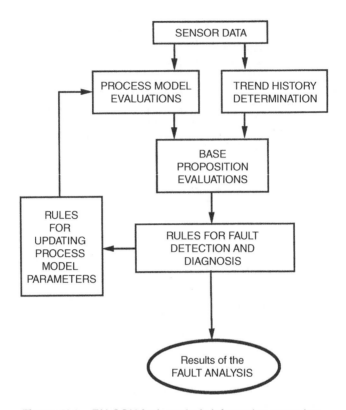

Figure 11.3 FALCON fault analysis inferencing procedure.

from the plant and the communications with the process data collection system and the human/machine interface. Except for these checks, all of the inference engine's conclusions were derived logically from the diagnostic knowledge base. The inferencing procedure used to reach those conclusions is outlined in Figure 11.3. The inference engine's logical structure and operation are described in greater detail by Dhurjati et al. (1987).

11.4.2 The Human/Machine Interface

Briefly, the purpose of the human/machine interface was to allow the process operators to interactively understand the reasoning process used by the inference engine to make its diagnoses. For each potential diagnosis, the program could also explain why the other possible process fault situations were not plausible. The human/machine interface consisted of the HP 2397A Color graphics terminal with touch screen capability and an interactive explanation program written in VAX FORTRAN-77 (over 4,000 lines of

FORTRAN code). The program communicated with the process operators via the touch-screen terminal. In this communication, it gave both the inference engine's conclusions and the process information used to reach those conclusions.

The database of the human/machine interface consisted mainly of predetermined explanation text inter-spliced with variable identifiers. Those identifiers got instantiated with current process information supplied by the inference engine once a fault situation was diagnosed. The predetermined explanation text was organized into a series of interconnected explanation trees, the roots of which were the initial fault announcements made to the process operators. These initial fault announcements were only intended to alert the operators and identify the fault situation present. As such, they did not give supporting evidence for the fault hypothesis tendered, nor did they explain the diagnostic reasoning used to derive that hypothesis. Such information was stored at lower levels in the explanation trees. As one progressed down through those trees, the level of technical detail given by the explanations progressively increased. The program thus allowed the operators to control the level of detail in the explanation they received while reviewing the fault analyzer's conclusions. The program worked in a similar fashion when explaining why other possible process fault situations were not plausible hypotheses. A much more detailed description of the development and operation of the human/machine interface and the lessons learned from its use at the plant were presented by Mooney et al. (1988).

11.4.3 The Dynamic Simulation Model

The main purpose of the dynamic simulation model was to act as a plant substitute during the development and testing of the FALCON System (Fickelscherer et al. 1986). This program was developed in an interactive simulation package available at the University of Delaware known as DELSIM. The model was written in FORTRAN-77. It contained 277 simultaneous ordinary differential equations, over 1050 variables, and over 300 parameters (over 7,300 lines of FORTRAN code). All of the parameters, except seven, were either the actual physical dimensions of the process equipment or the actual physical properties of various process compounds and mixtures. One of the seven parameters whose value was not equal to its actual process value was the LTC separator level controller proportional band that had to be detuned for stable operation of the simulation. The other remaining six parameters were estimates of the equivalent length of pipe used in the pressure drop calculations. On average, these had to be set to values five times larger than the theoretically predicted values in order to get the proper two-phase process and water recirculation flow rates (Olujic 1981). No other tuning of the dynamic simulation model was required.

The Fourth-Order Kutter-Meyerson integration algorithm with variable integration step size was used to solve the model. The average integration step size required was approximately 0.3 seconds. Solving the model took approximately three CPU minutes per minute of simulated process behavior while running under a UNIX operating system on a DEC VAX 11-780 computer. Although this comprehensive dynamic simulation model required major computational efforts to solve, it was definitely felt that the benefits derived from being able to accurately predict the actual process behavior over its entire range of possible operating conditions would be well worth such efforts. It directly allowed the University personnel to gain a detailed understanding of both normal and abnormal process behavior without spending an inordinate amount time at the plant and without requiring frequent conversations with the busy plant personnel.

The accuracy of the simulation for predicting the actual process behavior was verified via a three-step formal verification procedure (Fickelscherer et al. 1986, Fickelscherer 1990). From this protocol, it was determined that the dynamic simulation model was indeed a very accurate representation of the actual process system.

The main purpose of the dynamic simulation model was to act as a plant substitute during the initial development of the fault analyzer. Process fault situations would be simulated with the dynamic simulation model, and the results would be used to test the performance of the fault analyzer. The dynamic simulation would thus be both a powerful and a highly flexible fault analyzer development tool. The dynamic simulation was subsequently used extensively in both the development and verification of the FALCON System's current knowledge base. A total of over 260 process fault situations were simulated and used to test the fault analyzer's performance capabilities.

This need for process fault data was acute. Of all the process fault situations that the fault analyzer was expected to diagnose, actual plant data existed for only about 30% of them. And for all but a few of those fault situations, the actual fault's exact magnitude and rate of occurrence were unknown; they could only be estimated from the process data itself. Furthermore, the available process fault data typically existed for each such fault situation at just one set of the possible process operating conditions.

In the target adipic acid process, an actual process fault situation occurs approximately once a month. Moreover, the ones that do occur reoccur relatively more frequently than the vast majority of the other potential process faults (e.g., LTC frosting). In fact, most of the potential process fault situations actually occur either very infrequently or have never occurred. Consequently, this made obtaining actual process fault data for those faults impossible.

Dynamic simulation models represent the best means for creating this extremely useful fault data. In our case, all of the various possible process fault situations could be simulated at a variety of known magnitudes and rates of occurrence, and over the entire range of process operating conditions. Process fault situations that would be extremely hazardous if they occurred in the plant can be induced quite safely in the dynamic simulation.

Another benefit from creating the dynamic simulation model was derived from consolidating and codifying the various diverse sources of process knowledge that existed. This turned into a recursive procedure for learning about the actual process system's behavior during both normal and abnormal operation. The development of the dynamic simulation model required a highly detailed understanding of the process system's topology and normal operation. This understanding was obtained from: (i) a plant visit, (ii) process P&I Diagrams and operating manuals, (iii) FORTRAN programs for a steady-state process simulation model of the LTC recycle loop and the process system's automatic defrost control algorithm, (iv) extensively interviewing and corresponding with DuPont engineers, and (v) inspecting actual plant data. Taken together, these diverse information sources contained an enormous wealth of process knowledge that greatly improved our understanding of the adipic acid process system itself and its behavior during both normal and abnormal operation. Elicitation of this knowledge was critical for the eventual success of the FALCON Project.

Observing and correcting the behavior of the dynamic simulation model during its development taught us a great deal about the behavior of the target adipic acid process system. It also helped to accelerate us along the learning curve associated with thoroughly understanding both normal and abnormal process behavior. Furthermore, this highly detailed, fundamental understanding of process behavior made it possible for us to eventually develop a highly competent fault analyzer. Thus, having first developed the comprehensive dynamic simulation model was a major factor in the eventual success of the FALCON Project. The actual process behavior was just too complex for it to be adequately explored by any other means within the time constraints of the project.

Consequently, there were many invaluable benefits derived from having available the high fidelity dynamic simulation model of the adipic acid process during the development of the FALCON System. Obviously, however, it is impractical to develop such simulations for every process system for which one wanted to build a process fault analyzer. Many process systems are not understood well enough to enable such high fidelity dynamic models to be created. Even for those that are understood well enough, some maybe are still so complicated that any benefits derived from using the dynamic simulations would not justify the time and effort required to create them. Moreover, even for those process systems for which accurate dynamic simulations can be readily created, the inordinate amount of time and effort required to exhaustively verify the corresponding fault analyzers' performances using simulated fault data greatly reduces both those fault analyzers' cost effectiveness and their feasibility. Furthermore, such exhaustive validation strategies would further require that a high fidelity dynamic simulation be maintained along with the fault analyzer after any process modifications occurred. Such a situation actually occurred during the FALCON Project when the actual adipic acid process was modified. Its associated ramifications to the maintainability of the fault analyzer are described below.

Fortunately, in general, dynamic simulation models of the target process system are not necessary in order to develop competent and robust fault analyzers. As discussed in detail in Chapter 8, fault analyzers based on the Method of Minimal Evidence (MOME) do not require specific testing of the fault analyzer's underlying diagnostic logic. Rather, competent and robust performance requires only that the engineering models of normal process operation are all well-formulated.[1] This major advantage greatly simplifies the creation, verification, and maintenance of these programs, substantially reducing the necessary effort to develop competent and robust fault analyzers by many orders of magnitude over that required by the FALCON Project.

11.4.4 The Diagnostic Knowledge Base

The diagnostic knowledge base of the fielded FALCON System was capable of competently diagnosing approximately 160 potential process fault situations, 60 of which could be announced to the process operators. It was created by employing a model-based diagnostic strategy known as the Method of Minimal Evidence (Fickelscherer et al. 1987, Fickelscherer 1990, 1994). This knowledge base was written in Vax Common Lisp and contained approximately 800 diagnostic rules (over 10,000 lines of Common Lisp code). It was developed over a 3-year period in collaboration with DuPont. The knowledge base was tested on over 6,000 hours of both actual and simulated plant data to verify its capability to competently diagnose process fault situations in the adipic acid process system.

---

[1] The concept of well-formulated process models is defined in Chapter 8.

11.5 THE DERIVATION OF THE FALCON DIAGNOSTIC KNOWLEDGE BASE

A brief description of the development effort exerted to create the FALCON System's fielded knowledge base is given here. This includes the knowledge engineering protocols undertaken to study the solution of the fault analysis problem by human experts and how that led to the derivation of the MOME diagnostic strategy to automate process fault analysis (Fickelscherer 1990, 1994). A discussion of the lessons learned from FALCON's performance on plant data during its development was given by Dhurjati et al. (1987). The results of the online plant test were presented by Varrin and Dhurjati (1988).

The fielded FALCON System was the result of a 3-year development effort that occurred between December 1984 and January 1988. This development effort mirrored those of other knowledge-based systems reported in the literature.[2] It had four separate stages: (i) a rapid prototype development, (ii) the full system development, (iii) the full system verification, and (iv) a 10 -month online plant evaluation. The first two steps were done entirely at the University of Delaware (the University), the third was done jointly at the University and at DuPont's Engineering Department Headquarters (Louviers), and the fourth was performed at the DuPont adipic acid plant (Victoria), with University and Louviers personnel implementing the necessary knowledge base modifications.

11.5.1 First Rapid Prototype of the FALCON System KBS

An initial prototype of the FALCON System was rapidly developed as a proof of concept program. Its diagnostic strategy was based on qualitative physics (Chester et al. 1983). As discussed in Chapter 7, qualitative differential equations used to describe dynamic process operations are called confluences. Correctly describing physical systems with confluences is a procedure known as envisionment. The resulting fault analyzer was inherently ambiguous due to the loss of diagnostic process detail directly resulting from this envisionment. This diagnostic strategy was quickly abandoned as a methodology for obtaining the desired real-world program performance. It led to a formal search for a more robust diagnostic strategy.

---

[2]A large number of actual knowledge-based system applications are referenced in Chapters 2 and 10.

11.5.2   The Fielded FALCON System Development

The following describes the development effort required to create the fielded FALCON System.

11.5.2.1   *A Better Understanding of the Problem Domain*   Actual plant data collected during both normal operating conditions and some specific process fault situations became available to the University during 1985. This information source, more than any other, made us appreciate how difficult the task of building a competent, real-world fault analyzer was going to be. This program would have to correctly operate in an extremely complex domain, especially during non-normal process system operation. Being able to predict such operating behavior was essential in order to properly determine the required patterns of diagnostic evidence, regardless of the diagnostic strategy utilized, to be employed to correctly classify the current process situation.

As discussed previously, the fielded fault analyzer was developed after the comprehensive dynamic simulation model was completed. This was the University's primary tool to accurately predict process behavior under all conceivable situations. Once a verified simulation was available, numerous test cases were created in order to observe the behavior of the process in various operating states, especially in fault modes causing emergency process shutdowns. This greatly added to our understanding of how the process behaved during highly transient situations. Such situations were of interest because the fault analyzer would certainly encounter emergency process shutdowns online in the plant. Consequently, measures would have to be taken to ensure that the fault analyzer would react appropriately.

11.5.2.2   *Knowledge Engineering with the DuPont Engineers*
While establishing this improved understanding of the problem domain, a search began for a more suitable diagnostic strategy to use in the FALCON System. We met frequently with DuPont engineers and discussed the merits of the various possible strategies.

These meetings helped us rule out using methods related to fault tree analysis. Based on their experience, the DuPont engineers believed that in general developing and maintaining accurate fault trees would be too time consuming, too difficult, and too expensive to be practical.

For other practical reasons, we also temporarily ruled out trying to incorporate the dynamic simulation into the diagnostic strategy, either as a whole, in parts, or at reduced levels of modeling complexity. The reasons included (i) concerns about the real-time operation of the resulting fault analyzer, (ii) problems encountered in our attempts to develop a realistic and efficient diagnostic strategy based on using the dynamic simulation, (iii) the large amount of time and effort that would generally be required to develop and verify actual dynamic simulations accurate enough for this purpose, and (iv) the fact that if a high fidelity dynamic simulation model was used in this manner, it would also have to be maintained as part of the fault analyzer.

One diagnostic strategy discussed that did show promise was quantitative model-based reasoning. Such strategies base their diagnoses on the evaluation of fundamental conservation equations and process devise models. However, it appeared from our initial investigation that there were too few useful equations and models derivable for our process system to allow for the unique discrimination of any fault situations. This was partially due to the particular collection of process sensor variables then being currently monitored and partially due to the limited analysis being performed on the available models at that time.

11.5.2.3 *Protocol Experiments with the DuPont Engineers*   In order to get past this bottleneck in strategy selection, it was decided that we should more closely investigate the diagnostic strategies used by the actual process engineers. The FALCON Project was very fortunate to have access to one particular process engineer, Steve Matusevich, who had over 25 years of experience at Victoria. During the course of the project, he was our chief contact at the plant for answers to specific process questions. He also helped to develop and verify the dynamic process simulation, reviewed the human/machine interface explanations, and helped to develop, review, and verify the diagnostic methodology used in the knowledge base. The FALCON Project also had access to other DuPont engineers located at Louviers. These individuals were mostly consultants by profession and, as such, they tended to be specialists at solving specific engineering problems rather than being experts about the operation of any particular process system. It was decided to structure the investigation of diagnostic strategies in such a fashion that the effects of these different backgrounds on the engineers' approaches to performing fault analysis could be determined and compared. Two sets of protocol experiments were devised to accomplish this.

The first set of protocol experiments was designed to confront the engineers with simulated process fault situations and then monitor how they analyzed that data to derive plausible fault hypotheses. During their analysis, the engineers would be encouraged to describe what they were thinking about and, if required, to ask for additional process data. After a diagnosis was made, the engineers would be asked to summarize what they believed were the significant reasoning steps they had used to reach their conclusions. If their specific diagnosis was incorrect, the engineers would then be given the opportunity to uncover the mistakes present in their reasoning process. These experiments were called the "Mystery Fault Diagnosis" protocols.

The second set of protocol experiments was designed to simply ask the engineers what response patterns they would expect to observe in the process variables if a particular fault situation was occurring in the plant. These results would be compared to the corresponding results of dynamic simulation runs. Any discrepancies between the two results would be resolved through discussion. During their analysis, the engineers would be encouraged to explain their reasoning and to discriminate between significant and non-significant variable responses. This was especially encouraged for the response of process variables that were potentially ambiguous. These experiments were called the "Anticipated Fault Response" protocols.

The Mystery Fault Diagnosis protocols were first performed with Duncan Rowan and Tim Cole, two consultant engineers from Louviers. Both had a familiarity with the adipic acid process system, but did not consider themselves to be process experts. Without having much experience with the actual process system to guide them in their analysis, they tended to take a systematic and highly structured approach in order to reach their fault diagnoses. Their approach was based almost exclusively on evidence derived from the evaluation of conservation equations and process equipment models. Although it would sometimes take them a considerable amount of time to reach a conclusion, their diagnoses were generally correct.

Similar protocol experiments and interview meetings were held with Steve Matusevich. These interviews had the dual objectives to determine how he performed fault diagnosis and to collect his troubleshooting heuristics. From the protocol experiments, we found out that he relied heavily upon trend analyses, causal reasoning, and troubleshooting heuristics. Consequently, he usually could very quickly recognize and correctly identify process fault situations using just those diagnostic techniques. He would resort to evaluating conservation equations for additional evidence only when he was confronted with very difficult process fault situations, i.e., those situations in which his troubleshooting heuristics did not hold or in which the causal reasoning he used gave poor discrimination between plausible candidate fault hypotheses.

The Anticipated Fault Response Protocols were also conducted with Steve Matusevich and Duncan Rowan. From these experiments, it was learned that other than process system topology and an overall understanding of the process system operation, process experts use very little process specific knowledge when predicting process trends: they rely mostly on standard causal reasoning and engineering judgment. With respect to fault situations that could cause a potentially ambiguous response for a given process variable, the expert's judgment was very case specific. For cases in which the magnitude and/or the propagation rate of a disturbance could not be accurately determined, Steve Matusevich took the position that he would accept any of the possible responses of particular variables as long as they did not violate any possible causality argument. None-the-less, he was capable of correctly predicting the responses of particular variables in ambiguous situations once approximate estimates of the magnitude and rate of occurrence of the disturbance were known.

From these two sets of protocol experiments, we determined that comparable results could be obtained in many process fault situations by using either qualitative causal reasoning or quantitative model-based reasoning. However, in difficult process fault situations, the additional diagnostic information available in the quantitative models was required in order to derive unique fault hypotheses. Consequently, it was decided to use quantitative model-based reasoning as the primary diagnostic strategy to create the fielded fault analyzer.

11.5.2.4 *Primary Model Derivation Issues* The following three examples describe some of the real-world issues concerning the derivation of potential quantitative (a.k.a., Primary) models for performing fault analysis.

Example of Rate of Occurrence Limits on Model-Based Reasoning

Example 11.1   The following example will demonstrate the limits of diagnostic sensitivity on detecting and diagnosing faults with engineering models alone. Consider the DuPont adipic acid process shown in Figure 11.1A. An overall mass balance for this system can be derived as follows:

$$P1 \qquad 0 = \text{FKA-S} + \text{FNA-S} + \text{FA-S} - \text{FOG-S}$$
$$- \text{FPH-S} - K_{\text{SEP}}{}^* \left[d(\text{LSP-S}) / d(\text{time}) \right]$$

Now consider that there is a slow drift in the LTC separator level sensor LSP-S. The PI controller on this level would maintain its set-point, but the actual separator would either fill or drain depending on the actual direction of the drift. If this drift was below the diagnostic sensitivity of P3 above to detect, the fault would go unnoticed until additional evidence of the fault situation manifested itself. In the adipic acid process, this would be as follows.

Case 1: Actual LTC Separator Level Gets too Low   The FPM-S and FPH-S meter readings would become highly erratic as more and more off-gas becomes entrained in the recycle and product streams, respectively, cavitating the process pump.

Case 2: Actual LTC Separator Level Gets too High   The FOG-S meter reading would become highly erratic as more and more process liquid becomes entrained in the off-gas and hits the associated flow meter.

Example 11.1 demonstrates the need to augment the MOME logic with heuristics based on sensor trend analysis. These heuristics are intended to fill in the gaps in the model residual calculation sensitivities resulting from the normal background noise inherent in the various sensor measurements and process. In FALCONEER IV, all process variables and Performance Equations (see Chapter 8) can have Exponentially Weighted Moving Averages (EWMAs) continuously calculated to determine if those variables are in control or not (see Appendix 6.A). Erratic behavior of FPM–S, FPH-S, and/or FOG-S sensors would be easily detected by continuously performing these EWMA calculations.

Example of Data Sampling Limits on Model-Based Reasoning

Example 11.2   The following example will also demonstrate the limits of diagnostic sensitivity on detecting and diagnosing faults with engineering models alone. Again consider the adipic acid process shown in Figure 11.1B. The LTC process side exit temperature (i.e., TPO-S) is controlled by a cascaded PID controller scheme. The controller output on the master loop is calculated with the following equation:

$$TPO_{error} = TPO\text{-}S - TPO\text{-}SP$$

$$P2 \qquad 0 = TWI\text{-}SP - (K_p^* \, TPO_{error} + \int(K_i^* \, TPO_{error})d(\text{time})$$
$$+ K_d^*[d(TPO_{error}) / d(\text{time})])$$

Simulating the behavior of a PID feedback controller with P2 above (representing the most fundamental description of the PID controller's normal behavior) would be possible only if the controller's input variables (controlled temperature TPO-S, set-point TPO-SP, and controller constants K_p, K_i, and K_d) and output variables (set-point TWI-SP) were all being measured. In addition, the controlled variable would have to be sampled frequently enough so that the integral and derivative action of the controller could be closely approximated. Without an adequate sampling frequency, the agreement between the actual and simulated responses would be poor during highly transient process operating conditions. It would thus be difficult to accurately monitor whether or not the controller was operating properly.[3]

Example of Formulating Deep Knowledge from Heuristics

Example 11.3   A heuristic that was used by process operators of the DuPont adipic acid process (Figure. 11.1A) was that under normal circumstances, the temperature rise between the LTC separator and the LTC reactor (i.e., TSP-S–TPO-S) was normally about +3 °C. A rigorous steady-state reactor model based on the known kinetics of adipic acid formation and the treatment of the overhead pipe to the separator as a two-phase plug flow reactor and the separator itself as a CSTR reactor leads to the following Primary model:

$$P3 \quad 1.0 - X = ((cp_p{}^* \, (TSP\text{-}S - TPO\text{-}S) * exp$$
$$(-K * fact * k((TSP\text{-}S + TPO\text{-}S) \, / \, 2))) \, /$$
$$((FKA\text{-}S + FPM\text{-}S) \, / \, rho_p \,))$$
$$((1.0 + (k(TSP\text{-}S) * (LSP\text{-}S * A + B))) \, /$$
$$((FKA\text{-}S + FPM\text{-}S) \, / \, rhop))$$

where k (Temperature) = first-order reaction rate constant for that temperature; cp_p = process liquid heat capacity; rho_p = process liquid density; fact = fraction of overhead pipe containing process liquid

$$([FPM\text{-}S + FKA\text{-}S] \, / \, rhop) \, / \, (1.0 + (FOG\text{-}S^*C^* \, ((TPS\text{-}S + 273) \, /$$
$$(PSP\text{-}S + 14.7))))$$

K, A, B, C are process specific constants, and X = Overall TWKA conversion factor which can range from 0.0 to 1.0.

A value of 1.0 indicates agreement between the various measurements used in the calculation, especially the temperature rise of (TSP-S–TPO-S).

This model calculates the actual overall TWKA conversion factor very precisely during all steady-state process operating conditions based on the

---

[3]In the original FALCON KBS application, actual process operating data were sampled every 15 seconds. This interval was not frequent enough to accurately calculate the derivative term in the PID controller model, and thus, this primary model was not used by the KBS. However, primary models of all four PI controllers were included in the KBS.

underlying so-called deep knowledge (i.e., in this case, the engineering models of underlying reactor configuration).

The potential usefulness of a given Primary model for detecting assumption variable deviations should always be carefully considered during its derivation. For instance, although it is possible to derive models codifying experiential heuristics, the usefulness of such models' residuals for producing diagnostic evidence is usually very limited.

Firstly, all of the models' underlying modeling assumption variables may not be fully understood. As a general rule, qualitative process models (i.e., those based on experiential heuristics or by creating confluences through envisionment rather than employing first principles) require many more modeling assumptions, many of which are implicit, to specify the process states in which the model correctly describes normal process behavior. This observation has also been made by Koton (1985). Typically, the more implicit modeling assumptions required, the more likely it is that some of those assumptions will be overlooked. Such overlooked assumptions directly lead to misdiagnoses by the fault analyzer whenever they deviate significantly.

Secondly, even if all of the modeling assumption variables associated with a given qualitative process model are known, determining the actual mathematical relationship that exists between all these assumption variables is usually more difficult.

Finally, the resulting qualitative model's residual normal variance typically is also larger than the corresponding first principle model used to describe the same phenomenon. This directly reduces the best possible sensitivity of the resulting fault analyzer for those associated assumption variable deviations.

Consequently, to generalize the lessons of Examples 11.2 and 11.3, the following philosophy should be always followed when formulating Primary models of normal process behavior: with the only limits being the precise variables being measured and the frequency at which their corresponding measurements can be sampled, the most fundamental models of normal operating behavior should always be derived and evaluated by MOME. This brings to bear the best relationships (i.e., covers the most normal operating situations) which hold between the various assumption variables under normal operation. These models can be as complex as need be. The fault analyzer can then directly utilize as much as possible the same engineering knowledge used to originally design those processes in intelligently monitoring their daily operation.

11.5.2.5 Second Rapid Prototype of the FALCON System KBS
The second prototype of the FALCON System based on quantitative model-based reasoning was completed on February 1986. It was written in

Franz Lisp and was run on a DEC VAX 11-780 computer under a UNIX operating system. It could diagnose approximately 50 simulated process faults with a knowledge base that contained approximately 80 diagnostic rules (approximately 1000 lines of Franz Lisp code). The target faults being analyzed for had been selected by DuPont engineers as a realistic minimum set of possible process problems that DuPont felt the fault analyzer should be able to diagnose. This set became known as the *Minimum Fault Set*. The various process faults included in this set were failure of either process pump; malfunctions in all controllers, control valves, and control variable sensors; restrictions of feed and product flows; and loss of cooling in the LTC. By providing a minimum competency criterion against which its resulting performance would be judged, the Minimum Fault Set gave us a performance goal to focus on during the fault analyzer's development.[4]

The entire diagnostic methodology utilized by this prototype was reviewed by both DuPont and Foxboro engineers in order to check its soundness. Meeting their approval, the prototype was then tested with over 100 cases of simulated fault situations. This testing gave encouraging results and also helped to validate the logic of the model-based reasoning employed. Based on these testing results, it was decided to continue developing this prototype into a version that could be field-tested at the plant.

11.5.2.6 Fielded FALCON System's Knowledge Base Development The development of the fielded FALCON System's knowledge base from the prototype's knowledge base took place between March 1986 and mid-September 1986. It caused the knowledge base to increase from approximately 80 diagnostic rules (approximately 1000 lines of Franz Lisp code) to over 800 diagnostic rules (over 10,000 lines of Common Lisp code). This increase in knowledge base size was necessary to handle the complexity inherent in the actual process system behavior. The diagnostic strategy employed was improved upon as a result of the lessons learned, with it finally evolving into a more general diagnostic strategy known as MOME. By mid-September 1986, the fault analyzer was delivered to DuPont for an independent verification by Louviers engineers of its performance capabilities.

The enhancements described above made the fault analyzer capable of analyzing the plant for faults over the entire range of process production rates and operating temperatures, and during either steady- or unsteady-state operation. The entire FALCON System at that time was rewritten in VAX Common Lisp running under a VMS operating system on a DEC

---

[4]It set the intended scope of the fielded fault analyzer.

MicroVax II computer. Once this conversion was completed, a formal demonstration of the FALCON System was given at Louviers to DuPont and Foxboro personnel. At this demonstration, the FALCON System was shown to be capable of correctly diagnosing five plant data files that contained actual process fault situations.

11.5.2.7 *Process State Identification Used by the FALCON System*   After the formal FALCON System demonstration to DuPont and Foxboro personnel, a comprehensive attempt to formally verify the knowledge base commenced. A major problem that became more apparent during further testing with plant data concerned the process models being used to detect and diagnosis faults. There was a definite need to limit the conditions under which those various models were considered valid representations of the actual process behavior. It was evident that major process upsets such as pump failures and interlock activations invalidated many of the fundamental assumptions used during the development of those models. Consequently, it became necessary for the fault analyzer to always determine the current operating state of the process before the diagnostic rules were applied. Such determination would be required in order to determine which of the specific modeling assumptions were still valid and thus which of the process models would be appropriate in the fault analysis. In order to do this, process events such as process startups, shutdowns, and interlock activations needed to be explicitly monitored for by the FALCON System. The state logic typically required to thoroughly perform this type of analysis is presented in great detail in Appendix 5.A.[5]

Adding this capability for automatically determining current process state had several consequences. First, it allowed the FALCON System to be turned on when the process was either operating or shutdown: the fault analyzer automatically determined which state it was in. It also allowed the FALCON System to be run continuously, regardless of the plant state transitions that occur. More importantly, adding this capability also logically structured the entire knowledge base according to the patterns of evidence contained within the various diagnostic rules. This reduced the fault analysis of the incoming process data to an ordered, sequential search through the entire set of possible process faults. This search sequence in effect constituted a diagnostic priority hierarchy between the various possible process fault situations. The logical structure of this hierarchy within the FALCON System's knowledge base is illustrated in Figure 11.4. Discovering the

---

[5]This is the actual pseudo-code derived for the original FALCONEER program for monitoring the current operating state of the FMC electrolytic sodium persulfate process.

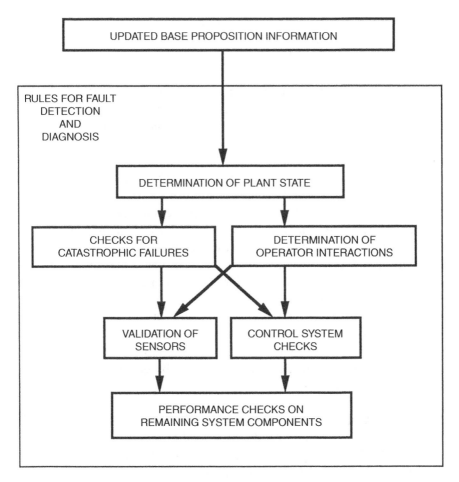

Figure 11.4 FALCON knowledge base search precedence hierarchy.

rationale behind it represented a major development in the model-based reasoning paradigm used as the primary diagnostic strategy.

These enhancements caused a substantial increase in the size and complexity of the knowledge base. The upgraded version of the FALCON System's knowledge base contained approximately 800 diagnostic rules (over 10,000 lines of Common Lisp code). It was capable of detecting, diagnosing, and announcing all of the 50 target process fault situations specified in the Minimum Fault Set, plus also an additional 10 sensor faults, along with about 100 additional fault situations which could be detected and diagnosed but which were not announced directly to the process operators, most of them diagnosing malfunctions in the interlock shutdown system. It was also capable of detecting, diagnosing, and announcing a few extremely dangerous fault situations that can occur during process startups. This updated version of the FALCON System's knowledge base was tested extensively with plant data throughout the Summer of 1986.

11.5.2.8 A Lesson Learned About Automating Fault Analysis

Incorporating the few diagnostic rules for the process startup fault situations demonstrated the true potential of rule-based diagnostic programs: it is not necessary to be able to diagnose all possible process fault situations in order to be able to derive substantial benefits from automated process fault analyzers. We were told by our domain expert, Steve Matusevich, that if any of the rules used to detect and diagnose the process startup faults ever fired even once, it could potentially eliminate a process accident whose overall cost could pay for the entire FALCON Project at least 10 times over (approximately $10 million US dollars). This estimate was based on the actual process damage and subsequent downtime caused by one of the aforementioned startup faults that caused an explosion. In a meeting at Louviers between University and DuPont engineers, it took less than 10 minutes to develop the rules required to diagnose those startup fault situations. Moreover, the resulting diagnostic rules very easily could have been implemented independently in a relatively simple FORTRAN program. Clearly, even limited attempts at automating process fault analysis subsequently can potentially have enormous benefits

11.5.2.9 Effects of the Process Modifications on the FALCON System

In mid-December 1986, the second formal demonstration of the entire FALCON System was given at Louviers to DuPont personnel. At that time, the FALCON System was shown to correctly diagnose three process fault situations that had occurred in plant since the previous formal demonstration 5 months earlier. The progress that had been made in the knowledge base verification at both the University and Louviers was assessed, and plans for its completion were discussed in light of a new development at Victoria.

This new development was that the cooling water system of the target adipic acid process system had been modified 2 weeks earlier. Consequently, the fault analyzer now had to be modified correspondingly and reverified. It was felt that this would be a good opportunity to examine just how much difficult it would take to maintain the knowledge base when such process modifications occurred. Such an understanding was useful to the engineers in further evaluating the costs, effort, and time required to maintain this fault analyzer. This certainly would not be the last time such process modification would occur. At Victoria, as in all modern chemical plants, changes occur on a periodic basis as their process systems evolve and/or are improved by their engineers.

The cooling water system of the adipic acid process had undergone modifications. One control loop was completely removed and another control signal was re-routed to automatically operate the cooling water return valve. The effects that these changes would have on the knowledge base were anticipated well in advance of the actual process modification and were outlined in a report submitted to DuPont during the Summer of 1986.

One purpose of this report was to request that the cooling water pressure sensor originally scheduled for removal when the modifications occurred be instead left in place. It had been determined that either this pressure sensor or, alternatively, the addition of a thermocouple to measure the makeup cooling water temperature would be required to maintain the fault analyzer's current level of diagnostic discrimination. If both measurements were available, it would also be possible to directly discriminate between faults in the existing two LTC cooling water thermocouples in all of their possible failure modes. Victoria engineers analyzed the situation and decided to leave the pressure sensor in place.

The other purpose of this report was to request that particular process information be sent to the University once the process modifications had actually occurred. This information was necessary in order to incorporate the appropriate changes in both the knowledge base and dynamic simulation model. This information included: (i) the new tuning constants for the cascaded PID controllers in the cooling water system, (ii) estimates for the expected ranges of fluctuation for the makeup cooling water temperature and feed pressure, and (iii) plant data covering a wide variety of process operating conditions. The plant data were mostly needed to derive a new valve curve correlation for the cooling water control valve.

Since the changes resulting from the process modification were all localized to the cooling water system, the knowledge base required only relatively minor modifications. Four potential process fault situations were eliminated by the removal of the one control loop, and only one new potential fault situation had to be analyzed for as a result of the new process configuration. The changes required to the knowledge base were the removal of about 20 diagnostic rules (approximately 160 lines of Common Lisp code) and the addition of about 5 diagnostic rules (approximately 40 lines of Common Lisp code). It took about 1 hour to determine which rules needed to be removed and to derive those rules that needed to be added. It took about 3 hours to completely implement those changes into the knowledge base. However, it took 3 weeks to formally verify that those changes were correct!!! And it would have taken much longer had the process side of the adipic acid process been affected by the process modifications. Fortunately, because the modifications had been restricted to only the cooling water system, we were able to determine that the diagnostic rules affected could not interact with the vast majority of the diagnostic rules within the knowledge base.

The reason it took so long to formally re-verify the knowledge base was a direct result of the approach we were using to perform it. In order to adequately test the changes in the knowledge base, we had to simulate 15 separate fault situations. The average simulation run would simulate about 2 hours of process time. This would take between 6 and 8 hours of CPU time to accomplish on one of the University's DEC VAX 11-780 computers. Since it was a time-sharing machine, it would take anywhere between 24 and 36 hours of actual elapsed time, depending on the machine load and the other inefficiencies associated with time-sharing, in order to complete the necessary CPU cycles. Once a given fault simulation was completed, it would take 30 minutes to test the fault analyzer with the resulting simulated fault data, and only about 1 minute to check the results. No changes were required in the knowledge base as a result of this concerted effort: the initial changes made were all correct.

The time delay encountered in the re-verification resulted from the fact that we were really performing two verifications at once. The first verification was of the various quantitative models directly affected by the plant modification. However, in this situation, there were very few such equations directly affected. Moreover, for those that were affected, the replacement models had been verified to be correct when they were derived. The second knowledge base verification being performed was of the patterns of diagnostic evidence used to diagnose the various process fault situations, i.e., the diagnostic rules. It was necessary to verify that these had been derived properly. However, if all of the modeling assumptions associated with the various models of normal process operation are known, along with each model's accuracy and its sensitivity to its modeling assumptions, it is a very straightforward and systematic procedure to derive the proper diagnostic rules. This led us to consider developing ways in which the procedure for deriving diagnostic rules based on MOME could be automated. It thus ultimately led to the creation of FALCONEER IV diagnostic rule compiler (Fickelscherer et al. 2005, Chester et al. 2008, Fickelscherer and Chester 2013).

11.5.3   The Fielded FALCON System's Performance Results

The FALCON System was tested during its development with 260 simulated fault situations and 500 hours of selected process data which contained 65 process operating events (e.g., emergency shutdowns, startups, and production changes), including 13 actual process fault situations. During its offline field test, the FALCON System monitored over 5,000 continuous hours of process data in real-time. These data included 22 process operating events, 8 of which were actual process fault situations. The FALCON System was then tested online for 3 months. DuPont independently rated its real-time performance at better than 95% correct responses during that test (Rowan 1992).

Although it preformed competently online, maintaining and improving the fielded FALCON System's diagnostic knowledge base (written in Common Lisp based data structures comprising of 20 Primary models and 800+ diagnostic rules) proved to be impractical for anyone other than the original developer (i.e., the University of Delaware) (Rowan and Taylor 1989). Being a research project, maintainability was not given as high a priority as was the FALCON System's performance with actual process data.[6] From a research viewpoint, generalizing the underlying logic of this model-based diagnostic strategy was paramount, allowing future such development project activities to be as streamlined as possible. This effort led directly to the formulation of MOME (Fickelscherer 1990, 1994).[7]

11.6 THE IDEAL FALCON SYSTEM

It became apparent during the FALCON Project that because of the nature of the problem being solved, the fault analyzer's development could never truly be finished: as discussed, incremental improvements would continue to be required periodically as the process system and its daily operation evolved. Consequently, we decided to characterize the performance of the Ideal FALCON System so that we could gauge how close we were to achieving such ideal performance. The following list defines our ideal performance criteria for automated process fault analyzers:

(1)   The fault analyzer will be able to correctly classify all possible significant single process operating events; i.e., it will not diagnose any of those events incorrectly, nor will it diagnose a fault situation when none exists.

---

[6]This is quite common to such development efforts. As Stock (1989) states that "The ad hoc procedure of architecture, design, and implementation associated with incremental development leads to an unwieldy, unstructured application that is extremely difficult to maintain, modify, integrate, and extend as new requirements become necessary."

[7]Jackson (1986) states that "This approach is very much in keeping with the tradition of 'standing back' from some working program to look at it from a higher level of abstraction, in order to see what has actually been learned from its implementation, and then performing a rational reconstruction which both extends the power of the original program and achieves its ends in a more principled way. There is much to be said for this as a research strategy in expert systems and artificial intelligence generally."

(2) The fault analyzer will be able to function over all possible operating rates and at any of the normal operating conditions while performing (1).

(3) The fault analyzer will correctly monitor both operator and control system initiated changes and will also correctly monitor all normal transients in the process state resulting from those changes.

(4) The fault analyzer will correctly monitor for faults prior to and during any emergency process interlock activations.

(5) The fault analyzer will be able to be turned on during any possible process state, automatically determine that state, and then monitor for all state transitions possible from that state.

(6) The fault analyzer will be able to monitor the incoming data and determine whether those data are unique, in the correct format and within the proper limits, and then adjust its analysis accordingly.

To summarize, these criteria state that the Ideal Fault Analyzer will operate properly under all circumstances and will correctly diagnose all single significant process faults that can occur during the production phase of process operations. Doing so mirrors an ideal concept of intelligence called rationality; i.e., a system is rational if it does the right thing (Russell and Norvig 1995). When this list of ideal criteria was compiled, it was believed that the FALCON System was indeed actually very close to being ideal. Minor problems still existed that would have to be dealt with in time, but all in all the basic development and verification of the fault analyzer had been completed. The experience and lessons learned from actually creating a competent real-world fault analyzer directly led to a conceptual algorithm capable enough to be a robust solution to the general automated process fault analysis problem.

11.7 USE OF THE KNOWLEDGE-BASED SYSTEM PARADIGM IN PROBLEM SOLVING

The true power of the knowledge-based systems' approach to problem solving is that it can be used to quickly automate solutions to poorly understood problems. This approach allows one to capture, organize, and directly utilize the diverse forms of domain and procedural knowledge required to solve those problems. The resulting solutions separate the domain-specific (i.e., application specific) knowledge from the underlying procedures or algorithms (i.e., generalized methods for solving problems) followed to obtain those solutions. Separating the two types of knowledge in this manner directly facilitates the study of poorly understood problems, eventually leading to generalized solutions for those problems. Once obtained, these generalized solutions can be implemented in a variety of ways, possibly even in conventional computer hardware and software.

This was the primary use of the knowledge-based systems' paradigm throughout the FALCON Project. As described above, this approach to problem solving was, therefore, not an end in itself, but rather it was a means to an end. It allowed us to rapidly prototype potential solutions to the problem of automated process fault analysis, intensely examine the limitations of those solutions, and then incrementally evolve to more robust solutions. Using this paradigm in our investigation directly led to the derivation of the MOME quantitative model-based diagnostic strategy, which then eventually resulted in the creation of FALCONEER IV (Fickelscherer et al. 2005, Chester et al. 2008, Fickelscherer and Chester 2013).

ACKNOWLEDGMENTS

We would first like to thank the other two key chief investigators at the University of Delaware on the original FALCON Project, the late Professors Prasad S. Dhurjati and David E. Lamb, along with many student research assistants, including Oliver J. Smith IV, George M. A'zary, Larry Kramer, Dave Mooney, Lisa Laffend, Kathy Cebulka, Apperson Johnson, and Bob Varrin, Jr. We would also like to thank DuPont and its employees: Duncan Rowan, Rick Taylor, John Hale, Robert Wagner, Bob Gardener, Tim Cole, and especially our domain expert, the late Steve Matusevich. At Foxboro Inc., we would like to thank Dick Shirley, Dave Fortin, and the late Terry Rooney. We would furthermore thank Oliver J. Smith IV and Duncan Rowan for reviewing earlier versions of this treatment; their comments about ways to improve it were invaluable.

REFERENCES

Chester, D.L., Lamb, D.E., and Dhurjati, P.S. (1983). An expert system approach to on-line alarm analysis in power and process plants. *Computers in Engineering ASME* 1: 345–351.

Chester, D.L., L. Daniels, R.J. Fickelscherer, and D.H. Lenz, United States Patent No.: US 7,451,003, "Method and System of Monitoring, Sensor Validation and Predictive Fault Analysis," 2008.

Dhurjati, P.S., Lamb, D.E., and Chester, D.L. (1987). Experience in the Development of an Expert System for Fault Diagnosis in a Commercial Scale Chemical Process. In: Proceedings of the First International Conference on Foundations of Computer Aided Process Operations (ed. G.V. Reklaitis and H.D. Spriggs), 589–626. New York: Elsevier Science Publishers Inc.

Fickelscherer, R. J., P. S. Dhurjati, D. E. Lamb, and D. L. Chester, "Role of Dynamic Simulation in the Construction of Expert Systems for Process Fault Diagnosis, Paper No. 51d, AIChE Spring National Meeting, 1986, New Orleans, LA.

Fickelscherer, R. J., P. S. Dhurjati, D. E. Lamb, and D. L. Chester, "The FALCON Project: Application of an Expert System to Process Fault Diagnosis," Paper No. 82a, AIChE Spring National Meeting, 1987, Houston, TX.

Fickelscherer, R. J., Automated Process Fault Analysis, Ph.D. Dissertation, Department of Chemical Engineering, University of Delaware, 1990.

Fickelscherer, R.J. (1994). A generalized approach to model-based process fault analysis. In: *Foundations of Computer-Aided Process Operations II* (ed. D.W.T. Rippin, J.C. Hale, and J.F. Davis), 451–456. Austin, TX: CACHE, Inc.

Fickelscherer, R.J., Lenz, D.H., and Chester, D.L. (2005). Fuzzy logic clarifies operations. *InTech* (October) 53–57.

Fickelscherer, R.J. and Chester, D.L. (2013). *Optimal Automated Process Fault Analysis*. New York: AIChE/John Wiley and Sons, Inc.

Jackson, P. (1986). *Introduction to Expert Systems*, 114. Reading, MA: Addison-Wesley Publishing Co., Inc.

Koton, P. A., "Empirical and Model-Based Reasoning in Expert Systems," Proceedings of the Ninth International Joint Conference on Artificial Intelligence, Los Angeles, CA, Vol. 1, Los Altos, CA, Morgan Kaufmann Publishers, Inc., 1985, pp. 297–299.

Lamb, D. E., Chester, D. L., Dhurjati, P. S., and Hale, J. C. "An Academic/Industry Project to Develop an Expert System for Chemical Process Fault Detection," AIChE Annual Meeting, Paper No. 70c, Chicago, Illinois, 1985.

Mooney, D.J., D. L. Chester, D. E. Lamb, and P. S. Dhurjati, "Design and Operation of the FALCON Interface", Proceedings of ISA/88, 1988, 747–758.

Olujic, Z. (1981). Compute friction factors fast for flow in pipes. *Chemical Engineering* 88 (25): 91–93.

Rowan, D.A. (1986). Chemical plant fault diagnosis using expert systems technology: a case study. *IFAC Kyoto Workshop on Fault Detection and Safety in Chemical Plants* 81–87.

Rowan, D.A. (1987). Using an expert system for fault diagnosis. *Control Engineering* 160–164.

Rowan, D.A. and Taylor, R.J. (1989). On-line fault diagnosis: FALCON project. *Artificial Intelligence Handbook, Instrument Society of America* 2: 379–399.

Rowan, D.A. (1992). Beyond *FALCON*: Industrial Applications of Knowledge-Based Systems. In: *Proceedings of the International Federation of Automatic Control Symposium* (ed. P.S. Dhurjati), 215–217. Delaware, USA: Newark.

Russell, S.J. and Norvig, P. (1995). *Artificial Intelligence – a Modern Approach*, 826–830. New York: Prentice Hall International.

Shirley, R. S., and D. A. Fortin, "Status Report: An Expert System to Aid Process Control," Proceedings of ISA/85, 1985, pp. 1463–1470.

Shirley, R.S. (1986). Status report 2: an expert system to aid process control. *Proceedings of the TAPPI Engineering Conference* 425–430.

Shirley, R.S. (1987a). Status report 3 with lessons: an expert system to aid process control. *Proceedings of the Annual Pulp and Paper Industry Technical Conference* 132–136.

Shirley, R.S. (1987b). Some lessons learned using expert systems for process control. *Proceedings of the American Control Conference* 2: 1342–1346.

Stock, M. (1989). *AI in Process Control*, 140. New York: McGraw-Hill Co., Inc.

Varrin, R. D., Jr., and P. S. Dhurjati, "Implementation of an Expert System for On-Line Fault Diagnosis in a Commercial Scale Chemical Process," Paper No. 21b, AIChE Spring National Meeting, 1988, New Orleans.

12

FAULT DIAGNOSTIC APPLICATION IMPLEMENTATION AND SUSTAINABILITY

Perry Nordh PEng.

Honeywell Inc., Calgary, Alberta, Canada

OVERVIEW

In industry, new technologies help drive better operation, more uptime, more efficiency, and yet we still have some of the age-old problems when these technologies are implemented. Why?

Applying new technology in the real world relies on machine and human interaction. Experience dictates that the implementation of new technology can be successful with careful attention to some basic principles that can apply to many advanced technologies.

While much less academically focused, this chapter explores application implementation strategies to ensure a new technology can be used effectively with long-term sustainability. Note that the principles here are not exclusive to monitoring and diagnostic software. Introducing any new technology

Artificial Intelligence in Process Fault Diagnosis: Methods for Plant Surveillance,
First Edition. Edited by Richard J. Fickelscherer.
© 2024 John Wiley & Sons, Inc. Published 2024 by John Wiley & Sons, Inc.

requires planning, setting appropriate expectations, and defining what success looks like.

CHAPTER HIGHLIGHTS

- Learning from the history of fault diagnosis applications.
- Time, a real-life challenge for engineering staff.
- Inputs, outputs, and instrumentation, the impact of accuracy and maintenance on technologies.
- Complexity of use for engineers, DCS technicians, and console operators.

NOMENCLATURE

APC	Advanced process control
Control loop	Feedback controller single in, single out (see PID)
Control performance assessment	Report card on the control application
DCS	Distributed control system
Deadtime	Time elapsed from control move to input response (PV)
Feedback control	Using current error to calculate control action
Minimum variance control	Aggressive control to minimize error
OP	Output
P&ID	Piping and instrument diagram
PID	Proportional, integral, derivative (control algorithm)
PLC	Programmable logic controller
PV	Process value
SP	Set point

12.1   KEY PRINCIPLES OF SUCCESSFULLY IMPLEMENTING NEW TECHNOLOGY

In many cases, applications introducing new technology meet with initial success, only to be abandoned later, often only 3–6 months into operation. Moving beyond the "how" of a new technology is critical to creating value through sustained operation. Not to take anything away from the complexity or the detailed work required, but any industrial application of an exciting new technology will require *people* to be able to understand, use, teach, and maintain it over time.

In addition, each person involved will have specific expectations of a new technology. Many aspects beyond the technical fault detection should be considered, such as:

- What are the design conditions (e.g., production rate, grades/products, ambient temp)?
- What it should look like (e.g., colors, graphics, consistency with other views)?
- How should notifications work (e.g., console alarm, SMS/text, email, reports, web view)?
- What defines success (e.g., false positives, false negatives)?
- What will need to be adjusted/corrected/maintained in the next week/ month/year?

In this chapter, a series of questions to ask about any potential technology implementation will be presented, along with an analysis of control performance monitoring to serve as an example of how a new technology can find organizational gaps, communication gaps, and potential pitfalls.

12.2   EXPECTATION OF ADVANCED TECHNOLOGY

Critical to the acceptance of any technology is setting expectations correctly. Do your homework and ask a few critical questions. The past can be a fantastic teacher.

- Has a similar application been tried before?
- If so, what made it a success or failure?
- Was the application still in use after 3 months? After 6 months?

Sometimes, a failure has nothing to do with the technology, but rather the organizational preparation and the expectations set for each group involved. Thus, a further division will help analyze these expectations into Who, What, and When. Note, the Why and the How questions are more than adequately addressed in other chapters in this book.

12.2.1   What Are the Expected Actions?

Implementing a fault detection algorithm or new data analytics approach generally enables a new alert or new information or new insight. It is important then to evaluate what the expected or potential actions that can be taken, and ask pertinent questions like:

- What do you expect the actions to be from this new insight/fault/alert?
- Is there just one course of action? Or multiple?
- If there is multiple – what order or what priority is required?
- Who is expected to take each action? (see the next section)

Going a couple steps beyond the technology results enables any technology implementer to evaluate the desired outcome from that new information and remove potential barriers to adoption of the technology.

12.2.2 Who Is the Audience?

The "who" questions are likely the most critical to the long-term sustainability of any technology. Any additional alert or information adds work to the person on the receiving end. In many cases that person is already extremely busy, so asking the following questions can help define the expected workflow:

- Who is going to be the primary person dealing with results from the algorithm?
- Can this primary person take an action directly based on the alert?
- If not, can this primary person (A) put the action into someone else's (B) queue?
- If in someone else's (B) queue – can the priority be adjusted by the primary person (A)?
- How many queues are involved? How many layers in the organization are required?
- What additional motivation is there (if any) to resolve the issue and or prioritize?

Obviously, this is not a definitive list of potential people issues when implementing a new technology. The key is determining who cares about the result and what can they do about it. Then, take advantage of that perspective when creating the implementation plan and training for the new application.

The primary objective should be to send the notification to someone who cares and has the organizational role to do something about it. Sending a notification to everyone often ends up in the assumption that someone else is going to look after it.

12.2.3 What Are the Failure Modes?

While the people in the workflow are critical, there are other aspects that can cause issues, spurious results, and even failures even though the underlying algorithm is performing perfectly. Real examples can range from instrumentation issues, data collection frequency, and other unforeseen operational and procedural problems. Some important questions to ask include the following:

- Where are the data coming from? Is the update frequency adequate for the algorithm? (note that data from a SCADA or satellite network can be relatively slow compared to on premise control systems)
- What inputs are required? Is the data from the process reliable? Are there potential instrumentation issues? Calibrations/blowdowns required? Is data reconciliation appropriate?
- Which inputs are critical? Is there a mechanism to shut off alerting/ calculations if any critical input is not available?
- Is there a specific operating range the algorithm for detection was designed to operate within? What happens outside that operating range?
- Is there anything else that could impact the reliability of the alert/ alarm? Note: this could be as simple as things like ambient temperature, feed quality of the raw material, or a delayed lab result.

The goal for any technology should be long-term deployment for long-term profitability. Using technology to drive long-term results will pay dividends over time, but does take careful planning to ensure that the implementation continues to deliver value.

12.2.4 When Is an Alert Expected and Valid?

Assume that all people and technical issues have been considered, the algorithm is generating the alert reliably, and all sensors are working properly, things are good, right? Not necessarily, timing is important; especially if the technology deals with an impending upset or an impending failure. If the alert/alarm comes only minutes or seconds before the upset or failure, there is no time to react or change course to avoid an incident.

Any predictive alarm or alert must give the operations team enough time to react.

Excellent data analytics and modeling will achieve a valid and repeatable result but could be a waste of effort if there is no time to change course and take corrective action.

Not all algorithms are predictive in nature, some are more historical and focus on diagnosis. In the upcoming example of regulatory control performance, the analysis is looking at historical data and diagnosing the issue, thus timing is less critical, but still important to remove variability and impending failures.

12.3 DEFINING SUCCESS

So, when are we "done"? Algorithm development can be a roller coaster of successes and failures, rework and retesting, negotiation and fixes, resulting in improvement over time. There is usually a strong temptation to rework/ redesign as more is discovered.

If no schedule has been defined, rework/redesign may be appropriate. In industry, however, this is rarely the case with budgeting restrictions, time restrictions for staff, and schedule limitations, a clear definition of what "success" means is critical to declaring victory.

Statistical proofs such as % correct, % false positive, and % false negatives are one way to look at algorithm results. Ensuring all team members have a clear understanding of the measurements that will be applied gives everyone visibility into progress, deficiencies, and allows contingency planning mid-project.

A simple rule can be applied. The care and maintenance of the application will have to be carried on after you are promoted. Have you documented it well enough? What level of education will that person need to carry on this work? Many applications fail shortly after being handed off to another person or another engineering team.

At this point, it is important to point out that the engineering team could have done all these things correctly, but the operations team still needs to accept the application. Do not ignore the effort to create user training for the person(s) who will receive the alarm/alerts. This is really part of the "people" issues, but cannot be ignored as it is a critical part of a sustainable application.

12.4 LEARNING FROM HISTORY

In this section, we will look at the example of regulatory control monitoring over multiple decades of research and development, including a definition of what regulatory control means, what failure modes are expected, and how the problem was dealt with over time, from a research project to a commercially available product. We will cover some background, some of the technical and workflow issues, and use these as an example of potential issues when implementing algorithmic fault detection Miller (2006), Nordh (2012).

In industry, the goal of production without any defects or loss of quality is often hampered by the reality of wear and tear, failures of sensors or equipment, or even inadequacy of the original design. History reveals specific challenges in implementation of these "warning lights," that if not addressed, will cause the failure of the best technology or algorithm.

Some early warning implementations have been successful, and some have not. Why?

To answer these questions, experience will help explain. We will examine the monitoring and diagnosis at the regulatory control layer within industrial plants as an example.

12.5   EXAMPLE: REGULATORY CONTROL LOOP MONITORING

The technology for using data to detect issues at the regulatory layer of industrial processes has existed since 1989. A seminal paper, written by Prof Tom Harris (1989), established a data-driven comparison for control loop performance based on the concept of minimum variance. The concept allows the performance to be calculated from normal operational data without a physical test. This was followed soon afterward with a paper on univariate performance assessment co-authored by Desborough and Harris (1992). These two papers formed the genesis of a new industrial software category, and the concepts have been utilized by multiple industrial software vendors to deliver engineering studies and commercial performance assessment reporting software Huang et al. (1997), Thornhill et al. (1999).

12.5.1   Regulatory Control Failure Modes

Detecting symptoms are easier than detecting the actual failure. That may seem intuitively obvious, but without doing a physical test on the equipment, the step of detecting the root cause is incredibly difficult. Understanding why requires a bit of background.

12.5.1.1   Background   Every industrial process on the planet depends on a base regulatory control layer comprised of feedback controllers often referred to as "control loops." Like transistors within electronics, control loops form the basis of industrial processes and it takes many control loops to make a final manufactured product to be sold. In almost all cases, this regulatory control layer is implemented on a distributed control system (DCS) or programmable logic controller (PLC). All DCS and PLC systems include a basic feedback controller called a PID (proportional, integral, derivative) controller.

Fundamentally, the goal of a PID controller (sometimes called a PID loop) is to control a process variable such as a flow or temperature, to a specific value or set point. When the process value (PV) does not equal the

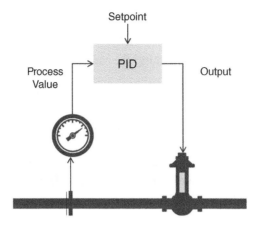

set point (SP), this results in an error (error = PV-SP, or, error = SP-PV, depending on the system). To resolve this error, a controller can move an output (OP), often to a valve, to change the process conditions and bring the PV closer to the SP. This is a simple definition of feedback control.

Moving the OP in the right direction requires a control calculation. A PID controller uses a combination of proportional control, integral control, and derivative control to calculate a control output move.

12.5.1.2 *Problem Scale*   Most refining complexes have between 1,500 and 3,000 PID controllers, and the largest petrochemical complexes can have upward of 5,000 PID controllers. Thus, using data-driven methods to automate fault detection within the regulatory control layer can save significant troubleshooting effort.

12.5.1.3 *Symptoms and Failures*   Unfortunately, any PID controller will happily do the wrong thing all day long, meaning that if there is a fault such as an erroneous process value, or a bad set point, or a sticky valve – the algorithm will still calculate, but may not control the process value properly. Often, a "problem" brought to the technical team is a visible symptom but may or may not point directly to the root cause.

Possible symptoms may include the following:

- Oscillation,
- Slow (sometimes called sluggish) response,
- Saturation (meaning the control calculation no longer influences the PV),
- Not holding set point,
- Other unexpected behaviors.

The symptoms exhibit visible/detectable expressions of the root–cause failures, which may include the following:

- Valve/control stiction (meaning the valve or control element does not move when asked to),
- Actuator damage (can look like valve stiction since the valve does not move properly),
- Under/over tuning (meaning the control moves calculated are too big or too small),
- Actuator tuning (can look like a tuning issue among other things),
- Process interactions (a disturbance from another part of the process),
- Instrument failure,
- Actuator failure,
- Along with many other potential issues (too many to list).

Thus, utilizing data to point out issues and faults holds great promise across the industrial world. The problem is, we do not always know exactly what we are looking for in the data, or what is important until after something bad has happened. Note that without data-driven methods, determining the real root cause may require visual inspection, process tests/steps, or even a shutdown of the process to enable inspection and repair of nonvisible components.

Utilizing the Harris/Desborough method referenced above provides an assessment approach that minimizes effort to detect issues and start diagnosis for regulatory PID controllers. Technically, the approach evaluates the predictability of the error signal. If the error signal trend is close to white noise, the controller is doing very well at rejecting disturbances, i.e., closer to minimum variance. Conversely, if the error signal trend is close to a sine wave, the controller is oscillating and could even be injecting disturbances to the process, i.e., furthest from minimum variance.

12.5.2 Expectations of Loop Monitoring

The promise of analytic fault detection for regulatory controls *without a plant test* was introduced with excitement and high expectations. Initial software releases were in many respects research projects with engineering interfaces that were difficult to understand without a deep control engineering background. In addition, the practical use of the software exposed weak points of the analytics.

12.5.2.1 *Technical Issue: Deadtime*   A key problem in this minimum variance comparison/analysis is deadtime. Deadtime is defined as the time from a control action, or output (OP) move, to a corresponding response on the process value (PV).

Proven over thousands of data sets, estimation of deadtime was found to be primarily correlated to the type of value being controlled. Therefore, deadtime could be estimated/assumed based on loop type without significant error. For example, many flow control loops exhibited consistent timing in response, generally responding quite quickly to OP changes within 1–2 minutes including deadtime. Conversely, loops controlling temperature had much longer response times, and thus, the type of value being controlled became a key configuration parameter in data collection.

12.5.2.2 *Technical Issue: Data Collection Frequency* Data collection frequency also became a key factor, again based on the response time of each loop. In order to capture the maximum possible information about a control loop, the collection frequency must be adjusted in relation to the time constant of the loop. This resulted in a specification for data collection based on loop type:

- Flow loops – 1-second data collection intervals.
- Pressure loops – 5-second intervals.
- Temperature loops – 30-second intervals.
- Level loops – 30-second intervals.
- Other loops – 30-second intervals.

The sampling theorem is clear, you must collect twice as fast as anything you want to see/detect. In other words, the collection frequency is vitally important to an accurate analysis.

12.5.2.3 *Technical Issue: Data Collection and Compression*
How much data do you need for any one analysis? A key question, especially if the analysis is to be done remotely or even on a different network level.

In the initial consulting work (before any commercial software was available), there were no practical guidelines. Industrial plants did have significant volumes of historical data, but often it was compressed and at 60 second intervals or longer. Six to eight months of this historical and compressed data were evaluated for any analytical content; however, multiple issues were found:

- Fast loops such as flow and pressure had no relevant data (data collection was too slow).
- Slower loops such as temperature and level had much of the dynamic content removed and resulted in a degraded analysis (no accurate dynamics left after compression).

Hence, new ways of collecting the data at the right frequency and without compression had to be invented.

12.5.2.4   *Technical Issue: Data Collection Volume*   In designing this new data collection mechanism, a key question arises: how much data do you need for any one analysis? The answer is especially important if the analysis is to be done remotely or even on a different network level.

Academically, various data set lengths had been tried, but again in the early stages of this technology development, the "best" data set length was unknown. Various trials were conducted, combining data set frequency and data set length to evaluate the results versus actual with thousands of manually classified loops, resulting in a criterion of 5,000 samples regardless of collection frequency.

Thus, for regulatory loop performance evaluation, ideally 10–15 time constants should be used (per loop type) to get an accurate result. In practice then, over a 24-hour period, there is generally no problem getting enough data for flow loops and pressure loops, but temperature and level need a longer time span to truly represent the actual performance.

12.5.2.5   *People Issue: Reporting and Diagnosing – Establishing Workflow*   As the technical accuracy of the analysis improved, it was clear that even though the developers and those with deep control engineering knowledge could interpret and act on the information, the need for interpretation itself became the issue.

Users wanted to know the "how" of fixing the problem, not just the statistics and analysis that pointed to the issue.

At the control engineering/instrumentation level, a step-by-step wizard approach was developed to help the user walk through the troubleshooting steps, augmenting the algorithmic results with the user's own process knowledge to derive the best action for process improvement. Note that some of these steps were not a direct "fix," but sometimes included a plant test to confirm the issue as detected by the algorithms.

12.5.2.6   *People Issue: Reporting and Diagnosing – Aggregating Data*   Multiple users also emerged causing multiple issues in how the data were reported. In the control engineering world, the users wanted more step-by-step approaches on fixing a problem (as above). In the management world, the users wanted to see a summary of performance across multiple areas and even across multiple facilities.

However, aggregating the data required much more than a simple average – especially given the multiple loop types, multiple possible performance ratings, and multiple levels of importance for each contributing element.

In the case of regulatory control, aggregating performance required a newly formulated index that calculated performance across multiple control elements, weighted by the importance of each (based on the impact to the final product/profit).

12.5.2.7 *People Issue: Time and Notifications – Push Versus Pull* Reporting an issue, however, does not fix anything; there is no fundamental change to the operation unless a person reads the alert, assesses the next action, and takes said action.

In the regulatory control monitoring market, the initial goal was to make problems visible. Although that goal was successfully accomplished, the actual level of control issues in industry did not benefit. Years after commercially software was available, a paper presented at the 6th AIChE International Conference on Chemical Process Control showed engineering workloads increasing with less and less time to review performance (Increasing Customer Value of Industrial Control Performance Monitoring-Honeywell's Experience) (Desborough and Miller 2001).

Expecting someone to "go and look" at a report is valid; however, depending on workload and how important the content is to that person's role will determine the effect.

Push notifications (email, text/SMS, process alarms) can often break this barrier; however, this too can be fraught with issues if there are too many alerts or inaccurate alerts. A control valve will happily do the wrong thing all day long, and a human will not tolerate inaccurate alerts for any length of time.

12.5.2.8 *People Issue: Sustainability* Any application that is implemented should be useful and maintainable. In the long-term feedback from regulatory loop monitoring customers, these main themes have emerged as feedback:

- Tell me what is changed (monitoring).
- Tell me what to do about it (diagnostics).
- Do not make the system itself a hassle (maintenance).

12.6 WHAT SUCCESS LOOKS LIKE

Thus far in this discussion, the focus has been on the technical and people aspects of implementing and commissioning a successful application. Of course, the reason these fault diagnosis applications are created is to derive benefits in time taken to discover and fix any issues which should deliver monetary benefits over time.

To be sure, any application that does not deliver monetary benefits over time (in some form) will be decommissioned.

Success then can be defined within multiple categories:

- Technical – delivering an accurate and timely fault alert/diagnosis.
- People – the person receiving the alert has the skill and responsibility to act.
- Benefit – the application delivers benefit over time, i.e., worth the upkeep/training/licensing.

Many companies have struggled in tracking "success" and even though these applications are delivering benefit, the real monetary impact has never been calculated.

12.7 EXAMPLE: SYSTEMATIC STEWARDSHIP

Managers must evaluate the use of resources and make hard decisions on what to keep and what to demise. To make these decisions, it is very helpful to have structured data showing the impact of each resource/application over time. This allows for easier decision making, but also can help justify additional expenditure on applications with consistent delivery of benefit.

Top tier companies have employed several techniques to help calculate and track benefits over time Kumar and MacGowan (2022). In all cases a regular, systematic review helps build a better picture of long-term performance. For example, after commissioning, evaluations might resemble the following:

- Weekly reviews – engineering, operations, technical specialists
  - Concerns/escalations
  - Knowledge transfer
  - Tracking actions week to week
- Monthly reviews – engineering, business team leadership, controls management, operations
  - Assessment of each application – uptime, performance metrics
    - Downtime assessment – if the application is down, why?
  - Valuation of application impact (in monetary terms, e.g., dollars or euros)
- Quarterly and annual reviews
  - Extending the time window of the monthly reviews to calculate value over the year
  - Allow for barrier mitigation with more investment and/or operational/ equipment/instrumentation changes

Tracking the final benefit can be difficult and even seen as unnecessary by the technical team, but if the technical team is able to show thousands or even millions of dollars or euros worth of benefits each year for a specific application, further work and further investment can easily be justified.

12.8 CONCLUSIONS

For the design of any application, we can then summarize a few key points that need to be considered.

12.8.1 Motivational Requirement

- The assets must be failure prone.
- Failure must have a significant impact.
- Must have asset domain expertise.

12.8.2 Setup Requirements

- Must detect asset-specific faults.
- Must require low configuration effort.
- Failures must be detected from available data.

12.8.3 Usage Requirements

- Must have low false positives.
- Must have low false negatives.
- Must provide clear path to action.
- Must show performance in context.
- Performance measurements must be traceable and transparent.

12.8.4 Sustainment and Continuous Improvement

- Must have a direct/regular feedback path for customers to identify problems with the software.

REFERENCES

Desborough, L. and Harris, T.J. (1992). Performance assessment measure for univariate feedback control. *The Canadian Journal of Chemical Engineering* 70 (1): 1186–1197.

Desborough, Lane & Randy Miller, Increasing Customer Value of Industrial Control Performance Monitoring -Honeywell's Experience, AIChE Symposium Series, 1 January 2002, presentation at the 6th AIChE International Conference on Chemical Process Control, Tucson, AZ, 2001.

Harris, T.J. (1989). Assessment of control loop performance. *The Canadian Journal of Chemical Engineering* 67: 856–861.

Huang, B., Shah, S., and Kwok, E.K. (1997). Good, bad or optimal? Performance assessment of multivariable processes. *Automatica* 33: 1175–1183.

Kumar, S. and MacGowan, J. (2022). *Syncrude Sustainment of APC Benefits*. Honeywell Users Group.

Miller, R. (2006). *Loop Scout History – A narrative by Randy Miller*. Honeywell Internal Publication.

Nordh, P. (2012). *Honeywell Technology Symposium – Remote Monitoring and Big Data Experience*. Honeywell Internal Presentation.

Thornhill, N., Oettinger, M., and Fedenczuk, P. (1999). Refinery-wide control loop performance assessment. *Journal of Process Control* 9 (2): 109–124.

13

PROCESS OPERATORS, ADVANCED PROCESS CONTROL, AND ARTIFICIAL INTELLIGENCE-BASED APPLICATIONS IN THE CONTROL ROOM

Rajan Rathinasabapathy[1], Atique Malik[2], and Richard J. Fickelscherer, PE[3]

[1]*Department of Chemical and Biomolecular Engineering, UCLA, currently with Technical Services - Process optimization, Phillips 66, Los Angeles, CA, USA*
[2]*AI Control LLC, Edwardsville, IL, USA*
[3]*Department of Chemical and Biological Engineering, State University of New York at Buffalo, Buffalo, New York, USA*

CHAPTER HIGHLIGHTS

- Enumeration of common pitfalls in the sustainability of Advanced Process Control (APC) applications.
- Elaboration of various issues to promote successful artificial intelligence (AI)-based applications in the control room.

Artificial Intelligence in Process Fault Diagnosis: Methods for Plant Surveillance,
First Edition. Edited by Richard J. Fickelscherer.
© 2024 John Wiley & Sons, Inc. Published 2024 by John Wiley & Sons, Inc.

- Emphasis on the role process operators play in ultimate acceptance and continued adoption of AI applications.
- Discussion on the role of support engineers and support programs to sustain benefits.

OVERVIEW

Advanced Process Control (APC) refers to a broad range of technologies and control methodologies implemented on top of basic process controls on Distributed Control Systems (DCSs) within the industrial Process Control Network. Its applications are primarily model-based, multivariable predictive controllers. APC automatically adjusts multiple single-loop controllers, typically every minute, to drive processes to their most optimal points within the unit constraint envelope. This can be thought of as the most knowledgeable operator optimizing the process every minute. The standard APC objective is economic benefit – optimizing the process unit to minimize the cost, maximizing unit rate and valuable products, achieving uniform product quality by reducing process variability, and running the process unit closer to its operating constraints in an energy efficient way. There are additional tangible and non-tangible benefits, better monitoring and management of equipment, safety, reliability, and environmental constraints across different operating conditions. Chemical process industries have a variety of artificial intelligence (AI)-based applications for different purposes. In our work, AI-based applications are closed-loop applications, developed using AI tools and deployed in the control room environment, where process operators are the ultimate end users. In the recent past, there have been APC implementations based on AI models developed using deep learning/machine learning technologies (Lahiri 2017). The subset of AI methods applied to closed-loop optimization is described in Meyn (2022) and Farsi and Liu (2023). The authors of these articles view AI-based applications as an extension of APC, especially for deployment in the control room, with the same end user and the same support engineer.

Historically, a successful and sustainable APC program has depended on a lot more factors other than the technology and implementation itself, especially the "human factor." In this chapter, we discuss our real-world experiences in the oil refining, petrochemical, and fertilizer industries in building and deploying traditional APC, including what it takes to be successful, the pitfalls to avoid, and how to ensure long-term value. We believe that the end-user engagement, system requirements, and factors affecting a sustainable APC program will also affect AI-based closed-loop deployments in a very similar way. The traditional APC challenges discussed in this chapter also apply to the ever-evolving, innovative AI-based applications that are now taking hold or will be used in the future in the process industry.

13.1 INTRODUCTION

What is AI exactly? The answer to this question has been continually attempted ever since the term originated in the 1950s. AI has, thus far, eluded a precise definition, with various attempts at doing so reflecting the thinking of particular humans making them based on their personal backgrounds and actual experience. John McCarthy, who originally coined the term in 1956, defined AI as "machines that can perform tasks that are characteristic of human intelligence" (Kissinger et al. 2021). Rich (1983a) defined AI as "The study of how to make computers do things which, at the moment, people are better at." She more specifically further stated: "AI is the study of techniques for solving exponentially hard problems in polynomial time by exploiting knowledge about the problem domain." (Rich 1983b). More formally, "AI is the attempt to solve N-P[1] Complete problems in polynomial time." More importantly, the paramount emphasis is on how the given AI algorithm performs in the real world on average (Rich 1983c). The bottom line here then is the resulting accuracy and usefulness of the AI program's answers rather than the actual time usually required to compute those answers.

Another view on AI is that AI researchers are trying to create a computer that "thinks" (Charniak and McDermott 1985). The emphasis of such AI is also fundamentally concerned with working programs: it is not committed to any particular way of producing those correct results. This is more in line with the following: "The goal of AI is to make computers more intelligent" (Mishkoff 1985). Randal Davis' similar take on such machine intelligence is that it should solve a problem in a reasonable fashion and not be concerned whether it models humans exactly (Mishkoff 1985). Russell and Norwig (1995) further believe that AI strives to "build" intelligent entities as well as understand them. Such capability is directly related to these programs' ultimate rationality, i.e., it is concerned with expected success given what has been perceived.

From the above discussion and perspective, continuing advances in AI can, therefore, be considered as part of the concerted, rapid evolution in Computer and Information Science. These advances continuously gain

---

[1]Nondeterministic polynomial time.

more relevance as their underlying algorithms and methodologies become more powerful and applicable to ever more diverse fields of human endeavor. This directly allows for more useful and novel applications. It is precisely in this vein of emphasis that the entire preceding treatment has viewed AI. Its application to automated process fault analysis is thus just one important subtopic within the far-reaching field of APC.

The various flavors of AI's application in automated process fault analysis efforts previously described throughout this book are consequently primarily the results of continuous advancement of modern human technology, especially those in computer software, hardware, networks, control theory, instrumentation, sensors, etc. The remainder of this concluding chapter discusses how all this rapidly maturing technology can be more readily applied and accepted by its ultimate human end users, the actual process operators. They are the ones who most truly expect to tangibly benefit from their committed acceptance and consequential reliance upon such cutting-edge applications.

The rest of this chapter is organized to discuss:

- History of sustainable APC
- Operators as ultimate end users
- APC Technology impact, design, and development considerations
- Need for a strong support program

13.2   HISTORY OF SUSTAINABLE APC

Traditionally, Advanced Process Control (APC) refers to multivariable model-based closed-loop control, implemented on top of basic process controls. It helps to optimize the process unit by driving process to its optimal point within the unit's constraint envelope. The typical APC objective is economic benefit, but there are numerous non-tangible benefits. APC implementation has been successful in the wider process industry over the past few decades and has played a key role in both meeting operational objectives and generating benefits in millions of dollars. Forbes et al. (2015) discusses various aspects of APC (Model Predictive Control) in industry, the challenges, opportunities, and maintenance requirements. Implementing APC is only just a start; unfortunately, not all APCs are sustained effectively to continuously provide the same maximum benefits year after year.

There are many possible reasons why APC benefits degrade over time: (i) APC is not properly monitored and maintained, (ii) evolving process equipment modifications, (iii) changes in operating strategy or ongoing process changes, (iv) alterations in feed rate quality or unit revamps, (v) instrumentation degradation, (vi) process deterioration (e.g., fouling), (vii) deterioration in regulatory controller performance, (viii) lack of local skilled support staff, (ix) lack of good understanding of the technology by operators, (x) complexity of the chemical process, (xi) linear versus non-linear processes or a narrow linearity envelope, (xii) unclear APC objective, (xiii) underlying APC design itself, (iv) particular APC technology used, (xv) poor process operator engagement and involvement, (xvi) ineffective process operator training, (xvii) insufficient vendor support, (xviii) narrow stake holder awareness, (xix) poor tracking of key performance indicators (KPIs), and (x) lack of management recognition, support, visibility of value, etc.

A combination of some of the above factors derails the benefits of APC initially achieved. Experience shows that APC benefit sustainment is a mix of art and science. As an example, just tracking APC uptime or specific variable uptime does not automatically equate to benefits. Sustainability of APC benefits requires a comprehensive and collaborative approach, involving multiple stakeholders working together. While some of the above factors could be subtle, they could also become the weakest sustainability link. There has been a lot of APC research over the decades about novel modeling techniques, modeling issues, and technology improvements; however, APC sustainability issues have not changed much in decades. The effort that is needed to make APC work discussed about three decades ago by Friedman (1992, 1997) is more or less the same even today. Li et al. (2011) discussed similar issues – a technology advancement not necessarily always results in plant performance improvement in the mineral processing industry. They described four pillars for success of APC: reliable instrumentation and associated control system, data processing and HMI,[2] control algorithms, and control room operators. Huang et al. (2021) discussed APC challenges in the pharma industry and mentioned "people" as the foundation pillar to enable long-term success and value delivery. In our view, the "human factor" – process operators are as important as or even more important than any other pillars for long-term value of advanced applications in the control room. In the rest of this chapter, we use the terminology "operators" to refer to the ultimate end user of these advanced applications (APC- or AI-based applications), typically who operate the process units from the DCS, and are also referred to as "process operators," "DCS operators," "control room operators," "board operators," etc.

---

[2]Human–machine interface.

13.3 OPERATORS AS ULTIMATE APC APPLICATION END USERS

Operators play a critical role in the success and sustainability of APC applications. They may initially be reluctant to accept APC for a variety of reasons, including anxiety about machines taking over their jobs, lack of confidence in a new system, prior disappointing experience, giving up control of the unit or changes to trained practices, and believing no further process improvement is possible. To ensure a successful APC program, it is essential to gain the operators' confidence, make them partners in the optimization process, and engage them from the very beginning, and acknowledging and quickly resolving any technical issues that arise.

Operator understanding and effective engagement with APC goes beyond training on a specific application. Operators need to be trained on the process, basic controls, broader APC concepts, troubleshooting tools, and underlying design strategies. A sustainable APC controller requires operators to be part of the solution, and the design should be accommodative, removing as many underlying barriers in their work and understanding as possible. If the APC design and operation is difficult to understand, it will be either misused or not fully utilized. If APC makes inexplicable process moves that operators cannot understand, they will likely drop variables from APC, causing unintended process moves or overcompensation. Other possible outcomes include clamping variable limits to a comfort zone, limiting optimization opportunities, simply turning APC off, or worse, ignoring APC. Even in situations with a simple and easy to understand APC design, it can be challenging to understand APC moves when there are multiple constraints that directly conflict with competing priorities and tradeoffs.

Negative outcomes can be greatly mitigated by seeking operator input throughout the APC life cycle and keeping everything simple. This can help build operator confidence and make them feel they are part of the solution. It also provides an opportunity for the APC designer to learn about what is important to the operators and address those needs in the design phase. Also, there are regular field activities that can affect APC, and often those activities are not tracked in DCS or history databases. DCS operators are aware of these activities and are the best to provide that knowledge in the design phase and mitigate impact when APC is online.

Li et al. (2011) discussed the pitfalls of inadequate knowledge of complex control systems and the impact on operators. A complex control system that is hard to learn and use will increase the mental workload of operators rather than being beneficial. AI-based application deployments in the control room environment will not be very different from APC applications; they need to be simple and easy to understand and use. Operators will be required as partners in any successful advanced application.

13.4 APC APPLICATION DESIGN CONSIDERATIONS

Operator trust on APC applications is built slowly over time, but it can be lost very quickly. Poor predictions caused by model errors (erroneous gain, incorrect time delay, or settling time), poor APC tuning, or priorities can cause unsteady process operation, eroding operator's confidence. Model errors could be due to various reasons – incorrect testing environment, not accounting for different disturbances such as seasons or day to night changes, test data intertwined with cause/effect data, mix up of steady state versus ramp behaviors up, etc. Moreover, the assumptions made today in modeling may not be applicable tomorrow. Extreme caution should, thus, always be taken when using automated adaptive model development tools. While these tools can sometime help fasten up model development and deployment, they are not perfect, certainly cannot replace an experienced engineer, or organize a proper test environment, nor account for all different scenarios. Unforeseen model errors arising from these tools also can directly erode operator's confidence. Similarly, extreme care needs to be exercised for APC targets set by external targets from calculations or third-party applications such as multi-unit optimization; these likewise should be reviewed periodically to revalidate all their original assumptions. Consequently, APC applications with multiple pages of variables make it complicated for the operator to use the application, even if it is a simple controller. Too many variables can sometimes lead the operators to ignore the application altogether.

Furthermore, performance of all process units typically degrades with time due to fouling of equipment, catalyst deactivation, valves sticking, etc. Unit turnarounds are meant to periodically rejuvenate these units to start a new operating cycle. Feedback on this and other changes to the APC models must thus occur continuously to make them reliable and remain consistent. In the case of the APC, the feedback occurs at each controller execution that happens in seconds or minutes. It accomplishes this high frequency feedback by adjusting the future predictions using the CV[3] feedback, but model mismatches should not be corrected or relied on feedback alone. The APC support engineer should determine if and when the models need to be

---

[3]Controlled variable.

updated. Any process engineering changes, instrument changes, instrument scaling, control valve changes, controller tuning and equipment upgrades, etc. will all affect models, and the models likely require updates. Also, the APC engineer should have a continuous dialog with economics and scheduling to keep the prices updated in the controller's optimizer. Most of these issues cannot be designed and should be addressed by APC revamps and/or tuning on an ongoing basis, using a strong support system.

One of the unique challenges of APC is the controller design itself. Is APC design an art or a science or both? For the same exact process, 10 different APC engineers, tasked with the same objectives, are very likely to implement 10 different APC controllers. Their uniqueness stems from the implementation engineer's knowledge, experience, and overall strategical approach to a control problem solution: there are consequently multiple ways to solve the same problem. Therefore, currently, there is not one standard way of designing an APC. Any associated standardization guidelines could be specific to company or sites or even engineer. For example, one engineer could prefer to model distillation tower temperature in APC and cascade reflux in DCS, whereas another could prefer to model reflux in APC directly. One could prefer to maximize product flow with product spec as a constraint versus another who prefers maximizing product spec with flow as constraint. These are likely to have similar optimization results, but the end user is thoroughly confused if they must operate a mix of APCs with such different design strategies. Such design uniqueness also makes it challenging for cross support APC when critical support engineers are unavailable. A standardized design approach within the plant/site would thus more likely greatly help sustainability.

Besides APC standardization, another area for building operator confidence in the given APC application is making sure operators are not caught off guard by unit emergencies or upsets. While it is common knowledge that APC is not designed for startup, shutdown, emergencies, or upset conditions, operators need a head start when they run into unit upset conditions. APC should be designed to turn itself off as soon as there are indications of a unit upset. (See Appendices 5A and 5B for process state identification logic and pseudo code.) There could also be variables in the unit that are not used in APC for optimization, but those same variables could be used in early detection of potential problems. For example, even if a heater fuel burner pressure is typically not used in APC for optimization or control, a low or high burner pressure is a cause of concern for the operator. The low/ high burner pressure values could be used to turn APC off. The turnoff threshold should be set to give the operator enough lead time to take appropriate action. Low furnace O_2 is another example. Designing with the operator's confidence in mind is very helpful.

Setting up APC economic objective functions to receive the optimization targets from third-party applications enhances closed-loop automation but takes the operator out of the loop. Without knowing the changes happening in the upstream/downstream units and the changing economics that dictates the target changes, the operator would not fully know why the targets changed or APC is behaving differently, ultimately reducing their involvement. Active communication with operators about upcoming objective function changes, and/or mandating the operator's approval for the calculated results, would help sanitize the optimization process and keep the operator engaged. In situations where MV/CV[4] limits are used for optimization, the relevant operators should own those limits and make all changes as needed.

Most actual industrial processes are characterized by non-linearity, complex properties and may also even be subject to frequent rate and/or quality changes. APC having models with long delay times or non-linear process behavior or frequent process changes that require frequent APC limit changes (optimization or control envelope) is subject to additional scrutiny and susceptibility. Complex processes do not have to lead to large, complicated APC applications with intricate engineering calculations ungraspable by operators. Such APC either will directly lead to an unsustainable application or else will require intense engineering support. To date, there has been extensive literature about small versus big controllers; either way, the simpler the better. Smaller, simpler controllers have a higher probability of being turned on by operators after a startup or unit upset without engineer's support. Breaking up big controllers into smaller sub-controllers is a very useful option and helps with easier understanding, support, and maintenance.

Regarding complex processes, their APC applications do not have to be designed to comprehensively control the entire process. In fact, it is highly recommended to leave all well-performing basic process control loops or loops that do not add any dollar value out of APC (i.e., level in a tower or accumulator). A comprehensive solution that tends to take away most of the operator's routine tasks is thus not recommended, as it erodes away the operator engagement. A smart APC application, i.e., one that models mostly the dominating dollar variables, is the preferred approach. Actively involving the operator in such programs' development typically leads to a long-term sustainable application. Operators should feel comfortable and view APC- or AI-based applications as a helper and not a plight. AI-based applications will face the same challenges in the design phase. Simple technology and simple applications with little obscurities are easy to understand and use, and hence they would be helpful.

---

[4]Manipulated and controlled variables.

13.5 APC DEVELOPMENT – INTERNAL VERSUS EXTERNAL EXPERTS

Over time, industrial processes change due to a variety of factors, such as equipment degradation, process modifications, feed quality changes, catalyst type changes, and control system changes. These changes can cause every APC application to become unique, even when the underlying process, priorities, and economics are all the same. A good support system is needed to keep APC up-to-date with these changes. This means that not all APC applications will have a safe list of variables to control and manipulate that stay the same over time.

The question of whether to build APC in-house or with external help is an important one. External experts can bring valuable knowledge and experience from other successful applications, but their knowledge of the local process and its history is limited. This can lead to APC that is not sustainable, as it cannot be easily adapted to changes in the process. In-house engineering resources have deep understanding of the local process and its history, which allows them to build more sustainable APC. Additionally, in-house staff can retain valuable intellectual expertise and use it for troubleshooting and investigations. This directly helps to build a customized APC application, without its design being stuck in time. In-house, experienced APC staff not only retains valuable intellectual expertise in APC, but this same experience is also utilizable in other process/operations troubleshooting and investigations, especially important in the age of graying staff (Ottewell 2015). AI-based online applications must address the issues surrounding process changes and ongoing design modifications to keep up with the changes. Local in-house experts, who are well versed with both process and AI applications, will be crucial in adapting the applications to ongoing process changes. A hybrid business model of external resources for designing and commissioning of APC and internal resources for long-term support and future enhancements is in practice in some places. The key to make this work requires internal resources being able to capture all knowledge and insights from external resources before the end of the project.

13.6 APC TECHNOLOGY

A simple, understandable technology is easier to learn and use, and it can lead to faster adoption and better results. APC has gained common acceptance in industry, and its success as a valuable control and optimization solution is all due to its simple underlying technology that is based on linear regression.

That technology has since been extended to non-linear problems where possible. Engineers and operators alike are now comfortable using this existing technology, as they can explain APC behavior using the model matrix, tuning, and strategies used in a simple language.

In the new, emerging era of AI-based modeling tools, neural-net-based deep learning technologies are now being applied as APC tools. While these AI-based tools have an advantage of building models quickly using historical data, the resulting models have complex structures and can be obscure because they are based on open-loop data (with inseparable cause and effect relationships in the historical data of the process). There could be times when the process moves by the application cannot be explained as the solutions are based on a complex model that cannot be easily interpreted yet. Both the operator and the support engineer could be confounded by the solutions provided by the black box-based technology. Operator confidence on APC is based on the ability to explain every solution, minute after minute. Without this ability, operator's confidence will eventually be eroded. One way to address this challenge is to develop transparent and interpretable machine learning techniques that can explain every solution in process engineering terms, without obscurity, which can increase trust in the tool and lead to effective use.

APC technology is a complex domain. APC engineers typically have strong backgrounds in mathematics and engineering. They also must possess unique skills of combining knowledge about process, unit operations, controls, planning and economics, with hands-on experience in DCS, data analytics, various modeling tools and good exposure to programming, computer networking, and IT. At the same time, they should effectively communicate with operators in a simple, welcoming, process language, without control, APC, or AI jargon. Such background and skills cannot be developed quickly or easily. For APC technology or AI-based technology, the authors foresee the support engineers will require the same type of background and skills. Colegrove (2020) mentions the different expertise needed to support a successful AI application. AI technology in the control room will be a long-term investment. Management commitment is subsequently essential to develop and retain these champions, with assured technical growth paths that do not lead to career stagnation or resource attrition. Qualified, experienced, and local engineers and a strong support culture will all play a vital role in sustainable AI applications. Currently, APC support engineer's availability in the control room, face-to-face with the operator as soon as it is required, helps retain the operator's confidence leading to long-term APC value. AI-based applications will require the same support model, through the life cycle of these applications.

Colegrove (2020) discusses the industry's need for tools that enable design and deployment of new technologies without having an advanced training in AI, data science, and allied fields and the need for new technology to address, develop, and introduce products from a human perspective that will address operational needs and integrates with other existing systems. Li et al. (2011) also discuss that a technology-centered approach without fundamental human elements issues will result in a technology not understood, not accepted, or used incorrectly. A technology and design that the operator can follow, understand, and effectively use is paramount.

13.7 APC SUPPORT

The APC life cycle is complex, starting from a conceptual phase to a design, a testing, an implementation, and a support phase, with support being the longest phase and the most important. This is because everything around APC is constantly changing, and it is important to ensure good support to maintain good performance. After its commissioning, APC reviews are typically limited to a small team. There are a lot of everyday changes at a given site that affect APC. These changes could be from economics (planning, scheduling, and economics group), process improvements, periodic process maintenance, safety and environmental changes, instrumentation changes, analyzer maintenance, operational changes, control system upgrades, software patches, new operator training, and unit testing. APC is not typically designed to handle all these changes that occur at a site. Therefore, it would be beneficial for the site support engineer to be engaged in the site's activities, to learn and educate others on activities that directly or indirectly impact APC. Some of these changes may require an APC modification. APC support is a team effort, built on collaboration between unit/process engineers, control system team, planners/schedulers, and importantly operators.

APC generates a plethora of data, and several years' worth of data can be stored in an APC local database and/or it can be easily transferred to a business database. This data is a gold mine for identifying issues and improvements. APC engineers have been analyzing such data for decades. Most APC software allows one to setup KPI reports and email on a scheduled basis. KPI tracking helps improve APC online time. Tracking APC on/off changes, MV/CV clamping, variables dropped from APC, and constant constraints can provide insights into process changes, issues with certain process regimes, feed quality changes, operating at the edge of linearity or wading into a non-tested process environment, seasonal changes, patterns with certain personnel, training issues, model fidelity issues, equipment constraint, and poor tuning, to name a few. However, at a given site, an experienced APC support engineer, being most likely the only one who can fully interpret that data properly, is too often time constrained. Management commitment is required to effectively address this problem by setting appropriate worker priorities.

Management of change procedures for new APC implementations and any major APC changes would be beneficial to track changes and train operators. During shift changes, operators should be encouraged to actively talk about APC, just like other hand over items. This should help make sure all APC or related issues are quickly brought to the support engineer's attention. Normally, the operators on each shift have their own ideas about where the operation limits of their given process should be and may change the feasible operating space for the control system by changing those limits. These changes, if not communicated between operators, can lead to serious issues. Li et al. (2011) discuss about communication issues at shift handover. Seamless communication also needs to happen between various groups; else, APC could still be optimizing for last week's demands.

APC software upgrades are not glitch free, and at times, the new software upgrade could change the controller behavior and keep the support engineer extremely busy, troubleshooting, testing, and separating process, control, and software issues. Version-to-version operator interface changes, especially feature changes, are not uncommon, adding strain on operators. Control system (DCS) upgrades, computer upgrades, and periodic patches can also lead into the same type of issues, eroding operator's confidence.

The APC support engineers play a critical role in the long-term benefits of APC applications. The sustainability of APC- or AI-based applications depends on the knowledge and skills of people involved in the creation and maintenance of existing applications. APC can quickly fall into disuse if not supported correctly or unsupported. Thus, the support requirements needed to maintain a comprehensive model of a major process unit require detailed process knowledge, various engineering skill sets, ability to work with operators and different site personnel, and some prior APC experience. A young, inexperienced engineer thus cannot be expected to learn to develop or even maintain the underlying models in a short time. The basic reality then is that, if APC is going to prove to be successful in the long term, its support requires the efforts of the company's very best engineers.

13.8 CONCLUSIONS

APC has been successfully used in the process industry for over three decades for closed-loop process optimization. A typical APC engineer has been a chemical/electrical/mechanical engineer, with a strong background in process, controls, modeling, mathematics, and computers. Process operators are the APC end users and have the most fundamental role in process operations, its basic control, and all actual APC usage. Ultimate APC success requires operators' total buy-in. APC is still just a technology used by humans, and so its design must take various underlying human factors into account to have a successful daily utilization and be sustainable. APC applications need continuous upkeep in terms of support, software upgrades, and modifications to process changes, and operators need continuous training. Without a support program in place with local resources, APC benefits deteriorate over time, or the applications are completely turned off.

AI-based applications that are planned to be used in the control room have many parallels to APC with respect to design, uniqueness of these applications, human interaction, human trust, and training, and will run into the same sustainability issues APC had historically. AI-based methods must address these issues in the design phase, meet operators' various operational requirements, and gain early trust of operators in order for them to be successful and sustainable. Only then will AI-based method gain the desired commitment for its continual use in order to potentially derive all subsequent possible tangible benefits. A continuous support program will be required with dedicated support engineers well versed with process, controls, data analytics, and AI, and it will need continuous investment. Undoubtedly, the future of ever evolving and improving AI-based applications will continue to be created by today's boldest control technology visionaries. These are truly very exciting times throughout the entire field of AI-based process applications.

REFERENCES

Charniak, E. and McDermott, D. (1985). *Introduction to Artificial Intelligence*, 1. Reading, MA: Addison-Wesley Publishing Co., Inc.

Colegrove, L. (2020). Artificial Intelligence in the chemical industry – why my industry puzzles over the vendors' struggles. *Journal of Advanced Manufacturing and Processing* 2.

Farsi, M. and Liu, J. (2023). *Model Based Reinforcement Learning*. IEEE Press.

Forbes, M.G., Patwardhan, R.S., Hamadah, H., and Gopaluni, B.R. (2015). Model Predictive Control in Industry: Challenges and Opportunities. *IFAC-PapersOnLine* 48–8: 531–538.

Friedman, Y.Z. (1992). Avoid advanced control project mistakes. *Hydrocarbon Processing* (October).

Friedman, Y.Z. (1997). Advanced Process Control: it takes effort to make it work. *Hydrocarbon Processing* (February).

Huang, J., O'Connor, T., and Ahmed, K. (2021). AIChE PD2M, Advanced Process Control workshop-moving APC forward in the pharmaceutical industry. *Journal of Advanced Manufacturing and Processing* 3 (1) (January): 10071–10090.

Kissinger, H.A., Schmidt, E., and Huttenlocher, D. (2021). *The Age of AI and Our Human Future, Little*, 56. New York: Brown and Co.

Lahiri, S.K. (2017). *Multivariable Predictive Control, Applications in Industry.* Wiley.

Li, X., McKee, D.J., Horberry, T., and Powell, M.S. (2011). The control room operator: The forgotten element in mineral process control. *Minerals Engineering* 24 (8): 894–902.

Meyn, S. (2022). *Control Systems and Reinforcement Learning.* Cambridge University Press.

Mishkoff, H.C. (1985). *Understanding Artificial Intelligence*, 4–20. Indianapolis, IN: Macmillan, Inc.

Ottewell, S. (2015). Plants Grapple with Graying Staff. *Chemical Processing* (July 29).

Rich, E. (1983a). *Artificial Intelligence*, 1. New York: McGraw-Hill, Inc.

Rich, E. (1983b). *Artificial Intelligence*, 37. New York: McGraw-Hill, Inc.

Rich, E. (1983c). *Artificial Intelligence*, 104. New York: McGraw-Hill, Inc.

Russell, S.J. and Norwig, P. (1995). *Artificial Intelligence: A Modern Approach*, 3–33. New York: Prentice Hall International, Inc.

Index

Artificial Intelligence in Process Fault Diagnosis: Methods for Plant Surveillance,
First Edition. Edited by Richard J. Fickelscherer.
© 2024 John Wiley & Sons, Inc. Published 2024 by John Wiley & Sons, Inc.

Printed and bound by CPI Group (UK) Ltd, Croydon, CR0 4YY

27/10/2024

14580473-0002